D1377167

Modern Astronomy

Modern Astronomy

D. Scott Birney

Associate Professor
of Astronomy
Wellesley College

Allyn and Bacon, Inc.
Boston London Sydney Toronto

Library of Congress Catalog Card No. 73-84855.

ISBN: 0-205-04241-4
Second printing . . . May, 1976

Contents

Preface

In the excitement of today's remarkable scientific and technological progress, astronomy has become an increasingly active and popular discipline. It seems that astronomy, the oldest of the sciences, provides something for nearly everyone. The non-scientist relates astronomy to the space program and to the remarkable discoveries within the solar system. Through the manned and unmanned space probes he feels that he is living in a great new age of discovery. At the other extreme, the theoretical physicist finds, in astronomy, problems that stretch the mind and the imagination. Stellar atmospheres and interiors are laboratories in which theories of the nature of matter are given their most severe tests. Somewhere in between, the college student obtains through astronomy a surprisingly comprehensive introduction to physical science. Here the student finds classical examples of the progress from observation to classification to theory. He sees how refined observations are followed by more complete theories. In numerous examples, these steps of the scientific method clearly and logically follow each other. Very few students who enroll in an introductory astronomy course will go on to become professional astronomers, but many will gain through astronomy an appreciation of the logical methods that are common to all scientific investigations. It is the author's intention that his book will facilitate this basic understanding for the students who use it. It is with this in mind that I have tried to emphasize the observational background and clearly distinguish between the observed and the inferred features of the universe.

At all levels in education today there is a growing emphasis on self-paced study and independence on the part of the student. This represents a welcome change in emphasis away from a concept that stressed the memorization of facts. Part of the new emphasis is on the student's understanding of basic principles and his ability to apply those principles in interpreting other data. It is this approach that I have tried to maintain throughout this book. Wherever possible we begin with observations and show how these are understood in today's world. It is hoped that the student will see something of the real beauty of astronomy as he begins to realize the complex manner in which perseverance, ingenuity, and chance have been used by astronomers. Sometimes, as in the case of pulsars, a few observations have opened an entirely new field of study. At other times a slow accumulation of data has been necessary before astronomers could recognize some important new point. The study of binary stars and the mass-luminosity relation is an example of this type of effort.

Those instructors who wish to supplement a text with their own material or with other readings should find that this book may be used as the basic text in a two-semester course. The book is intended, however, chiefly for use in a one-semester or one-quarter course. The type of course for which this book is designed need not require any very extensive mathematical or scientific background. Algebra and trigonometry have been used in a few places where a simple mathematical explanation helps to make a point clear, and new physical principles are introduced as necessary. The mathematically insecure student need not panic at the sight of each equation, since an understanding of the main points does not usually depend completely upon the mathematics. Any use of algebra and trigonometry will simply reinforce the training which the student has had in high school.

Finally, it is hoped that the current frontiers of astronomy are covered fully. Some modern advances are well beyond the scope of this text. Others, especially in the areas of the solar system and the structure of our galaxy, are highly pertinent and find a logical place in this discussion.

Sincere thanks are due to a great many students and professors who used the first edition of this book and who offered numerous suggestions for updating and improving it to produce this second edition. I am especially grateful to my Wellesley colleagues Sally Hill and Steve Adler, who have been most helpful and encouraging at all stages in the revision of this book.

Modern Astronomy

minolta

Appearance of the Sky

Few of us today have stood in a large open space far from city lights on a clear, moonless night and watched the great spectacle of the stars and the Milky Way. This view of the night sky is a memorable and inspiring sight, just as it has been to men throughout history. Until fairly recent times the study of astronomy usually began with this view of the night sky. The student learned the patterns of the stars and followed the motions of the moon and planets. He mapped the details of the Milky Way circling the heavens, and he watched the changes in position of the constellations from one season to the next. In the daytime he may have noted the changes in the sun's daily path across the sky throughout the year. Today, however, astronomy has become more complex, and many astronomers see rather little of the actual sky. The means of collecting data on the stars have become quite varied, and the analysis of these data is often lengthy and difficult. The astronomer must try to apply the most advanced engineering and technology in his instrumentation, and he must try to relate his knowledge of physics, chemistry, and geology to the data. In most situations, mathematical and statistical methods provide the tools by which the data can be understood and by which new theories can be formulated.

◄ **Figure 1.1.** A modern planetarium projector. The appearance of the sky may be realistically displayed on a domed ceiling with an instrument of this type. (Minolta Corporation photograph.)

Nevertheless, the basic goal of astronomers today is much the same as it was in very early times. We seek to explore and to understand the physical universe. In each period of history, however, we have only partially understood our observations. The answers to the old questions have come with great effort, and these answers have often led to a new series of unanswered questions.

Through the years the methods of astronomy have changed to meet the changing problems. From an ancient era when men puzzled over the irregularities in the motions of the planets, we have reached a point at which we seek to explain each minute feature and change in the spectra of the stars. The advent of the discipline we know now as "astrophysics" has made it possible for us to make reasonable statements about the diameter, temperature, composition, mass, and atmospheric pressure of stars which are revealed only as dimensionless points of light in our telescopes. Today, astronomy has become the application of meager but hard-won data to questions of the evolution of stars and the nature of the universe. In spite of the great change in emphasis caused by astrophysics, the most basic concepts still afford the logical starting place in the study of astronomy. The initial experiences of the student studying the night sky are the same as they have always been in this, the oldest of sciences. Today's astronomy student builds upon foundations laid thousands of years ago.

1.1 Constellations

In the night sky certain prominent groups of bright stars attract our attention. These groups are the **constellations,** which have been recognized and named independently by people in many cultures. The names used today for the more conspicuous constellations in the skies of the northern hemisphere are the Latin forms of names which in many cases may be traced back to the Greek and Babylonian civilizations. Such names reflect the occupations and customs of those early times. Thus, we see Aquarius, the Water Carrier, and Aries, the Ram, in the skies of autumn; Orion, the Hunter, in winter; and Sagittarius, the Archer, in summer. Near the north pole another group of constellations bear the names of figures in Homer's *Odyssey*: Perseus, Andromeda, Cassiopeia, and Cepheus. In a few cases, a star group represents some familiar object of a more modern day. To many of us, the Big Dipper does look like a water or milk dipper. In England, however, the same star group is called Charles' Wagon or the Plow. Much to the disappointment of most casual observers, few of the constellations bear any resemblance whatever to the object for which they were named. Such resemblance is certainly not necessary even though more descriptive names for the actual patterns and shapes might make learning the constellations easier for the student. The old names still live on, forming a rich and romantic link with the very distant past.

For the most part, constellations in themselves have no physical significance. They are merely chance patterns projected onto what appears to us as a spherical sky. Astronomers find it convenient, nevertheless, to use the names of the constellations when referring to particular areas of the sky or to individual stars. These names have become just as significant in discussions of the sky as the names of cities, states, and nations in discussions of the spherical earth. In 1922 astronomers from all over the world agreed upon specific boundaries between the eighty-eight constellations which cover the entire sky.

1.2 Use of Star Charts

Whether the constellations are studied by the individual student or as a class project, each student must learn to recognize groups for himself. Star charts for the identification of the brighter stars have been provided as an aid for those who wish to learn the constellations. It is likely that at first the student will find it difficult to make the transfer from the printed page to the actual sky, but once a few key patterns have been found, the difficulties are somewhat lessened.

It is fortunate that there are increasing numbers of students who have access to a planetarium (see Figure 1.1) because a well planned planetarium program can give students a rapid introduction to the sky. The lecturer can point to individual stars with a flashlight or superimpose mythological figures on the star patterns to reinforce later recognition. Later, when students view the real night sky, they will be well prepared to pick out the more prominent constellations and look for some of the fainter stars.

A good starting place for observers in the northern hemisphere is the region near the north celestial pole. With an unobstructed view of the northern horizon, an observer may look for the seven conspicuous stars which form the Big Dipper. The orientation of this group of stars with respect to the horizon will depend upon the hour of the night and the season of the year. The Big Dipper actually consists of the brighter stars in the constellation Ursa Major and may readily be identified in Figure 1.2. The observer facing northward should turn the diagram so that the name of the current month is at the top. The figure is drawn to show the appearance of the northern sky at approximately 8:30 PM. For observations made later than 8:30 PM the diagram should be rotated counterclockwise by about 15° per hour.

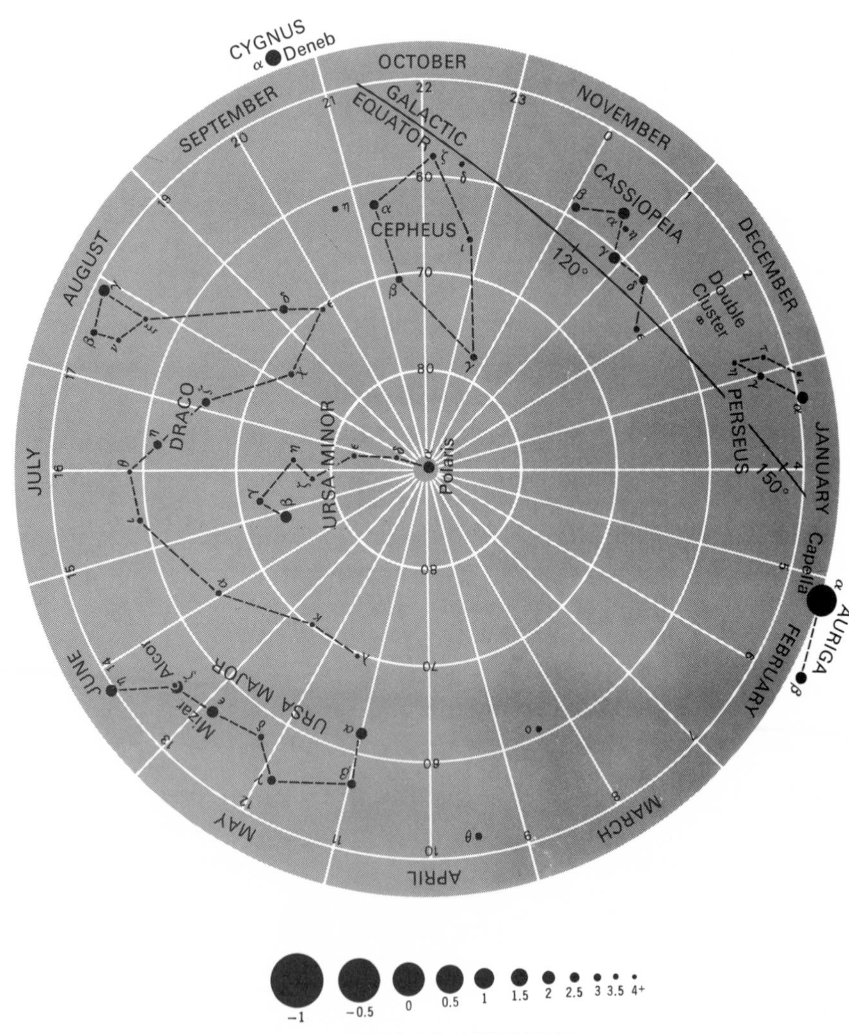

Figure 1.2. Constellations near the north celestial pole. Face north and hold the book so that the appropriate month is at the top. (From Stanley P. Wyatt, *Principles of Astronomy*, 2nd ed. p. 86. © Copyright 1971 by Allyn and Bacon, Inc.)

The last two stars in the bowl of the Big Dipper are marked on Figure 1.2 with the Greek letters α and β, and these stars are commonly referred to as "the pointers." The line of these two stars extended in the direction from β to α passes very near Polaris, the North Star. Another line passing from the handle of the Big Dipper through Polaris points toward the conspicuous constellation, Cassiopeia. On a reasonably dark night an observer can usually begin by finding Cassiopeia, the Big Dipper, and Polaris and then look for the fainter stars in Cepheus, Draco, and Ursa Minor. Having identified the stars in these northern constellations, one may proceed in the same manner to match the other star charts to the sky. This is especially easy in the winter months, when the bright stars of Orion and its surrounding constellations are plainly visible (see Figure 1.4).*

1.3 Daily Motions

Because of the earth's eastward rotation on its axis, all of the celestial bodies appear to rise in the east, move steadily across the sky, and set in the west. This daily motion of the sun is, of course, quite obvious to everyone. To a lesser degree we notice the similar motion of the moon. If we take time to do so we may easily see the daily motion of the stars across the sky as well. The stars which we see in the eastern sky in the early evening will be high in the sky by midnight and will be in the western sky by early morning.

1.4 Annual Motions

Throughout the year the appearance of the sky changes continuously. By late spring, the brilliant star groups so outstanding in the winter sky of the northern hemisphere are low in the west in the early evening and have been replaced by Leo, the Lion, and Boötes, the Bear Hunter. These, in time, give way in late summer and fall to Cygnus, the Swan, Aquila, the Eagle, and Lyra, the Harp. The Great Square in Pegasus dominates the skies later in the fall and presages the return of Orion and the winter skies. As

*In order to match the charts in Figures 1.3 to 1.6 to the sky, the book should be held directly overhead, with the top of the page toward the north. A southern hemisphere observer can use Figure 1.7 in a manner analogous to that described above for Figure 1.2.

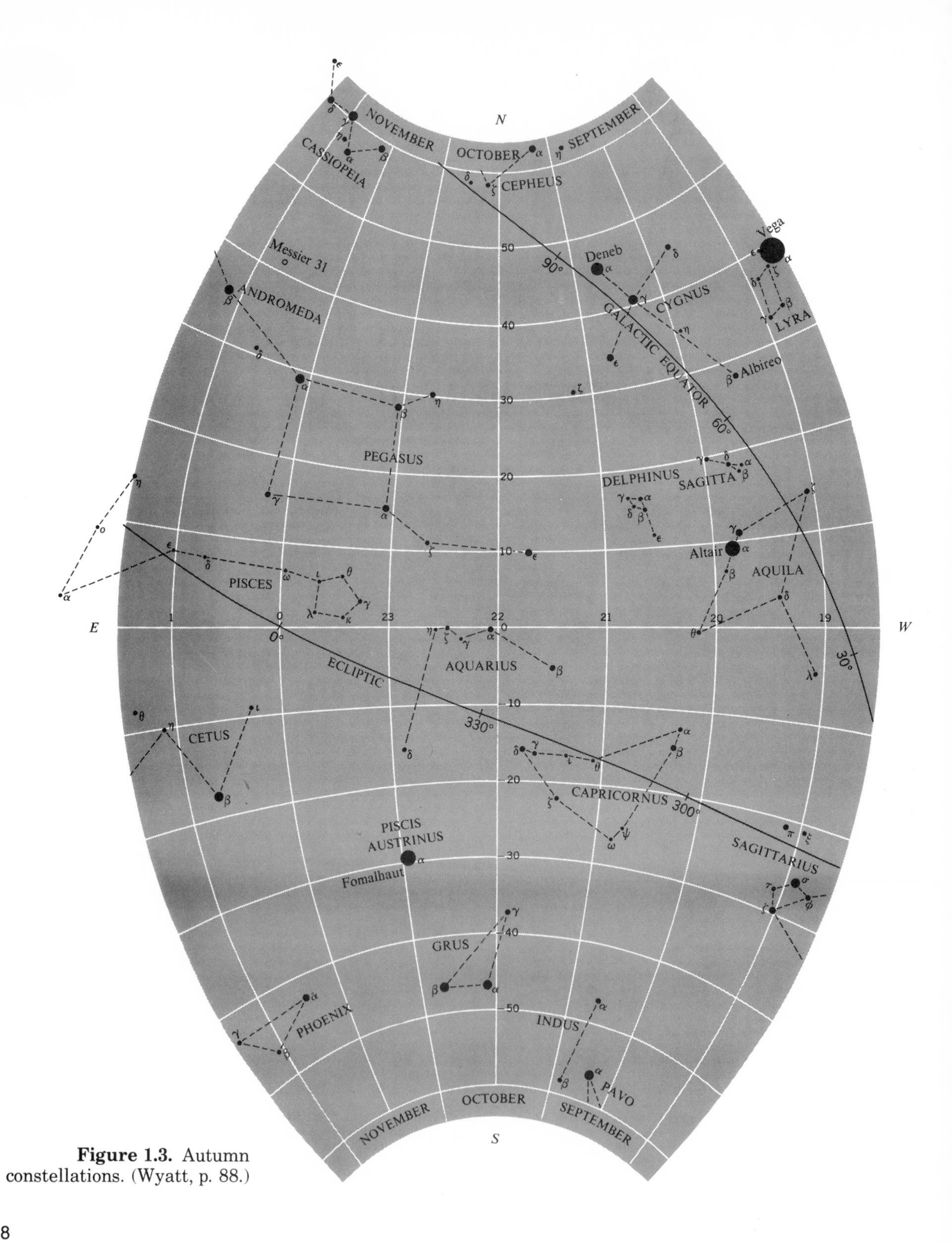

Figure 1.3. Autumn constellations. (Wyatt, p. 88.)

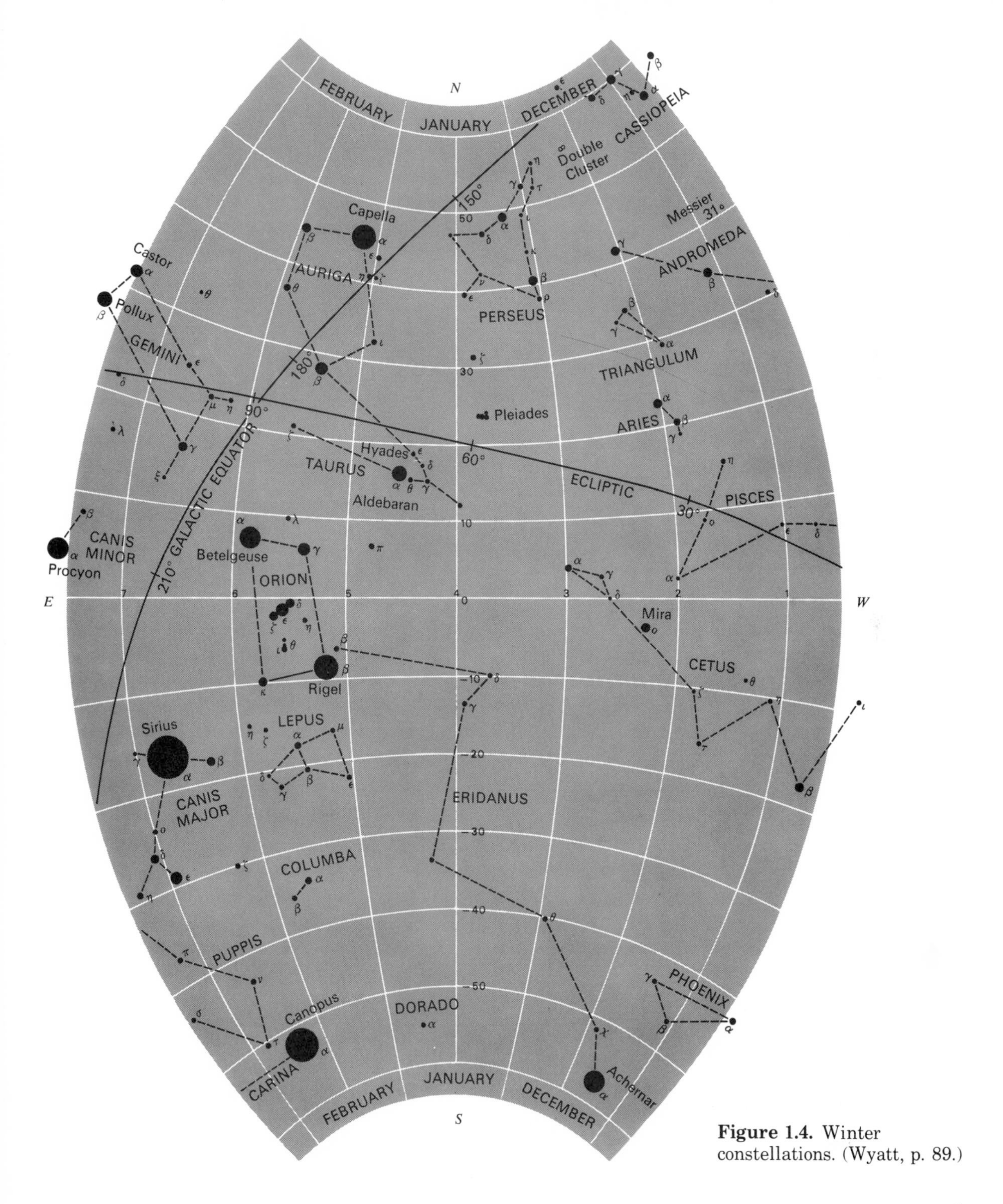

Figure 1.4. Winter constellations. (Wyatt, p. 89.)

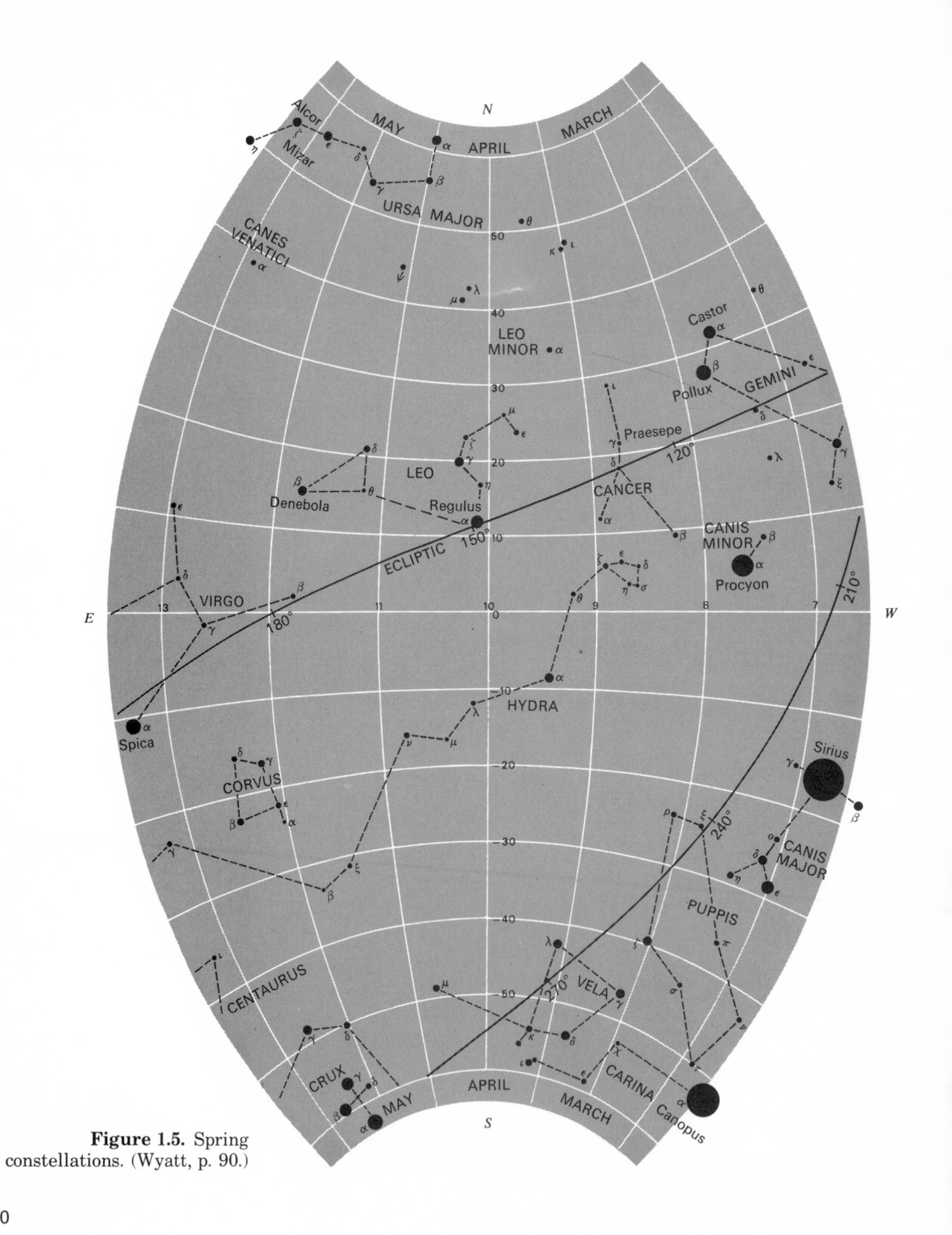

Figure 1.5. Spring constellations. (Wyatt, p. 90.)

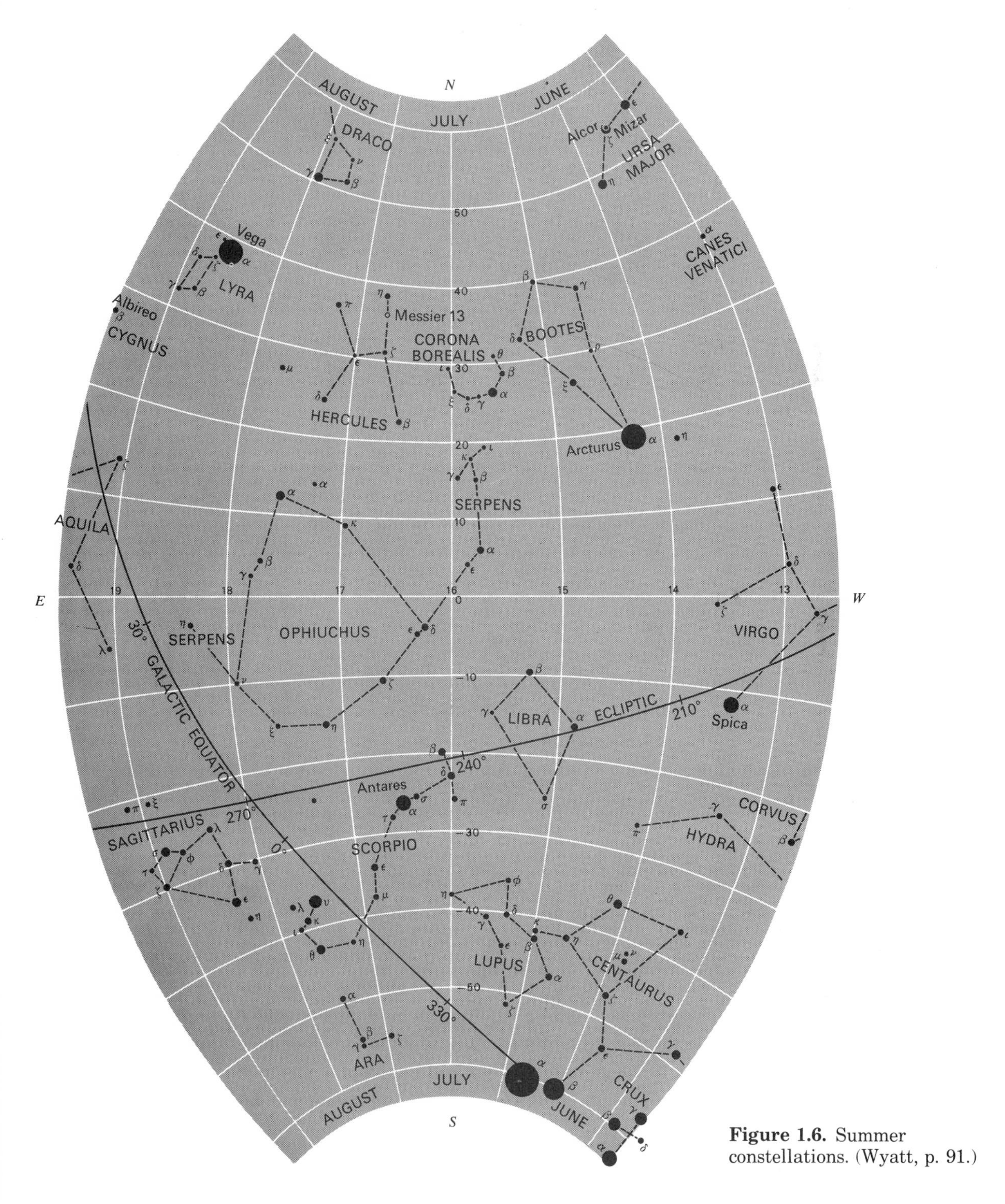

Figure 1.6. Summer constellations. (Wyatt, p. 91.)

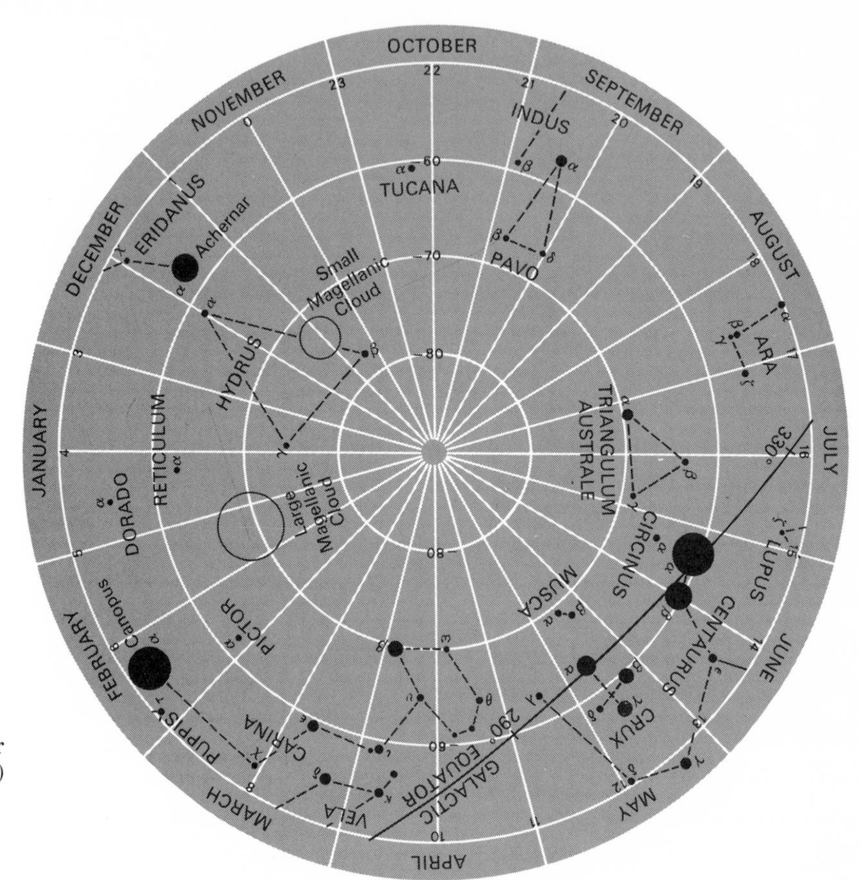

Figure 1.7. South polar constellations. (Wyatt, p. 87.)

weeks pass, one may notice that the constellations appear to move slowly and progressively westward. Stars that are high in the sky at sunset at the beginning of a month will be noticeably further west at the same hour at the end of the month. Careful observation will reveal that each star rises and sets four minutes earlier each day than it did the day before.

This annual march of the constellations results, of course, from the earth's revolution around the sun in the course of a year (Figure 1.8). As the earth moves progressively around its orbit, the stars which are seen overhead at midnight change continuously. Since the earth moves 360° around the sun in 365 days, it moves through an angle of approximately 1° per day. If the stars in the direction of the sun could be seen during the daylight hours, an observer would notice that the sun appears to move progressively

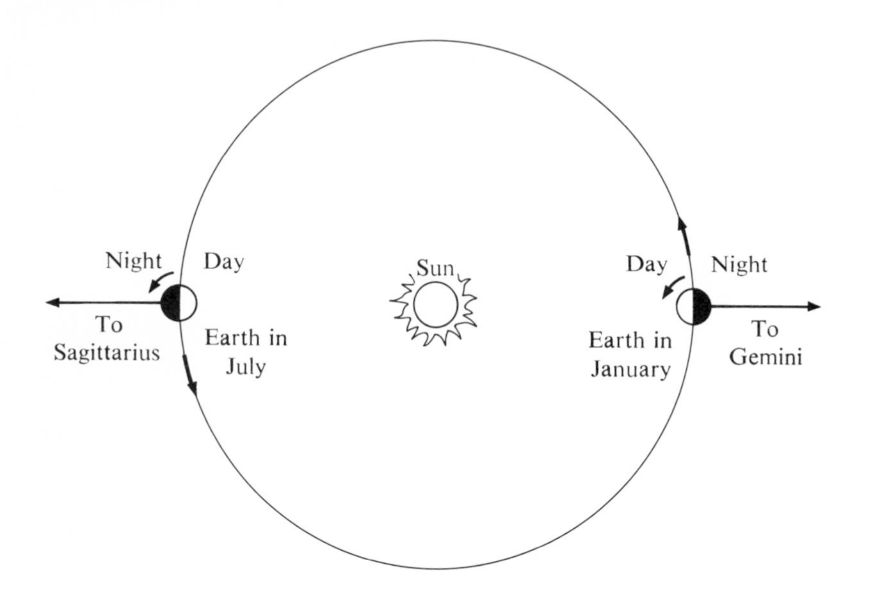

Figure 1.8. Because of the earth's motion around the sun, an observer sees different constellations overhead in each season. Gemini is on the meridian at midnight in January. Sagittarius is on the meridian at midnight in July.

eastward from day to day against the background patterns of the stars.

It should be remembered that the same changes in the appearance of the sky could result just as well if the earth stood still and the sun moved around it. The actual observational proofs which convince us that the earth moves rather than the sun will be discussed later.

1.5 The Celestial Sphere

Wherever we are on the earth the sky has the appearance of a dome covering the earth. At night the stars seem to be points of light set on the inside of this spherical surface. We recognize today that no such surface exists, but in many problems astronomers find it convenient to refer to the **celestial sphere** as if it did exist. Ancient peoples thought about the sky as a sphere, and seem to have been strongly convinced of its reality. More important, they recognized that this sphere, real or not, is extremely large with respect to the earth. It is so large, in fact, that two observers in different locations looking toward the same star are looking along parallel lines. The earth is a mere speck, and it is difficult to show both the earth and the celestial sphere together in a meaningful sketch.

As mentioned above, the visible portion of the vast celestial

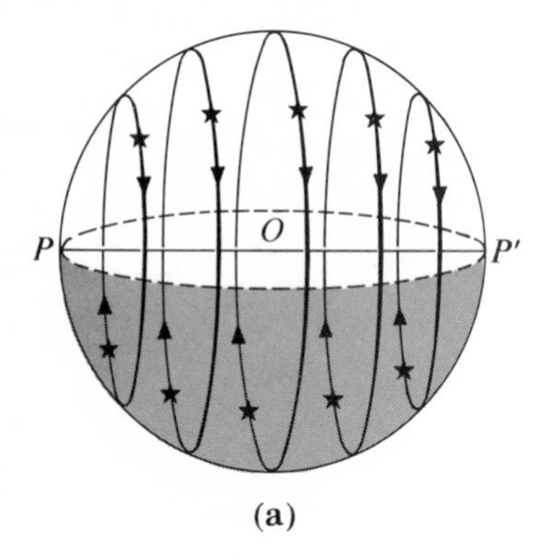

(a)

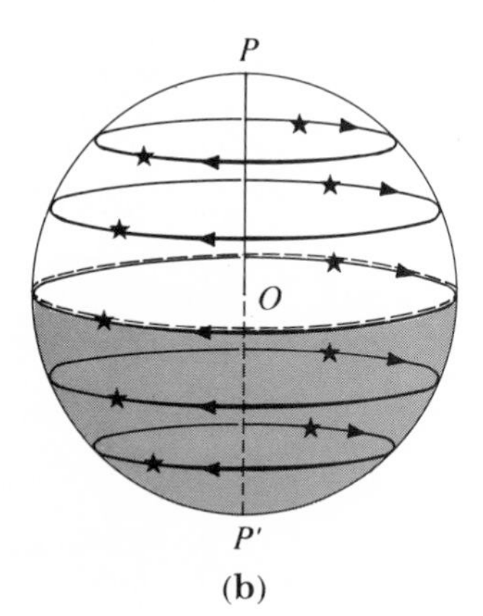

(b)

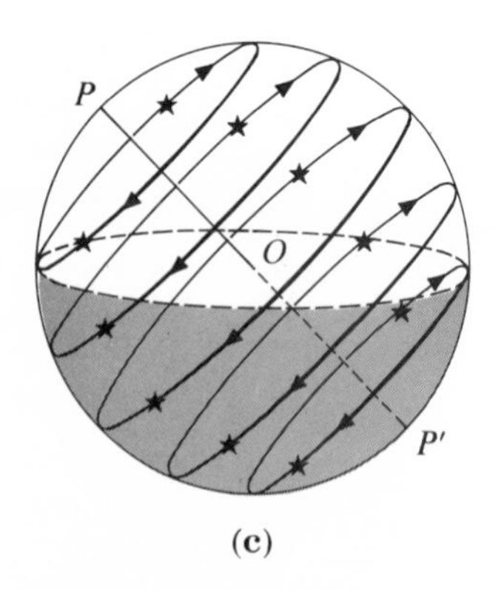

(c)

Figure 1.9. Apparent motions of the stars for observers at the equator, at the north pole, and at a latitude of 45°N.

sphere appears as a dome above the observer's head. The lower edge of this dome is the observer's horizon, which may be defined as being 90° from the zenith, the point directly overhead. Consideration of Figure 1.9 suggests that the appearance of this dome of the sky must depend upon the observer's location on the earth. Likewise, the apparent daily motions of the stars, planets, and sun due to the rotation of the earth will depend on the position of the observer on the earth.

In Figure 1.9a the observer at *O* is on the earth's equator. The horizon in the sketch is parallel to the earth's axis, and, due to the extremely large size of the celestial sphere, the observer looking northward would see the north celestial pole on the horizon. The observer at the north pole of the earth would see the north pole of the sky directly above his head, or 90° above the horizon (Figure 1.9b). In other latitudes the north pole of the sky is above the horizon by some intermediate angle (Figure 1.9c). The exact size of this angle between the celestial pole and horizon is equal in degrees to the observer's latitude, as indicated in Figure 1.10. The Greeks had long been aware that stars in the northern part of the sky appeared to be higher in the sky as one travelled to the colder northern lands. Homer mentioned a northern land where the Great Bear no longer bathes in the ocean; that is, the Great Bear, or Big Dipper, does not disappear below the horizon as it does in the latitude of Greece. The Greeks correctly reasoned that the changing heights of the pole with changes in latitude meant that the earth was round. Changes in latitude on a flat earth would not result in such changes in altitude of the pole (Figure 1.12).

Keeping in mind that the daily motions of the stars result from the earth's rotation on its axis, it is clear that the celestial sphere seems to rotate about the earth. The observer anywhere on the earth sees the stars move in circles around the celestial poles. Thus, along with the changes in appearance of the celestial sphere in different latitudes go changes in the apparent motions of the stars with respect to the observer's horizon. At the equator, the plane of the horizon is parallel to the earth's axis, and the stars will rise in the east at right angles to the horizon and set in the west at right angles to the horizon. In the course of any twenty-four-hour period each of the stars will be above the horizon for a total of twelve hours. At the north pole the horizon is at right angles to the earth's axis and the stars, consequently, move parallel to the horizon. The stars will neither rise nor set, and only the stars north of the celestial equator will ever be visible. At the intermediate latitudes in which most of us live, the stars rise and set at some angle to the horizon (Figure 1.11). Stars near the north celestial pole, such as the Big Dipper, will appear to move in circles around the pole and never go below the horizon. Stars near the south celestial pole never become visible to northern hemisphere observers.

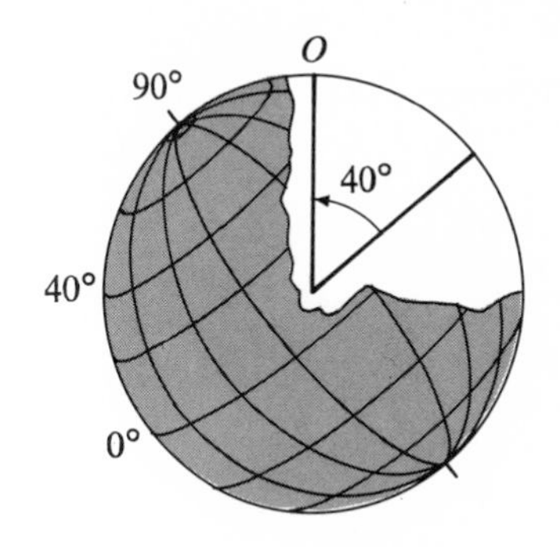

1.6 Planets

Against this background of the moving sky certain bright objects are seen to change their positions among the stars from night to night and from week to week. These moving bodies were recognized by ancient sky watchers, who described them as *wanderers*. Today we know them as **planets,** a transliteration of the Greek word πλανήτης, which means *wandering*. Only five wanderers, Mercury, Venus, Mars, Jupiter, and Saturn, were recognized by the ancients, but these five objects became of great concern to men for several thousand years. The motions of the planets became the basis of astrology, the practice of trying to understand the influence of the planets on men's lives.

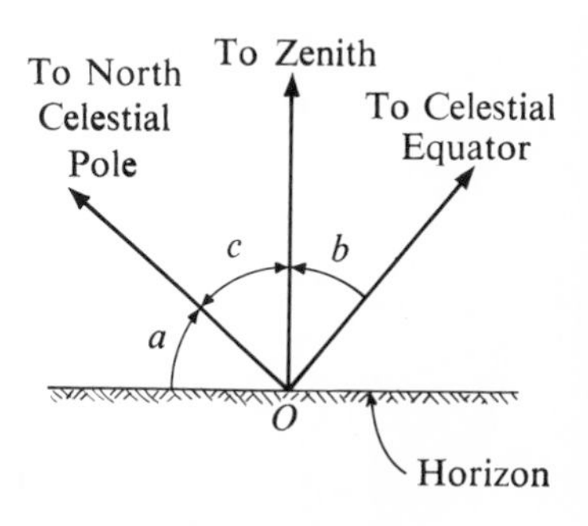

$\angle b$ = observer's latitude

$\angle a$ = altitude of north celestial pole above the horizon

$\angle a + \angle c = 90°$

$\angle b + \angle c = 90°$

$\angle a = \angle b$

Figure 1.10. At any location on the earth's surface, the altitude of the north celestial pole is equal to the observer's latitude.

Figure 1.11. Motions of the stars with respect to the eastern horizon. (Wellesley College photograph.)

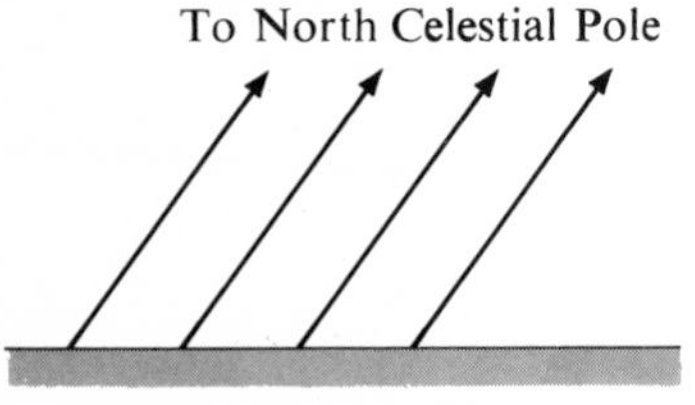

Figure 1.12. On a flat earth the altitude of the pole would be the same for all observers.

Even though today astrology has been discredited by scientists, it is recognized that in important eras in history, astrology gave men a practical reason for detailed observation and thoughtful consideration of planetary motions. Moreover, since astrology required the prediction of future positions of the planets in their background constellations, much effort was devoted to explanations of the mechanisms by which the motions of the planets were controlled. If the mechanical nature of the universe could be well understood, astrologers felt they could predict with ease and accuracy the future positions of the planets and the favorable dates for births, marriages, battles, and other important events of daily life. One peculiar characteristic of the planets caused the astrologers' search to become extremely complex and difficult: retrograde motion.

1.7 Retrograde Motion

In the course of their normal revolutions around the sun the planets move in the same direction as the earth moves in its orbit. Thus, the planets normally progress eastward among the background stars. Occasionally a planet such as Jupiter will appear to stop its eastward or direct motion, move westward for a time, stop again and finally resume its eastward motion. At such times the planet's path will often describe a loop as seen in Figure 1.13. This westward motion is called **retrograde motion.** Jupiter's motion may be retrograde for a period of several months at a time, and retrograde motion will occur at intervals of one year and 34 days. All of the planets show retrograde motion in an amount decreasing with the increase in their distances from the sun. It was retrograde motion which proved to be such a difficult problem for ancient astronomers and astrologers and caused the world systems of the early thinkers to become so complex.

1.8 Morning and Evening Stars

Less troublesome than the problem of retrograde motion was the fact that Mercury and Venus are seen only as morning "stars" or evening "stars." During certain periods, for example, Venus, the more easily observed of the two, is seen low in the western sky at sunset. As the weeks pass, Venus appears progressively higher in the sky at sunset. The situation reverses itself after a time and Venus is then seen progressively lower in the sky at sunset. Later

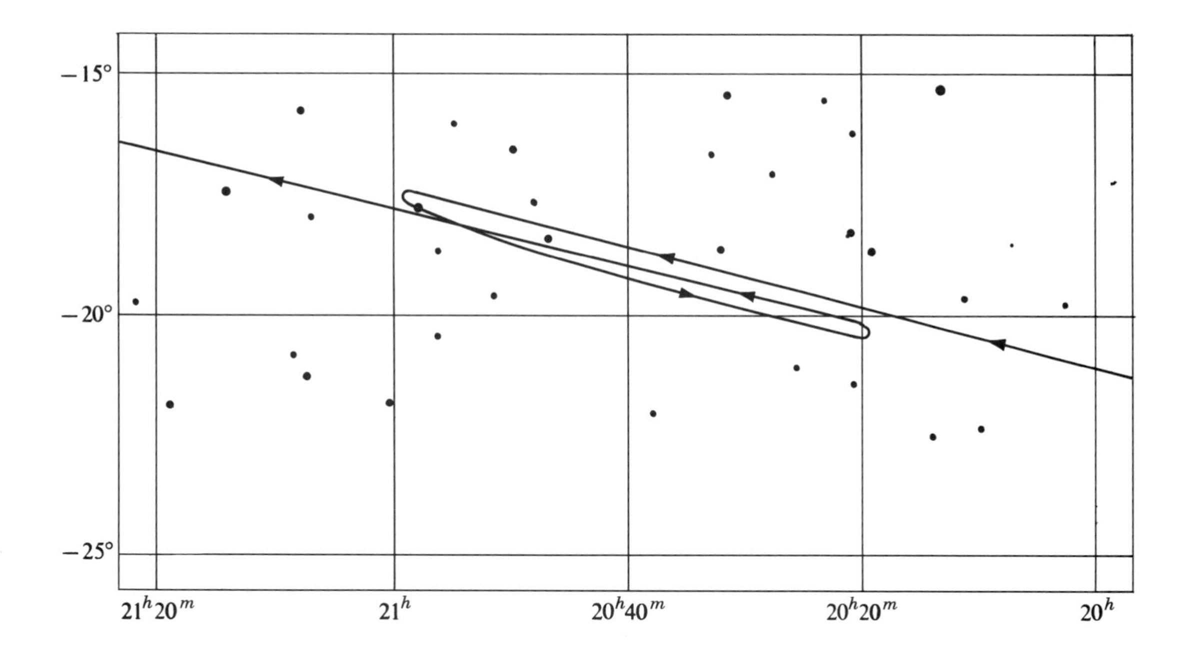

Figure 1.13. Apparent motion of Jupiter from March 1, 1973 to January 1, 1974. The motion is retrograde from June 1, 1973 to September 28, 1973. The background stars are in the constellation Capricorn.

on, one must look toward the east just before sunrise to see Venus low in the sky. Again as the weeks pass, Venus is progressively higher in the sky before sunrise; then progressively lower until it ceases to be seen as a morning star. This limited movement means that the angle measured from the earth between the directions to the sun and Venus is never larger than 48°. This angle is referred to as the **greatest elongation** of Venus. Venus may never be seen in the sky late at night. The same conditions apply with respect to Mercury, but in this case the greatest elongation is 28°.

Any model universe must, then, explain retrograde motion and the fact that Mercury and Venus are never far from the sun in the sky. In addition, a satisfactory model must also permit the prediction of the positions of the planets among the stars for any time in the future.

1.9 The Ptolemaic World System

In the ancient world one of the great seats of learning was the famous library at Alexandria. For hundreds of years men from all over the known world came there to teach and to learn. Much of the collection of perhaps 400,000 volumes was burned accidentally when Julius Caesar was fighting in Alexandria in 47 B.C. The rest of the library was eventually dispersed and lost. The last director of this famous institution was Claudius Ptolemy, who during his lifetime produced what was to become the astronomical authority for more than one thousand years. Ptolemy's work was titled *The Thirteen Books of the Mathematical Composition*, but it has been best known through the years by the title of its translation into Arabic, *Almagest*, meaning *The Greatest.*

Almagest contained Ptolemy's synthesis of the astronomy and astrology of his own era as well as that of hundreds of years before his time. Most important, however, was Ptolemy's own original scheme or mechanism for explaining retrograde motion and predicting future positions of the planets. The main points of Ptolemy's world system were:

1. The earth is round, stationary, and very small compared with the immense sphere of the sky.
2. The stars are fixed points on the inside of the great celestial sphere.
3. Day and night result from the daily rotation of the entire celestial system around the fixed earth.
4. Planets move around small circular paths, called **epicycles,** and the centers of these epicycles move around the earth on other circular paths, called **deferents.**

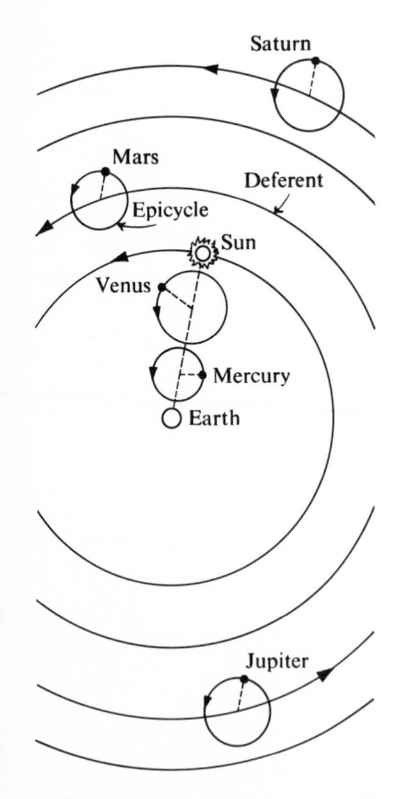

Figure 1.14. Ptolemy's world system.

Examining Figure 1.14, one may see how Ptolemy sought to explain retrograde motion. A planet such as Mars moves around its epicycle at some uniform rate of speed. The center of the epicycle moves along the circular deferent at a uniform speed. Thus, when moving on the inner side of the epicycle the planet would actually be moving in the retrograde direction, that is, from east to west in the sky as seen from the earth. The planet would, therefore, be a little closer to the earth during retrograde motion. This theory was in agreement with the observed fact that Mars is actually slightly brighter during the retrograde part of its orbit than at other times.

Perhaps the most remarkable thing about this system was that Ptolemy had laboriously determined the sizes of the circles and the rates of speed along the circles which would result in motions matching closely the observed motions of the planets in the sky. As a result, priests and astrologers were able to use Ptolemy's system to predict approximate future positions of the planets.

Ptolemy lived in the second century A.D. After his time the civilized world settled slowly into the medieval period. In most of Europe, life became a struggle against the elements and the conquering barbarians with little time for study of any sort. Meanwhile, across North Africa and the Eastern Mediterranean lands from the Atlantic Ocean to India the Arabs conquered everything and spread the Moslem faith. Among them were a number of mathematicians and astronomers who eventually translated *Almagest* into Arabic. It was from the Arabs who had conquered and occupied Spain that Ptolemy's *Almagest* finally made its way to Europe.

The Ptolemaic system had served the Arabs very well for hundreds of years. Many Arabs made careful and interesting observations (for example, they measured the size of the earth), but through the years they maintained the basic concept of the Ptolemaic system. At last, with the increase in travel and trade, Europe had begun to emerge. By 1300 universities had been built in Italy, France, and England, and the science, philosophy, and poetry of ancient Greece were rediscovered. In adopting the Ptolemaic system, however, the Church scholars refused to accept the notion that the earth was round. Anyone, they reasoned, could see that it was flat.

1.10 Copernicus

Into this emerging world Nicolas Copernicus was born in Poland in 1473. He studied extensively in Europe and particularly in Italy, with his interests leaning to mathematics, astronomy, and Greek language and writings. He eventually returned to Poland and entered the church as an exceptionally well-educated man. Copernicus maintained his interest in astronomy throughout his life, searching for some means of simplifying the complexities of the Ptolemaic system. His ideas gradually took shape and in 1543, the year in which he died, he set them forth in a small book *De Revolutionibus Orbium Cœlestium.* Copernicus was not an observer as some of the Greeks and Arabs had been. He was perhaps best described as a Greek scholar, and his ideas represent, in part at least, a synthesis of some ideas from the ancient Greeks.

Copernicus' great contribution, as set forth in his book, was that he removed the earth from the center of the universe and replaced it with the sun. Great simplifications in the explanations of retrograde motion were now possible. The earth was simply one of six planets in circular orbits around the sun. The revolution periods of these six planets were progressively longer for planets farther from the sun. Therefore, the earth moved about the sun more rapidly than Mars and more slowly than Venus. It was now easy to see, as shown in Figure 1.15, that as the earth passed Mars going around the sun, Mars would appear to stop its eastward motion among the stars, briefly move westward, and then resume its eastward motion once more. To Copernicus, day and night were the result of the rotation of a spherical earth on its axis. Being inside of the earth's orbit, Mercury and Venus quite naturally could not

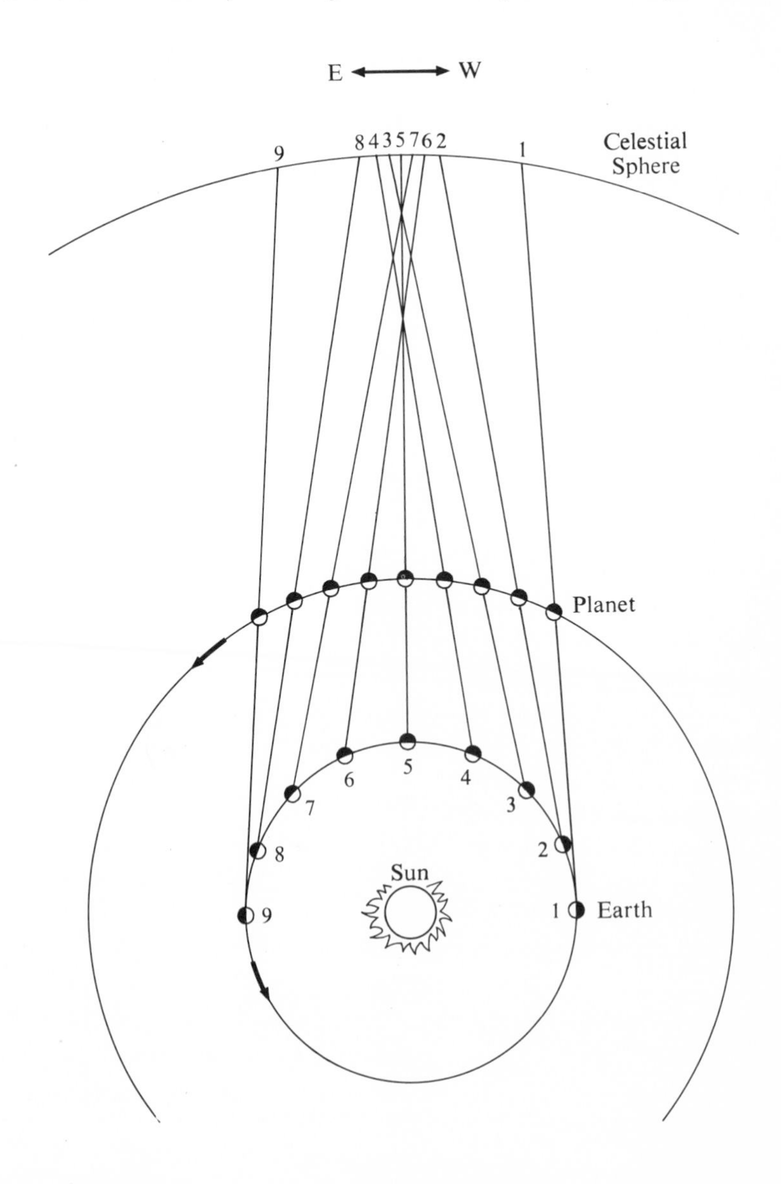

Figure 1.15. Retrograde motion in the Copernican system. As the earth moves from point 1 to point 9, the planet is projected against the celestial sphere at points 1 through 9. The planet moves eastward until it reaches point 4. It moves westward to point 6, and then resumes its eastward motion.

be seen at midnight when the observer was turned away from the sun. These two planets were seen, therefore, only as morning and evening "stars."

In spite of its great simplicity and success in explaining observed phenomena, the Copernican system was by no means very readily accepted by Copernicus' contemporaries. The Ptolemaic system had the authority of more than a thousand years of use and was the accepted system in the Church. More than mere simplicity was needed for the new concepts to win acceptance. No crucial observations had yet been made which would enable a thoughtful man to make a choice of one system over the other.

The first step toward the required proof did not come until 1675, when observations of the satellites of Jupiter by the Danish astronomer Römer led not only to a measurement of the velocity of light but to a proof of the earth's orbital motion as well. The first indirect proof of the earth's rotation had been the discovery about 1670 that the earth was not round but was flattened in the direction through the poles. It was argued that this flattening was a direct result of rotation. A nonrotating body would assume a perfectly spherical shape. More convincing proofs of both rotation and revolution were not actually achieved until after 1830. As late as 1778 it was stated in defense of rotation that the sun and certain of the planets were observed to rotate, and therefore it was very likely that the earth did also. In the light of the lack of strong observational evidence at the time, it is not surprising that the Copernican world system did not replace the Ptolemaic system for a great many years.

QUESTIONS

1. Describe the apparent daily motions of stars with respect to the horizon for observers at the equator, at the north pole, and in mid-northern latitudes.
2. What was the original significance of the word *planet?*
3. Why do we see a different group of constellations in the summer than in the winter?
4. Why is the concept of a celestial sphere meaningful even though we know that it does not exist in reality?
5. Why were retrograde motions troublesome to Ptolemy and his predecessors?
6. Describe the contribution of Copernicus to the problem of the motions of the planets.

Figure 2.1. Examples of some early astronomical instruments. Top: a ring dial for finding the sun's position; bottom: an astrolabe used for measuring the positions of stars. (Photographs from the Huggins Collection, Whitin Observatory, Wellesley College.)

The Earth as a Planet

The true nature of the earth as one of several planets circling the sun could not be positively proved by either the crude naked-eye observations or the philosophical reasoning of the ancient astronomers. Refined techniques and precise observations were needed, and these became possible only with the gradual development of new skills, first in metalworking and then in optics.

During the centuries between Ptolemy and Copernicus, astronomers and astrologers devised a few simple instruments to aid them in their task of comparing observed positions of planets with positions predicted from Ptolemy's epicycles and deferents. These instruments, crude as they may seem today, expanded the technology of their day. Ptolemy describes the armillary sphere which he used to measure positions of the sun and moon with respect to the celestial equator. Arabian astronomers later made extensive use of the astrolabe for measuring altitudes of the heavenly bodies above the horizon. In the medieval era the cross staff, quadrant, and sextant were devised for the measurement of angles between two celestial bodies or between such bodies and the horizon. Some of these instruments are pictured in Figure 2.1. All were in use long before the invention of the telescope, but none was capable of very precise angular measures.

2.1 Tycho Brahe

The design and use of pre-telescopic instruments were advanced to their highest level by a Danish nobleman, Tycho Brahe (1546–1601). At Uraniborg, his elaborate and highly successful observatory on the island of Hven near Copenhagen, Tycho refined the designs of older instruments and invented new ones. He worked continually to increase the accuracy of his observations, and with the great fixed quadrant shown in Figure 2.2 was able to measure positions of objects in the sky to within 10″ (10 seconds of arc). He was an ingenious and colorful man. Going about his work with a full complement of attendants and students and dressed in colorful robes, Tycho was probably the basis for our classical concept of the medieval court astronomer. (The peasants of Hven were further awed by his solid-gold nose. In a student duel he had lost the end of his nose and he had chosen to have his artificial one made of some more elaborate material than flesh-colored wax.)

Not only did Tycho construct the best instruments of his day, but he saw that they were effectively used. He observed and analyzed systematically, accumulating vast quantities of data. Eventually Tycho produced an accurate catalogue of the positions of the bright stars, proved that comets were not phenomena in the earth's atmosphere as Aristotle had said, and made accurate observations of the paths of the sun, moon, and planets among the stars.

Tycho's observatory flourished and grew from its founding in 1576 until 1597, when he left Denmark after a series of mishaps with the new king and various nobles and high officials whom he had offended with his autocratic attitudes. Eventually Tycho and one of his assistants arrived in Prague with a few of his more portable instruments and the extremely valuable records of planetary observations. In 1599 he became court astrologer to the Emperor Rudolph II, and remained in this capacity until his death in 1601.

Tycho was, of course, well aware of the works of Ptolemy and Copernicus, and he felt sure that his own observations were of sufficient accuracy to enable him to choose correctly one system over the other. In particular he looked for small systematic changes in the positions of stars. These small annual changes could prove that the earth really revolved around the sun. Unable to find such motions, Tycho invented for himself another earth-centered system which attempted something of a compromise between the Ptolemaic and Copernican theories. Though the details of his world

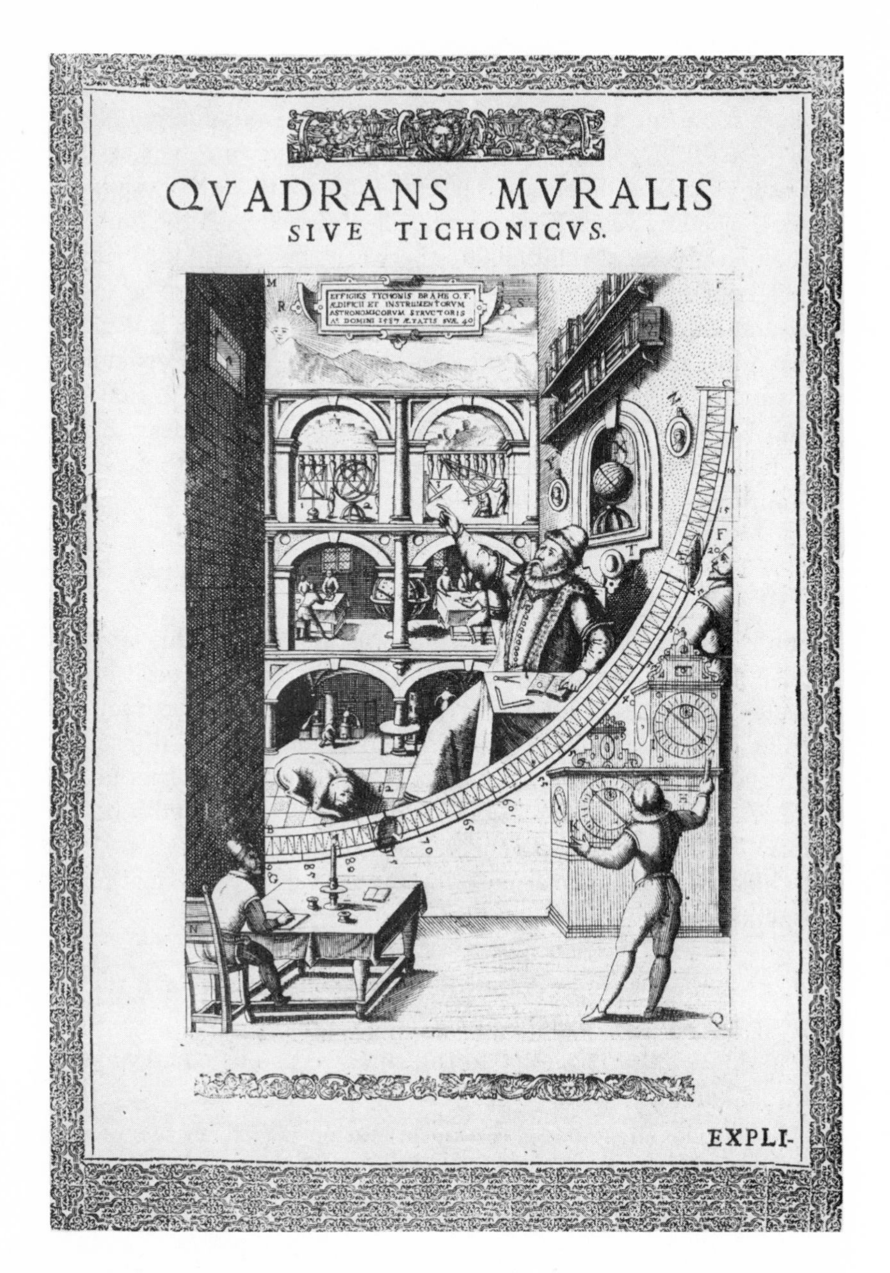

Figure 2.2. The Mural Quadrant at the observatory of Tycho Brahe. The observer at the right is sighting through the small opening in the wall at the left in order to measure the altitude of a body on the meridian. Other instruments used by Tycho are seen in the background. (Reproduced, from a text of 1598, by the New York Public Library.)

system are of little importance today, Tycho tried to fit a theory to his precise observations. He was, at least, scientifically correct in his attempts to interpret his own carefully made observations. We remember him today for his ingenuity, precision, and dedication to the accumulation of the best possible observational data.

2.2 Kepler

Tycho's successor as court astrologer in Prague was Johann Kepler (1571–1630), a young German mathematician who had assisted with the computation of tables of planetary positions in Tycho's last years. For many years Kepler had been obsessed with the problems of the mathematical relationships between the sizes of the orbits of the planets. A mystic and a strong believer in astrology, he sought long and hard for the keys which would enable man to unlock the unity and harmony of the universe. From childhood, Kepler's life had been relatively unhappy. He was never to succeed in that task which he had been so anxious to complete. Nevertheless, Kepler did make a lasting contribution of greater significance that even he himself understood.

2.3 Kepler's Laws

Tycho Brahe's great legacy to Kepler was the record of his long and precise observations of coordinates of stars and the positions of planets among the stars. With great faith in their accuracy and with great appreciation of their value, Kepler occupied himself for many years with efforts to find out just how these observations were related to the true motions of the planets. Eventually he was able to state the following three laws (the first two laws in 1609, the third in 1618), which describe quite precisely the motions of the planets:

1. The planets travel in elliptical orbits about the sun, and the sun is at one focus of the ellipse.*
2. The line joining a planet to the sun sweeps out equal areas in equal periods of time.
3. The ratio of the squares of the revolution periods of two planets is equal to the ratio of the cubes of their mean distances from the sun.

Kepler derived these three important laws through great perseverance and trial and error. He was fortunate that Tycho's most extensive observations had been of the planet Mars, for Mars' orbit is more eccentric than the orbits of the other bright planets (with

*See the Appendix for a discussion of the ellipse and other conic sections.

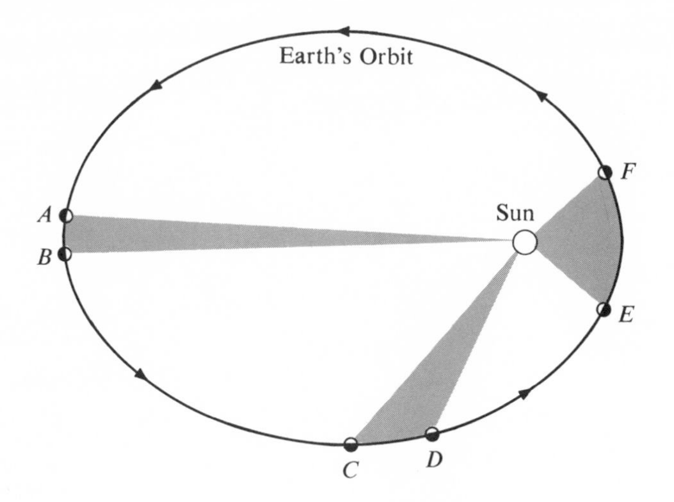

Figure 2.3. Kepler's laws state that the orbits of the earth and the other planets are ellipses, with the sun at one focus of the ellipse. The three shaded areas are equal in area and the planet covers the distances *AB*, *CD*, and *EF* in equal times.

the exception of Mercury). Therefore, in the orbit of Mars, ellipticity is more easily detected than in the orbit of any other planet.

Astronomers over the years have made use of Kepler's laws in many ways, and, as will be shown later in this chapter, Isaac Newton was able to show that there is a solid mathematical basis for them. In Kepler's own time, however, the laws freed men from the philosophical constraints which had come down from the Greeks before Ptolemy. The Greeks had held that the heavens were perfection and that the motions of the heavenly bodies must therefore be perfect. Perfect motion meant uniform motion in a circular orbit. Kepler's first law stated that the orbits were ellipses rather than circles, and the second law stated that the orbital velocity of a planet must vary, being the greatest when the planet is actually closest to the sun. From Figure 2.3 it may be seen that if the law of equal areas is to hold, the planet must move more rapidly in its orbit, when it is closest to the sun, at perihelion, than when it is farthest from the sun, at aphelion.

Kepler's third law, often referred to as the harmonic law, made it possible for astronomers to construct a scale model of the solar system without knowing its actual dimensions. From careful observations of the motions of the planets and in particular the intervals between oppositions (i.e., the times when the sun and an outer planet are in opposite directions in the sky), the **synodic period** of a planet could be found. The synodic period is the interval between successive oppositions and is indicated in Figure 2.4 for a planet in an orbit larger than that of the earth. With respect to the stars, the

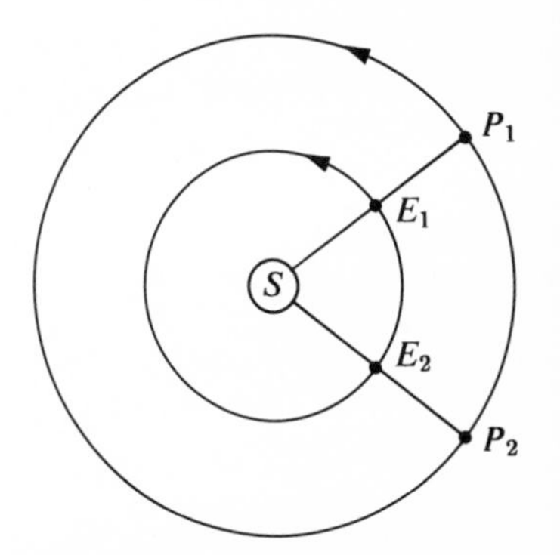

Figure 2.4. Synodic period of planet, *P*, is the time interval from one opposition of the planet to the next opposition. A planet is said to be in opposition when it is in the opposite side of the sky from the sun.

earth makes more than one complete passage around the sun in the time from one opposition to the next, so the synodic periods of the outer planets are always longer than one earth year. From a long series of observations the synodic period of a planet may be known with precision, and from this the **sidereal period** is calculated. (See Section 5.1.) It is the sidereal period which is to be used in Kepler's third law. For Mercury and Venus the synodic periods are determined from the intervals between conjunctions. Conjunctions are the moments when the sun and planet are in the same direction in the sky as seen from the earth.

If we let P_p and P_e be the periods of some planet and the earth, respectively, and D_p and D_e be their mean distances from the sun, Kepler's third law is written

$$\frac{P_p^2}{P_e^2}=\frac{D_p^3}{D_e^3}$$

This equation reduces to

$$P_p^2=D_p^3$$

when the period, P_p, is measured in years and the distance from the sun, D_p, is measured in terms of the mean distance of the earth from the sun. For convenience in measuring distances within the solar system, astronomers refer to the mean earth-sun distance as the **astronomical unit** (abbreviated A.U.). Since the periods may be accurately calculated from observations, the second equation makes possible the computation of the orbital radii of all the planets in astronomical units. Thus, even without knowing the distance from the earth to the sun in kilometers, it is possible to produce a scale model of the solar system. Table 2.1 lists the revolution periods and distances from the sun for all of the planets.

Kepler's works were published in a number of separate books. One of these, the extensive table of planetary positions which had actually been initiated by Tycho Brahe, found widespread use and acceptance throughout Europe. Other works, however, in which Kepler actually set forth his laws of planetary motion circulated very slowly. Even Galileo, with whom Kepler maintained a regular correspondence, seems to have been unaware of Kepler's laws as late as 1632.

Table 2.1. Planetary revolution periods and distances from the sun.

Planet	Revolution Period (years)	Mean Distance from the Sun (A.U.)
Mercury	0.24	0.38
Venus	0.62	0.72
Earth	1.00	1.00
Mars	1.88	1.52
Jupiter	11.86	5.20
Saturn	29.46	9.54
Uranus	84.01	19.19
Neptune	164.78	30.07
Pluto	248.42	39.44

2.4 Galileo

One of the truly brilliant men of Kepler's era was the great Italian scholar Galileo Galilei (1564–1642), to whom we give credit for the first use of the telescope in astronomy and for important work in mechanics, the study of the forces acting on bodies in motion. His straightforward reasoning and his abandonment of the restrictions imposed by the authority of the Church in secular matters set a worthy example for succeeding investigators, but eventually caused him great suffering. It is likely that his accomplishments could have been even greater if he had chosen to be a bit more diplomatic in his relations with the Pope and others in high places.

No claim is made that Galileo invented the telescope, but there is no question that he was the first to make significant use of it in astronomy. His telescopic discoveries, of which the first was in 1609, contributed greatly to the growing support for the heliocentric world system of Copernicus. Some of Galileo's more important discoveries are listed below:

1. The moon, which ought to be perfect, is covered with mountains and other irregularities.
2. Venus displays a cycle of phases similar to the phases of the moon, which are explainable only in terms of Venus' motion around the sun rather than around the earth.
3. The planets appear as round discs through the telescope, whereas

the stars are only dimensionless points, just as they appear without a telescope.

4. Four lesser bodies revolve in orbits about Jupiter and thus the earth is clearly not the center of all motion in the universe.
5. Saturn has peculiar appendages and is for this reason unique among the planets.*
6. The Milky Way is composed of countless stars too faint to be resolved as individuals with the unaided eye.
7. The sun, long held to be perfect like the moon, is sometimes marred by the appearance of dark spots on its surface.

Galileo encountered considerable difficulty in convincing his contemporaries of the reality of his discoveries. In many cases, men whom he wished to convince simply refused to look through his telescope. Others who did look through the telescope were unable to see all that Galileo himself was able to see. His telescopes were very poorly made and not solidly mounted, and it took some practice to learn to make effective use of them. Furthermore, the strong bonds of tradition and authority were not easily broken in a world in which rational analysis was a new concept understood and practiced by only a handful of men. Year by year, however, the combined ideas of Copernicus and Kepler, reinforced by the observations of Galileo, became better known and more widely accepted. Long before there were any actual proofs, most scholars in seventeenth-century Europe had begun to accept the idea that the spherical earth was a planet in orbit about the sun.

2.5 Newton's Law of Universal Gravitation

During the first half of the seventeenth century much thought was devoted to the idea of motion and to what the most basic kind of motion must be. Galileo had been among the first to make serious inquiry into the matter, and on the basis of his work some of his younger contemporaries were able to grasp the concept that uniform motion in a straight line is the natural thing. Any departure from uniform straight-line motion requires the action of an outside force applied in such a way as to change the direction of the motion or the velocity of the body in its direct course.

*The "appendages" were Saturn's rings seen with poor definition in Galileo's telescopes.

These early thoughts on force and motion were developed by the great English scientist Isaac Newton (1642–1727) into his well-known laws of motion:

1. A body in motion tends to remain in motion in a straight line unless acted upon by some outside force.
2. When a force acts on a body, the motion of the body is changed in the direction in which the force acts. In other words, a force produces an acceleration.
3. For every action there is always an equal and opposite reaction. That is, when a force is exerted on a body, the body exerts a similar force in the opposite direction.

Reasoning from these laws, Newton considered the motion of the moon around the earth. For the moon to remain in its orbit, some force must be continuously acting on the moon. Further, this force must be directed toward the earth. Newton sought to identify this force and to describe it in a quantitative way. He was able to do this by means of his law of universal gravitation. Newton stated that all bodies in the universe attract each other, and that the force of this attraction is proportional to the product of the masses of the two bodies and inversely proportional to the square of the distance between them or

$$F = G\frac{m_1 m_2}{r^2}$$

where F is the attractive force, G is a universal constant, m_1 and m_2 are the masses of the two bodies, and r is the distance between them. (The numerical value of G depends upon the units in which mass, distance, and force are expressed. See Appendix Table A.4.)

With the aid of calculus one can show, as Newton himself did, that Kepler's laws of planetary motion may be derived as a natural consequence of the law of universal gravitation. From this derivation one finds that the orbit of a body moving under the gravitational attraction of a second body may be an ellipse, a circle, a parabola, or a hyperbola. It also follows from this derivation that the correct form of the third law is actually

$$\frac{P_1^2\,(M_\odot + m_1)}{P_2^2\,(M_\odot + m_2)} = \frac{D_1^3}{D_2^3}$$

Here $M_\odot$ is the mass of the sun, m_1 is the mass of one planet, and m_2 is the mass of a second planet. Masses for the planets orbiting the sun are in all cases insignificant when compared with the mass of the sun, and the ratio $(M_\odot + m_1)/(M_\odot + m_2)$ is very nearly equal to unity. For this reason Kepler's third law, in which mass does not appear, gives an accurate description of the planetary motions. We shall see in later chapters a number of interesting applications which employ Kepler's third law as modified by Newton in order to determine the actual masses of various celestial bodies.

Newton received immediate and widespread recognition for his great contributions and several of his contemporaries saw ready application of his work. One such was Edmund Halley, who applied the law of universal gravitation to observations of comets and showed that the orbits of some comets are highly elongated ellipses along which the comet returns to the neighborhood of the sun at regular intervals.

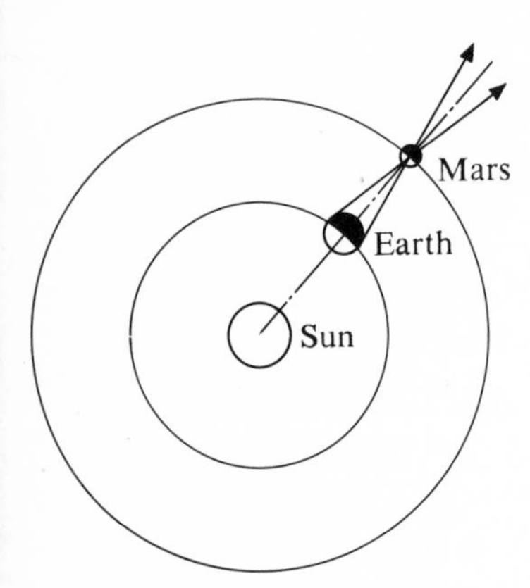

Figure 2.5. Mars at opposition. When the faster-moving earth overtakes and passes Mars, the sun and Mars are in opposite directions in the sky. At such times the distance from the earth to Mars is a minimum and may be measured by direct trigonometric means.

2.6 The Scale of the Solar System

The distance from the earth to the sun in kilometers, or some other familiar unit, is the kind of fundamental datum which interested all men regardless of whether they subscribed to the Copernican or the Ptolemaic world system. The sun is too far away for direct triangulation techniques even with the earth's 12,750-km diameter as a base line. We have already mentioned that Kepler's laws provided a scale model of the solar system based only on the revolution periods of the planets. Seventeenth-century scholars realized that measurement of a distance in kilometers between any two planets could provide the true scale of this model. When a planet such as Mars is at opposition, as in Figure 2.5, its distance from the earth is small enough that triangulation using the earth's diameter as a base line can be used to find the distance in kilometers between the earth and that planet. (See Figure 2.6.)

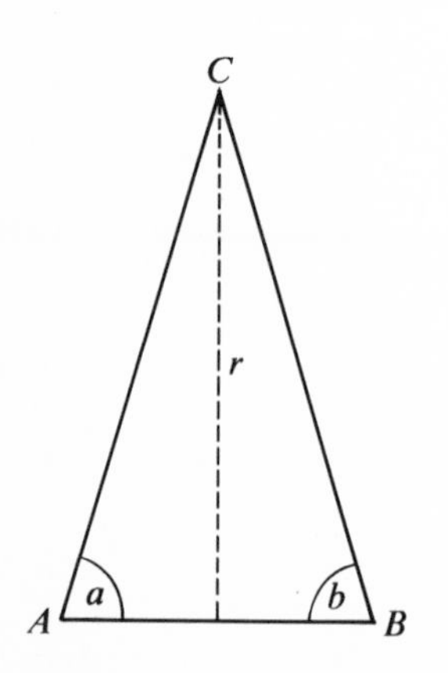

Figure 2.6. Triangulation. If the angles a and b are known and if the length of side AB is known, then the distance r may be calculated.

Historically this is the method that was actually applied in 1672 when simultaneous observations of Mars were made at Paris and at the French colony of Cayenne in French Guiana. The computed distance was approximately 0.4 A.U. and the measured distance was approximately 60,000,000 km. Thus

$$0.4 \text{ A.U.} = 60{,}000{,}000 \text{ km}$$

and the value of one astronomical unit is therefore approximately 150,000,000 km. The distance measured this way was not particularly accurate and, as a matter of fact, the earth-sun distance is still being refined today. Many techniques have been tried through the years. The most recently developed methods have employed radar waves reflected from Venus and the analysis of radio signals transmitted from spacecraft in orbits around the sun. All of the methods, however, are essentially similar in that the length of some distance known in astronomical units from orbital studies is measured in kilometers by some direct method. Referring again to Table 2.1, we need only multiply the orbital radii in astronomical units by 150,000,000 km per A.U. in order to find the radii in kilometers of all of the planetary orbits. Astronomers usually prefer to compute from the observations a quantity called the **solar parallax,** which is the angle which would be subtended by the equatorial radius of the earth if it could be seen from the sun. The most refined value for the solar parallax is 8.79415 arc sec ± 0.00005.

2.7 Proofs of the Earth's Revolution

Many men had speculated that the celestial sphere was only an illusion and that the stars were probably scattered randomly through space. Some stars should thus be closer to the solar system than others, and if the earth in truth revolved around the sun annually, these closer stars should be observed to show **parallax** or small periodic changes in position with respect to the more distant background (see Figure 2.7). The same effect may be seen when an object held at arm's length is viewed first with one eye and then with the other. The closer object seems to shift back and forth with respect to the distant background. Tycho Brahe was aware that many men before him had been reluctant to accept the sun-centered system simply because they were unable to detect the regular parallactic motions of stars. Recognizing that the observed effect would necessarily be very small, Tycho, nevertheless, felt that his equipment and techniques should be precise enough to reveal stellar parallax. When he could not detect this motion, he felt unable to accept the Copernican system. Even Tycho had overestimated the size of this very small effect, and it was not until 1838 that the parallax of a star was first measured. This was done by F. W. Bessel using rather specialized telescopic techniques. For the

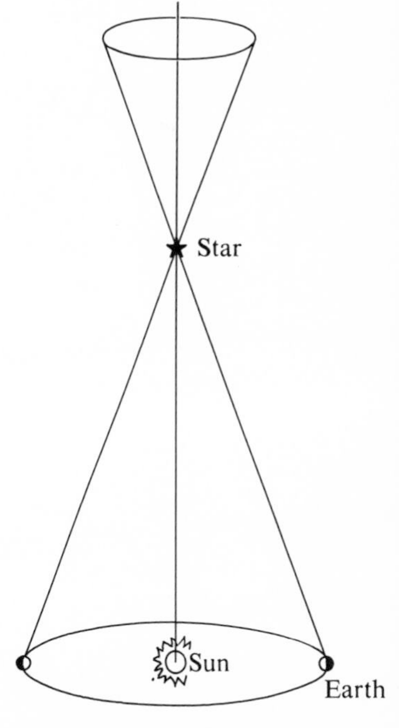

Figure 2.7. Parallax of a star as a proof of the earth's revolution. For the nearest stars this effect is extremely small.

nearest star, Alpha Centauri (visible in the skies of the southern hemisphere), the parallax angle is less than one second of arc—comparable to the angle between the front and rear bumpers of a full-sized automobile seen from a distance of 600 km. Such small angles are extremely difficult to measure with any degree of certainty, but the measurement of parallaxes provides the only direct means by which the distances to stars may be determined.

Since the earth's orbit is nearly circular, nearby stars at right angles to the plane of the earth's orbit appear to move in nearly circular paths against the background of more distant stars. Looking toward stars in the plane of the earth's orbit, the observer sees parallactic motion back and forth on a straight line, and looking in other directions in space, the observer sees parallactic motion along small elliptical paths.

A second proof of the earth's orbital motion was actually discovered accidentally in an early attempt to measure parallax. In 1726, while looking for systematic parallax effects in the positions of *nearby* stars, the English astronomer James Bradley noted systematic effects in the positions of *all* stars. With respect to his solidly mounted telescope the stars moved in little ellipses in the sky with a period of one year. Bradley correctly explained the stars' motions as a result of the combined motion of the earth in its orbit and the light traveling with a finite velocity. This **aberration** of starlight is exactly the same as an effect noted by anyone who has ever driven an automobile in a heavy snowstorm (Figure 2.8). If there is no wind, the snowflakes fall nearly straight down. As the car starts to move, however, the snowflakes appear to be coming from a point somewhere ahead of the car. When the driver of the car turns a corner the snowflakes appear to be coming from a new direction still ahead of the car. If the driver makes a complete circuit around a block and returns to his starting place, the snowflakes will have

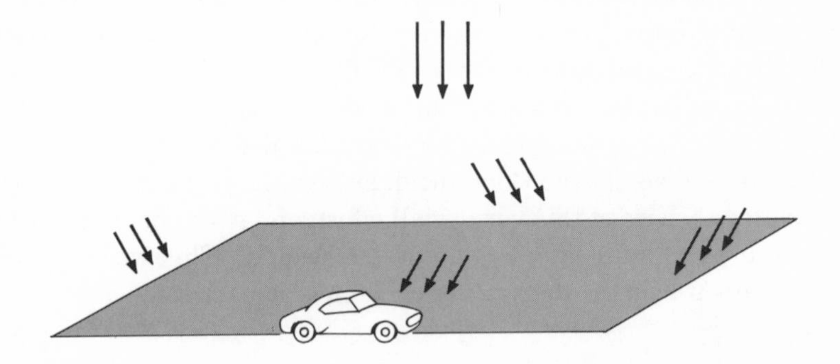

Figure 2.8. Aberration results from the combined motion of the car and the falling snow. Regardless of the direction in which the car moves, the snow always appears to come from a direction ahead of the car. The same effect is seen for starlight due to the earth's orbital motion.

appeared to come at him from four different directions. From the stopped car the driver sees the snowflakes coming straight down once more.

It is, of course, the motion of the car which causes the apparent change in the direction in which the snowflakes seem to be falling. The effect will be greater if the speed of the car is increased. Similarly the effect will be lessened if the snowflakes fall at a faster rate. If a is the angle through which the snowflakes seem to have been deflected, u the velocity of the car, and v the velocity of the snowflakes, then the expression

$$\tan a = \frac{u}{v}$$

shows the manner in which the angle a depends upon the two velocities (Figure 2.9). See Appendix A.1.

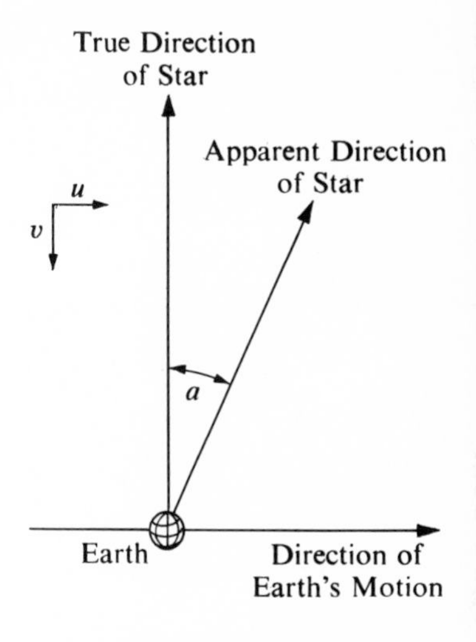

Figure 2.9. Aberration causes a change in the apparent direction to a star.

The same simple relation holds also for the aberration of starlight resulting from the earth's orbital motion if we let u be the earth's orbital velocity and v be the velocity of light. The observed value for a is 20.5 seconds of arc and is in good agreement with the value calculated from the accepted values for the earth's orbital velocity and the speed of light. (It should be noted here that observed aberration may be used in the above equation with a laboratory value for the speed of light to compute the earth's orbital velocity. With the length of the year known, the circumference of the earth's orbit easily follows. This method may be employed to find the earth-sun distance.)

2.8 Proofs of the Earth's Rotation

In Chapter 1 it was mentioned that the apparent daily motions of the sun, moon, and stars result from the earth's daily rotation on its axis. This explanation as an alternative to the daily rotation of the vast celestial sphere had occurred to men like Ptolemy, but they had rejected it for several reasons which may seem rather absurd to us today. They argued that if some body as large as the earth rotated so rapidly, it would surely fly apart. And secondly, they argued that if a ball were thrown straight up into the air, it would fall to the ground somewhat to the west of the point from which it was thrown. (A contemporary of Galileo helped to disprove this argument by having a rider on a speeding horse throw a ball straight upward and catch it again.) In spite of the fact that arguments such as these against rotation were not very convincing, actual proofs of rotation were not demonstrated until the first half of the nineteenth century. Within a span of only a few years, French physicists laid the groundwork for several convincing proofs.

Chronologically, the first proof of the earth's rotation resulted from the effort of French scientists to measure the size of the earth. Proceeding much as Eratosthenes had in Egypt many years earlier, surveyors measured the length of a long arc along a meridian of longitude. Latitudes in degrees were found from astronomical observations at many points along the arc. On a spherical earth there should always be the same number of kilometers between any two points one degree of latitude apart on a meridian. The surveyors found, however, that as they advanced northward the length of a degree of latitude became longer (Figure 2.10). This meant that the earth was flattened through the poles, that is, its polar diameter must be shorter than its equatorial diameter. Knowing that a body like the earth ought to assume a perfectly spherical shape, the French scientists reasoned that the earth could only be flattened by rotation. Rotation would cause the equator to become larger, and the polar axis would then become smaller.

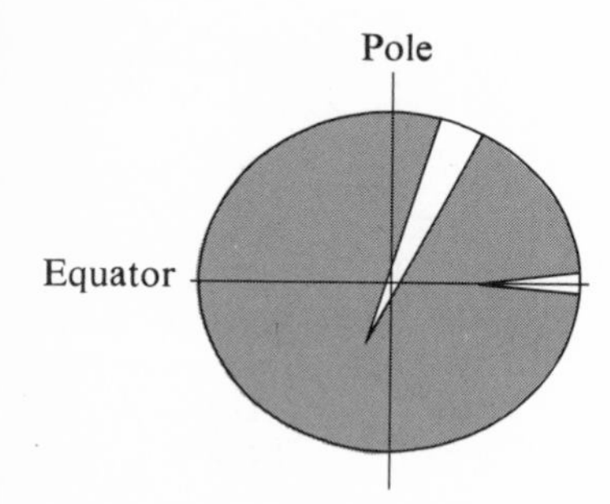

Figure 2.10. Because of the earth's flattened shape, the radius of curvature of the surface is greater in high latitudes than it is near the equator. Differences in latitude may be determined from differences in the direction of the vertical at various points.

As a result of the work of Newton it was known that the plane of oscillation of a free-swinging pendulum should remain fixed if left undisturbed. In 1851 J. B. L. Foucault constructed a pendulum in the Pantheon in Paris by hanging a heavy iron ball on the end of a 60-meter-long wire attached to the top of the domed roof. When this pendulum was allowed to swing in a true plane, it was seen that the plane did not remain fixed with respect to the floor. A pin on the bottom of the weight made a series of marks on the floor indicating that the plane of oscillation of the pendulum was moving in a clockwise direction. Foucault correctly reasoned that the floor of the room was actually turning counterclockwise under the pendulum as a result of the earth's rotation.

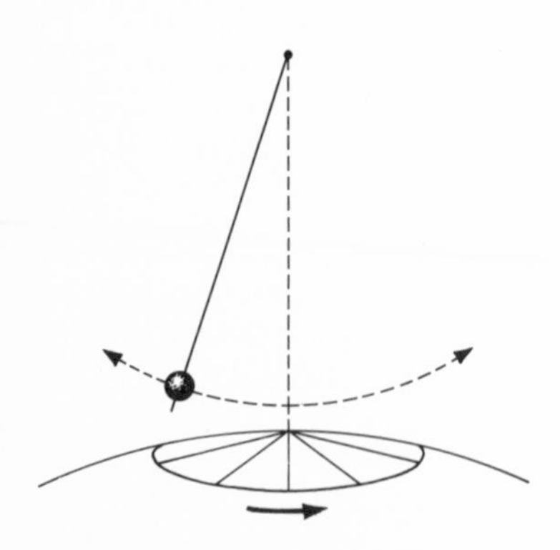

Figure 2.11. Foucault pendulum. The pendulum swings continuously in one plane as the earth rotates beneath it. The sketch depicts the experiment as it would appear if performed at the north pole.

This demonstration is usually more understandable if one imagines the Foucault pendulum experiment performed at the north pole of the earth (Figure 2.11). In such a case the constancy of the plane of oscillation would be clearly indicated by the fact that if the pendulum were started swinging toward some particular star it would continue to swing toward that star. Changes in the position of the plane noted with respect to points on the earth below the pendulum would clearly be the result of the earth's rotation. In twenty-four hours the earth's surface under the pendulum would have turned through a full 360°.

In Foucault's original experiment performed at mid-latitudes, the time required for one complete turn of the room under the pendulum was thirty-two hours. At the equator the time is infinite and the demonstration will not work at all. Detailed analysis shows that the time required at any point on the earth is given by

$$t = \frac{24 \text{ hours}}{\sin \phi}$$

where ϕ is the latitude of the point at which the experiment is performed.

In 1835 another French physicist and mathematician, G. G. Coriolis, did some important theoretical work on the motion of objects set in motion relative to a moving surface. Shortly afterward, other men made use of Coriolis' analysis to explain certain well-known natural phenomena as proofs of the earth's rotation. They showed convincingly that the pattern of trade winds and the rotation of hurricanes and tornadoes were results of the **Coriolis effect.**

The general direction of trade winds is indicated in Figure 2.12. Their explanation lies in the probable atmospheric circulation combined with the rotating earth. If warm air near the equator rises from the surface, air on either side of the equator will move toward the area of lower pressure near the equator. Other similar surface motions of air result from the descent of cooler air at other latitudes. This air circulation is also indicated in Figure 2.12. An

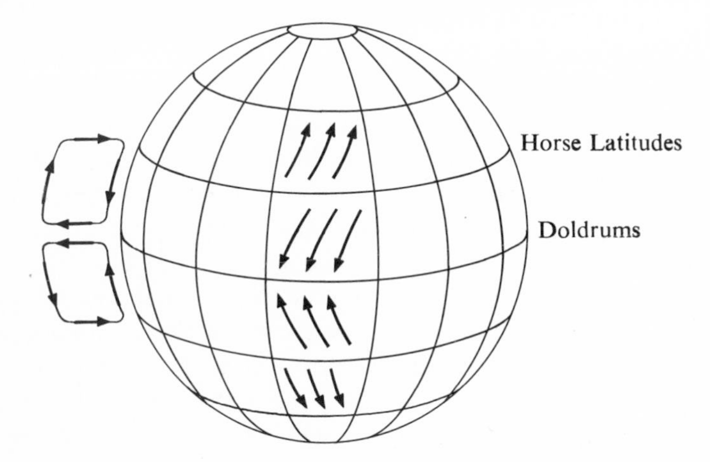

Figure 2.12. The pattern of trade winds in the Atlantic Ocean and the possible circulation pattern of warm air rising near the equator, cooling and descending at higher latitudes.

observer in the northern hemisphere looking toward the equator would expect the air to move due south. Instead, the trade winds seem to have been deflected to the right. In higher latitudes air moving northward likewise seems to have been deflected to the right. In the southern hemisphere similar deflections from the north-south line seem to be to the left.

In order to understand the Coriolis effect as an explanation of the trade winds and as a proof of rotation, consider two ships some distance apart in the northern hemisphere. As the earth rotates both ships are carried toward the east moving all the way around a large circle in twenty-four hours. The ship nearest the equator moves along a somewhat larger circle than the other one and must, therefore, have the greater eastward velocity of the two. Suppose now that ship A nearer the equator tries to hit B with a long-range shell. The shell travels northward with the muzzle velocity of the gun. It also has the original eastward velocity of the ship, however, so the shell moves toward the northeast. The target B is moved eastward also by the earth's rotation, but during the time of flight of the shell it does not move eastward as far as the shell. The result is that the gunner on ship A has the feeling that his shell has in some way been deflected to the right of the target B toward which he aimed. A gunner on ship B who now attempts to return the fire notices the same effect. His shell also seems to be deflected toward the right and hits west of A because A has a greater eastward velocity than that which has been imparted to the shell by the eastward velocity of B. Thus any object moving this way should be deflected toward the right in the northern hemisphere. In the southern hemisphere, the same analysis may be applied. It is readily seen in this case, however, that the deflection is toward the left.

In this manner the Coriolis effect explains the pattern of trade winds and offers a proof that the earth rotates on its axis. Other weather phenomena resulting from the Coriolis effect are the rotating windstorms such as tornadoes, hurricanes, cyclones, and typhoons. When a serious low-pressure area develops somewhere on the earth, air moves toward it along the earth's surface. The moving air is deflected to the right in the northern hemisphere with the result that a counterclockwise rotation develops. In the southern hemisphere such cyclonic storms rotate in the clockwise direction. It has generally been held also that the rotation in the drain of a wash basin is a small-scale manifestation of the same effect.

Somewhat similar to the Coriolis effect in its causes is another effect known as the **eastward deviation of falling bodies.** Greek philosophers had reasoned that if the earth rotates, the top of a tower should be moving eastward with a greater velocity than the base of the tower. Therefore, if an object were dropped from the tower, it should always land slightly to the east of the point directly below its origin. Ancients who actually tried this experiment noticed no deviation because the effect they were looking for was much too small to be detected without very precise measuring equipment under carefully controlled conditions. In modern times this experiment has been performed successfully in a deep mine shaft which was relatively free of disturbing air currents and where precise measurements were possible.

The earth's rotation may also be proven by observations of an earth satellite in a polar orbit (Figure 2.13). The orientation of the orbit plane will remain relatively fixed in space. On each successive pass overhead, however, the observer on the earth's surface sees the satellite farther toward the west. The earth must be rotating inside the satellite orbit. The communications satellites in "synchronous" orbits offer another proof. These satellites are placed in orbit directly over the equator, and the size of the orbit is chosen so that the satellite's orbital period will be 24 hours. If the satellite did not have a large eastward velocity, it would fall to the earth. So if some specific point on the earth is to remain directly below the satellite, the earth must be rotating toward the east.

Figure 2.13. A satellite in a polar orbit crosses the equator at points progressively farther west on successive revolutions around the earth. The plane of the satellite's orbit changes very slowly as the earth rotates inside the orbit.

2.9 The Sun's Apparent Path in the Sky

Anyone who cares to do so can make simple observations to show that the altitude of the sun above the horizon at noon varies throughout the year. An observer at latitude forty degrees north will see the sun only $26\frac{1}{2}°$ above the horizon at noon in late December and $73\frac{1}{2}°$ above the horizon at noon in late June. At noon on the twenty-first of March and the twenty-first of September, the same observer will see the sun 50° above his horizon. Earlier in this chapter it was mentioned that Ptolemy had used an armillary sphere to follow these motions of the sun with some degree of precision. In addition to the sun's north-south motion its eastward motion against the background stars may also be shown from simple observations. Although we cannot see the stars behind the sun during the day, we know from the changing nighttime pattern of constellations that the sun must move eastward regularly from day to

day. The apparent path of the sun as it moves throughout the year is called the **ecliptic,** and it forms a great circle* on the celestial sphere. The planets, in the course of their motions around the sun, are never far from the ecliptic. A band extending 9° on either side of the ecliptic and including all possible positions of Mercury, Venus, Mars, Jupiter, and Saturn is known as the zodiac. Twelve constellations along this belt are known as the signs of the zodiac.

Along the ecliptic four points have special significance and have been given indvidual names. These points are the equinoxes and the solstices. The **vernal equinox** is the point at which the sun moving northward crosses the equator. As we will see, this is a very important reference point on the celestial sphere. Directly opposite the vernal equinox on the sky is the **autumnal equinox,** where the sun moving southward crosses the celestial equator. The point at which the sun is at its greatest distance south of the equator is the **winter solstice.** The point at which the sun is at its greatest distance north of the equator is the **summer solstice.** These four points have been indicated in Figure 2.14. It is customary to specify the beginnings of the four seasons as the moments when the sun arrives at each of these four points in its annual journey along the ecliptic.

When we are able to understand and accept the proofs of the earth's orbital motion, we have no problem in understanding why the sun appears to move as it does along the ecliptic. In fact, the observed angle of 23½° between the plane of the earth's equator and the plane of the ecliptic indicates to us that the earth's equator is tilted 23½° out of the plane of its orbit. In Figure 2.15 the earth is shown in its summer and winter positions with respect to the sun (for a northern hemisphere observer). These are marked *S* and *W*, respectively. The earth's axis is shown tilted 23½° away from the perpendicular to the orbit plane. Even though the earth's winter position is 300,000,000 km from its summer position, the imaginary celestial sphere is so large that the earth's axis points toward the same point on the sky, namely the north celestial pole, throughout the entire year. In Figure 2.15a, representing the earth in summer, an observer at 40°N latitude is shown at *O*. The directions toward the celestial equator and the sun are indicated by arrows *E* and *S*. Clearly the observer at *O* sees the sun north of the equator.

*A great circle is formed by the intersection of a plane with a sphere in such a way that the plane includes the center of the sphere.

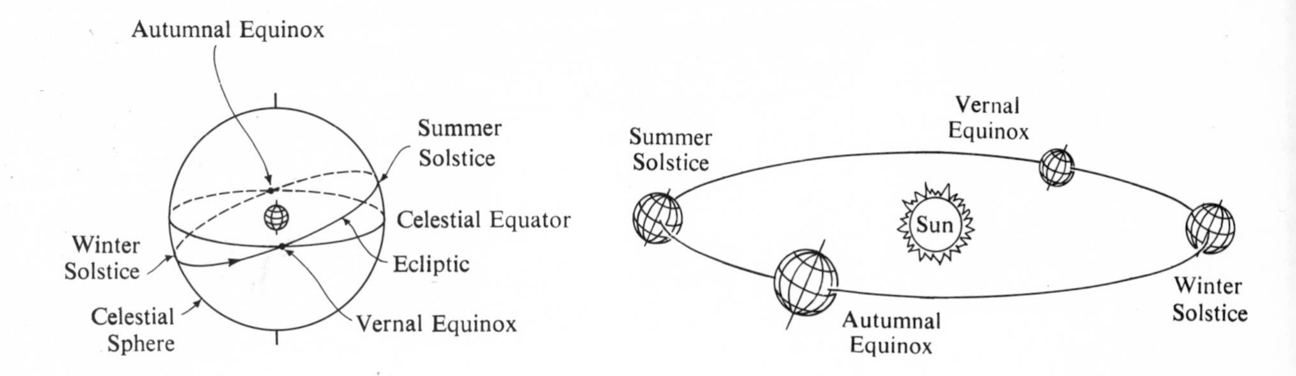

Figure 2.14. The sun's position on the celestial sphere changes as a result of the earth's orbital motion. As a result, the sun appears to move along the ecliptic.

The winter situation is indicated in Figure 2.15b, and the sun now appears to be south of the celestial equator. Obviously then, as the earth progresses around the sun the inclination of its axis causes the sun to appear to move north and south with respect to the celestial equator and eastward with respect to the constellations.

2.10 Seasons

In both Figures 2.15a and 2.15b the observer at midnight is at O', and he is looking toward opposite sides of the celestial sphere in the two cases. A line crossing the earth at right angles to the direction to the sun separates the day and night halves of the earth. As the earth rotates the observer is carried progressively from O to O' and back to O in twenty-four hours. In the earth's summer position the observer is on the sunward side of the day-night line for a longer time than he will be on the dark side. The reverse is true when the earth is in its winter position. Thus the duration of daylight varies throughout the year. It should be noted on Figures 2.15a and 2.15b also that this effect becomes progressively greater as the observer moves farther and farther from the equator.

With reference to the same figures, a tangent to the earth at O represents the observer's horizon, and it may be noted that the angle between the horizon and the direction to the sun becomes larger as the earth moves from its winter position to its summer position. As stated earlier, this angle may vary from $26\frac{1}{2}°$ to $73\frac{1}{2}°$ for an observer at 40°N latitude. When the sun is high in the observer's sky, the energy in a beam of unit cross section is concentrated into a considerably smaller area than when the sun is low in the sky (Figure 2.16).

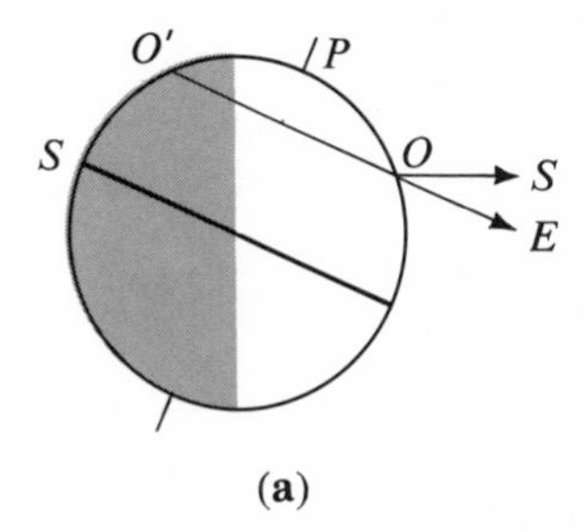

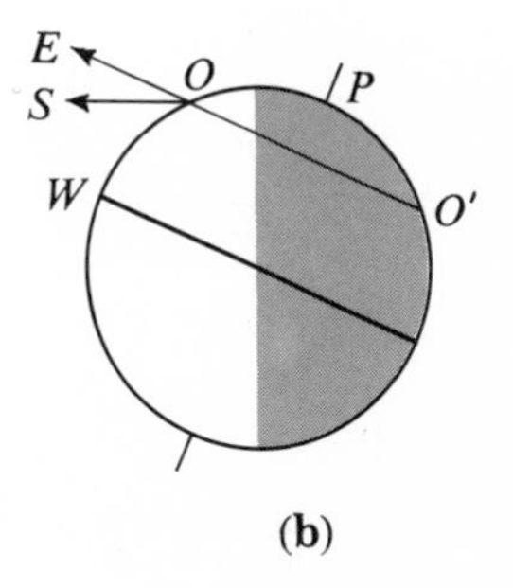

Figure 2.15. Because of the earth's orbital motion, the sun moves north and south of the celestial equator.

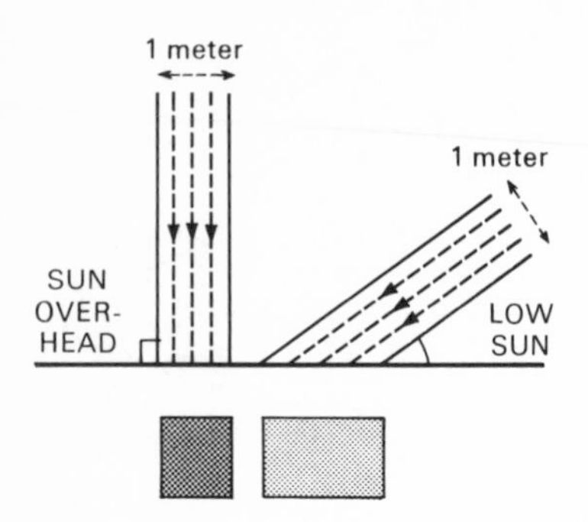

Figure 2.16. Energy in a beam of sunlight is concentrated in a smaller area if the beam strikes the surface vertically than if it strikes the surface at an angle. (Wyatt, p. 58.)

These two factors, the altitude of the sun at noon and the duration of daylight, are the principal factors which bring about the annual cycle of seasons on the earth. During the long days of summer the ground and the air are efficiently warmed by the sun. More solar energy is absorbed during the day than is radiated away at night, and the general level of heat in a given region increases. Both factors result, as we have said, from the fact that the earth's equator does not lie in the plane of its orbit. The cycle of seasons is therefore a direct consequence of the earth's orbital revolution.

2.11 Sidereal Time

Another consequence of the earth's annual revolution around the sun is related to the methods by which we reckon time. We are accustomed to measuring the rotation of the earth with respect to the sun. The interval from one noon to the next defines our twenty-four-hour day. As may be noted in Figure 2.17, however, this is not really the true rotation period of the earth. Since the earth's revolution period is just over 365 days, the earth moves slightly less than one degree along its orbit each day. The result is that after one noon passage of the sun, the earth must make slightly more than one rotation on its axis before the sun passes overhead a second time. The true rotation period of the earth may, however, be noted by choosing as a reference some point such as the vernal equinox on the celestial sphere. Time reckoned this way on the true rotation period of the earth with respect to the stars is referred to as **sidereal time** or star time. The vernal equinox is the reference point for the measurement of sidereal time, and it is customary to count sidereal hours from 1 to 24 rather than from 1 to

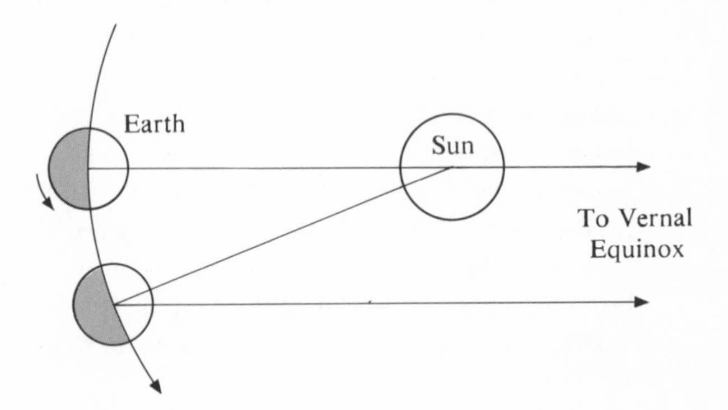

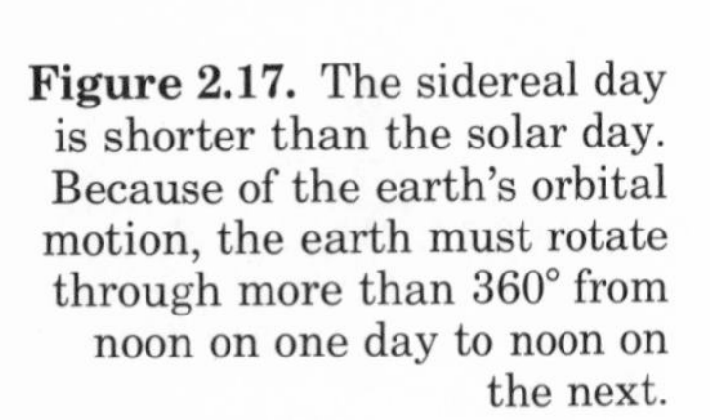

Figure 2.17. The sidereal day is shorter than the solar day. Because of the earth's orbital motion, the earth must rotate through more than 360° from noon on one day to noon on the next.

12. Our conventional **mean solar time** is based on the rotation of the earth with respect to the sun. Examination of Figure 2.17 shows that the earth must rotate through about one degree more than a complete rotation in order to complete a solar day. The solar day is therefore longer than the sidereal day. Since 360° is equivalent to twenty-four hours, 1° is equivalent to four minutes of time. Thus the solar day is approximately four minutes longer than the sidereal day. Because of the angle between the ecliptic and the celestial equator and because of the varying velocity of the earth in its orbit, the length of the solar day (from noon to noon) is not constant through the year. We adjust our clocks, however, to run at a mean or average solar rate.

QUESTIONS

1. Describe fully the great contribution of Tycho Brahe to astronomy and discuss the significance of his contributions.
2. Show how the apparent paths of the planets can be plotted from a series of precise measures of the angle between the directions to the planet and to selected stars on the celestial sphere.
3. State Kepler's three laws of planetary motion.
4. Which of Kepler's laws implies that the orbital speed of a planet must vary continuously?
5. How can Kepler's laws be used to derive a scale model of the solar system?
6. How did the observations made by Galileo with his crude telescopes tend to support the heliocentric system of Copernicus?
7. How did the work of Isaac Newton tend to support the works of Kepler and Copernicus?
8. What sort of observations are required in order to establish the scale of the solar-system model determined from Kepler's laws?
9. Describe fully one proof of the earth's revolution. Name one other proof.
10. Describe fully one proof of the earth's rotation. Name two other proofs.
11. Show with a sketch how the orbital motion of the earth causes the sun to appear to move on the celestial sphere.
12. What is the basic factor causing the cycle of seasons on the earth?

Figure 3.1. A spectacular solar halo caused by refraction of sunlight by ice crystals in high, thin clouds. (Photographed in Alaska on April 27, 1966 by Theodore Loder.)

3

Light

Since most of the astronomer's data concerning the sun, moon, planets, stars, and galaxies are derived from an analysis of the light received on earth from these objects, an understanding of some of the basic characteristics of light is important. The physicist recognizes light as electromagnetic radiation to which the human eye is sensitive. He has confirmed the nature of this radiation in a number of ways, and has related the visible radiation to the invisible ultraviolet, infrared, x-ray, and radio radiations. There are several ways in which we have come to think of light. We speak of light rays, light waves, and photons (discrete "particles" of light). The most meaningful concept to use depends upon the discussion at hand, and we shall have occasion in this chapter and in Chapter 4 to consider light in each of these three ways.

3.1 The Velocity of Light

Long before Galileo, men had speculated on the exact nature of light and the manner in which light traveled through the air. Much effort was devoted to attempts to measure the velocity with which light traveled from source to observer. Galileo and an assistant tried some crude experiments with lanterns and proved only that the velocity of light was very great. Some philosophers had even speculated that the velocity of light was infinite, that is, that light traveled instantaneously from source to observer.

The first determination of a velocity reasonably close to today's accepted value was obtained unexpectedly from a peculiarity noted in a routine series of observations of the satellites of Jupiter. In the middle of the seventeenth century, astronomers at the Paris Observatory had been attempting to make an accurate determination of the revolution periods of Jupiter's four inner satellites. In order to do this they noted the moment that a satellite disappeared into Jupiter's shadow, and timed the interval during which the satellite reappeared, passed in front of Jupiter, and disappeared into the shadow again. For the innermost satellite, Io, this period was about one day and eighteen hours, or 1.75 days. The first observations had been made when Jupiter was at opposition or on the opposite side of the sky from the sun. (Jupiter is at opposition when the earth is at position *A* in Figure 3.2. At such times Jupiter is high in our sky at midnight and is thus easy to observe.) Later on, observations of eclipses of the same satellite were made when the earth was near point *B* in Figure 3.2, and Jupiter was nearly in conjunction with the sun. The moments at which the eclipses then occurred were always late on the basis of the revolution period established when Jupiter was at opposition.

With reference to Figure 3.2, the preliminary revolution period of Io was determined when the earth was at position *A* in its orbit. On the basis of this value for the period it was possible to calculate the moments when future eclipses should occur. For example, if an eclipse had been observed to occur at midnight on the first of June and the preliminary period was 1.75 days, then in an interval of 175 days the satellite should have made 100 revolutions. If the period is really 1.75 days, an observer should see the satellite disappear into the shadow at midnight on November 23. (A correction must, of course, be made because of the orbital motion of Jupiter. For the sake of simplicity, however, this has been ignored here.) As the earth moved on toward point *B*, however, the eclipses occurred progressively later than their predicted times. After the earth had passed point *B*, the amount by which the eclipses were late became less and less until when the earth had returned to point *A*, the eclipses once more occurred at the times predicted on the basis of the preliminary period.

Ole Römer, a promising Danish astronomer, had been brought to the Paris Observatory under the auspices of the newly formed

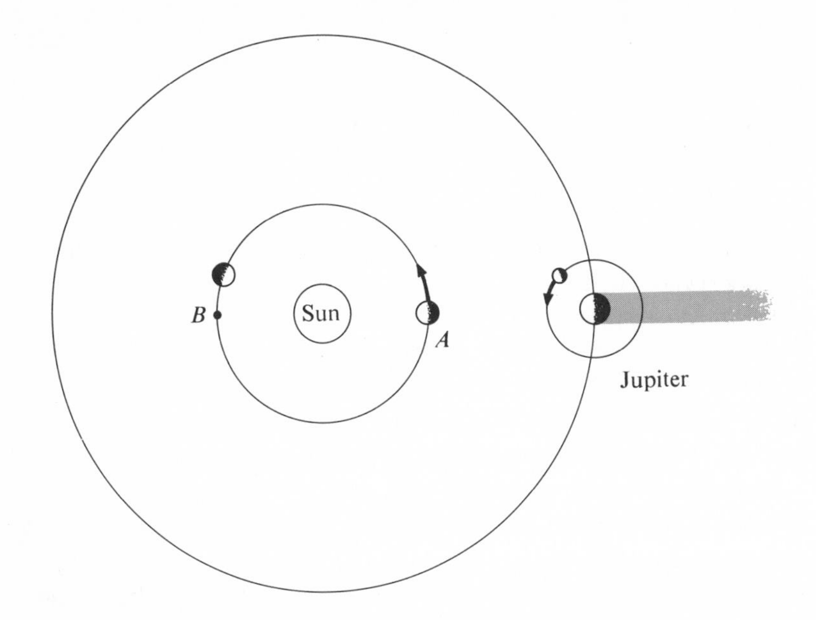

Figure 3.2. As the earth moves from A to B, the satellite appears to emerge from the shadow at moments progressively later than the predicted times.

French Academy. It was Römer who, in 1675, explained the delay in eclipse times as due to the longer time required for the light from the satellite to reach the earth in one position *(B)* than in another *(A)*. Knowing the maximum amount by which the eclipses were delayed and knowing the value of the earth-sun distance found from observations of Mars (Chapter 2), Römer then calculated the velocity of light. If the maximum delay (i.e., the time required by light to travel the diameter of earth's orbit) were about 1000 seconds (16 minutes), and if the earth-sun distance were 150,000,000 km, it followed that the velocity of light must be 300,000 km/sec. (With binoculars and an accurate clock an enterprising student can repeat these observations for himself.) The actual velocity of light determined by Römer was a good approximation to the modern value (299,792.462 km/sec) found from later laboratory methods. Great precision in his result was, of course, impossible since the earth-sun distance was not well known at that time. It is the method, however, which is interesting and significant.

Römer's determination of the velocity of light may be offered as another proof of the earth's revolution. Under the Ptolemaic system the distance of Jupiter from the earth would remain nearly constant throughout the year and no delay in the eclipse times should occur. Acceptance of Römer's explanation and computation of the velocity of light implies previous acceptance of the concept that the earth revolves around the sun.

3.2 Reflection and Refraction

Besides the speed with which light travels, the direction in which it travels and the means by which this direction can be changed are of importance. In this connection it is convenient to speak of light **rays** and to assume that these rays are simply straight lines along which the light travels. In elementary physics courses we learn that the direction of a light ray may be changed in either of two ways: by reflection or by refraction. Some spectacular reflection and refraction effects in nature may be seen in Figure 3.1.

Reflection occurs when light strikes any surface, with the amount of light and the colors of light reflected depending upon the nature of the surface. The shiny silvered surface of a mirror reflects nearly all of the light incident upon it. On the other hand, a dull black surface such as an asphalt highway reflects very little light. Other surfaces reflect and absorb selectively. Thus we say an object is green if it reflects green light and absorbs all of the other colors.

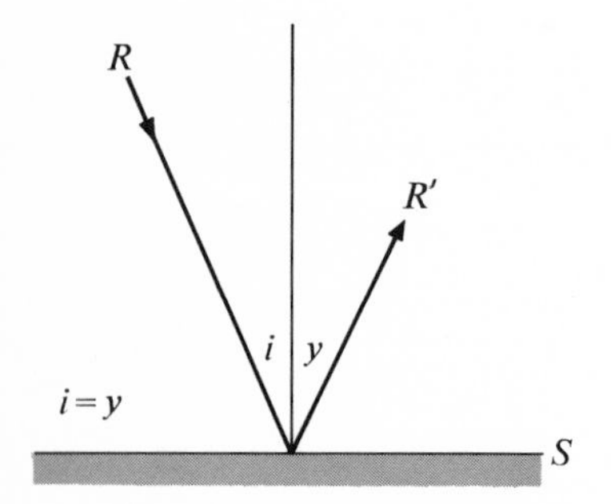

Figure 3.3. Reflection of a light ray from a plane surface.

When a light ray strikes a smooth surface, the ray will be reflected away from that surface in a specific direction, as indicated in Figure 3.3. The direction in which the light ray R strikes the surface S is specified by the angle i between the ray itself and the direction normal to the surface. Angle i is referred to as the **angle of incidence.** The angle y at which the reflected ray R' leaves the surface is known as the **angle of reflection.** The angle of reflection is always equal to the angle of incidence and the reflected ray always lies in the same plane as the incident ray and the normal.

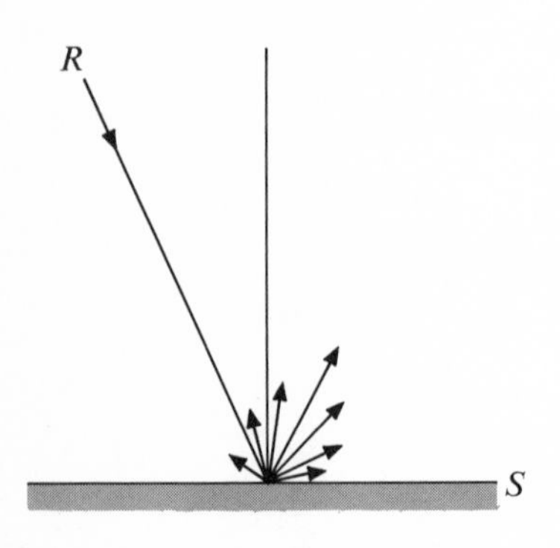

Figure 3.4. Diffuse reflection.

Most surfaces do not reflect light in quite the manner described above. A sheet of white paper, for instance, is better described as a **diffuse reflector.** It does not reflect an image as does a mirror, but it scatters the incident light in all directions. The sheet of white paper will be visible to a person looking at it from any direction even though all of the incident light may be reaching the paper in a single beam of parallel light, as in Figure 3.4. The arrows representing the reflected rays indicate that light leaves the surface in all directions. The longer arrows indicate that a greater percentage of light is scattered in the same general direction as the incident ray. The detailed nature of a surface determines the degree to which the incident light is scattered in the forward direction. The surfaces of the moon and planets are diffuse reflectors, and astronomers have been able to speculate on the fine structure of the moon's surface from the manner in which it reflects and scatters light.

The direction of a ray of light is said to be changed by **refraction** when the light ray passes from one transparent medium such as air into another transparent medium such as water or glass. As in the case of reflection, the refracted ray behaves in a certain predictable way. In Figure 3.5 the direction of the incident ray R has been altered as it passed from its original medium into a second medium at surface S. The amount by which the incident ray is bent will depend on the **index of refraction** of each of the two media. The letters N_1 and N_2 in Figure 3.5 indicate the indices of refraction of the media above and below the surface S. The exact amount by which the ray will be bent is given by Snell's law:

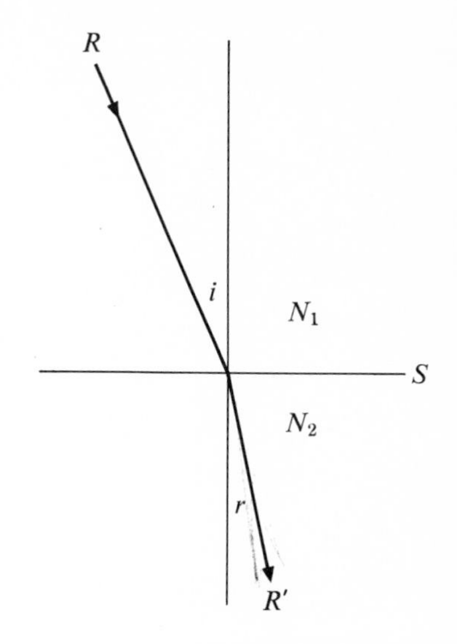

Figure 3.5. Refraction of a light ray passing from a less dense medium into a more dense medium.

$$N_1 \sin i = N_2 \sin r$$

where r, as shown in Figure 3.5, is the angle between the refracted ray and the normal. The index of refraction N for any medium is actually dependent upon the velocity of light in that medium. The higher the number, the slower the velocity of light. Thus the index of refraction of one type of glass is about 1.5; for water, 1.3; and for air, only slightly larger than 1.0. The index of refraction of air depends strongly upon the temperature, pressure, and water vapor content of the air. Another important characteristic of transparent media is that the exact value of the index of refraction really depends upon the color of the incident light. In glass, for example, the index of refraction for red light is less than the index for blue light. In the glass, the blue light travels more slowly than the red light. This makes it possible to use a glass prism to separate the individual colors in a beam of white light.

The common lenses used in eyeglasses, microscopes, telescopes, and other optical devices simply refract light in some desired manner. In the case of refracting telescopes, the function is usually to bring parallel rays of light to a single focus behind the lens.

3.3 Waves and Photons

A number of experiments involving interference and other phenomena suggest that light travels in **waves,** and the concept of light

Figure 3.6. Graphic representation of wave motion. ►

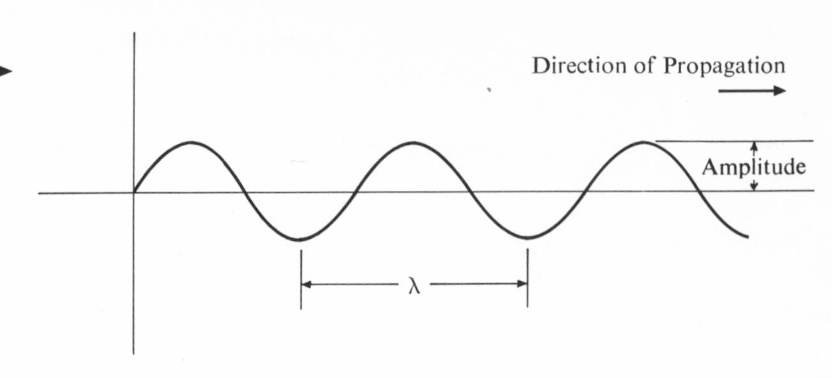

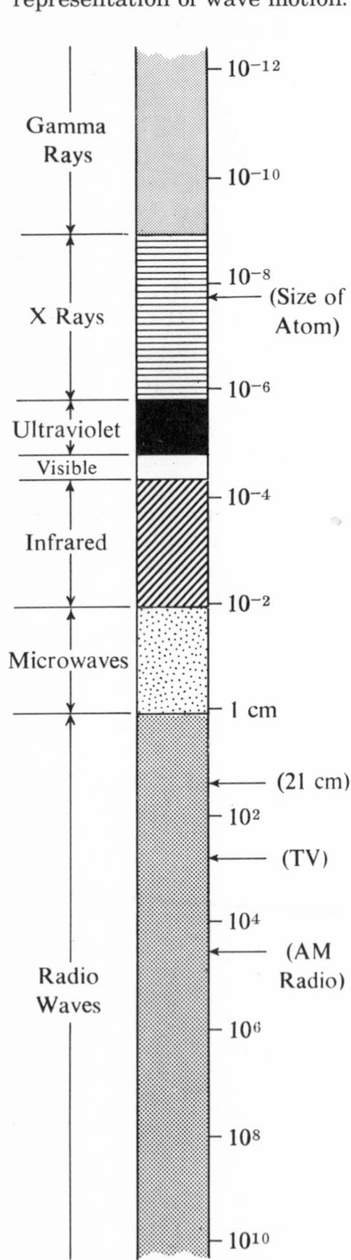

Figure 3.7. The electromagnetic spectrum from gamma rays to radio waves.

waves finds many applications. Wave phenomena imply periodicity, and Figure 3.6 shows one common representation of such periodicity. Light is a form of energy; the curve represents the variation in the amount of energy passing a given point as a function of time. The distance between crests on this curve is called the wavelength and is usually represented by the Greek letter λ (lambda).

Wavelengths of light are conveniently measured in angstrom units and range from about 3400 Å in the violet to about 7000 Å in the deep red. In more familiar units of length, one angstrom unit is equal to 0.000 000 01 cm (10^{-8} cm). Waves shorter and longer than those of this range do not stimulate the retina of the eye and are therefore invisible. Such waves in the ultraviolet and infrared, respectively, are detected to a limited extent by photography. Figure 3.7 represents a large portion of the electromagnetic spectrum and indicates the radiation processes and detection methods which apply to radiation of a wide range of wavelengths. It should be noted that the wavelengths representing visible light cover an extremely small part of the complete electromagnetic spectrum. Regardless of wavelength, all of these radiations travel at the same velocity, namely 300,000 km/sec, in the vacuum of space.

Other physical experiments give rise to the concept that light travels in discrete particles called **photons.** The amount of energy represented by a photon depends on the radiation process from which the photon originated. Photons representing a certain quantity of energy give the eye the sensation of blue, while photons representing less energy give the eye the sensation of red. To relate waves and photons we may then say that the greater the amount of energy possessed by a photon, the shorter is the wavelength of the detectable radiation. Blue light is therefore associated with short wavelengths and high-energy photons. Red light is associated with longer wavelengths and less energetic photons. Photons of extremely low energies are detected as radio signals.

3.4 Optical Spectra

It was stated earlier in this chapter that the index of refraction in transparent media is greater for blue light (short wavelengths) than for red light (longer wavelengths). Refraction may therefore be used to separate light of various colors travelling together in a single beam. Newton showed many years ago that an ordinary beam of white light was actually just such a mixture of light of all colors from violet to red. In Figure 3.8 a ray of white light is dispersed by a simple glass prism into a **spectrum** consisting of a regular progression of colors such as that seen in a rainbow. The spectroscope, the instrument used to disperse light into spectra, will be discussed in the next chapter.

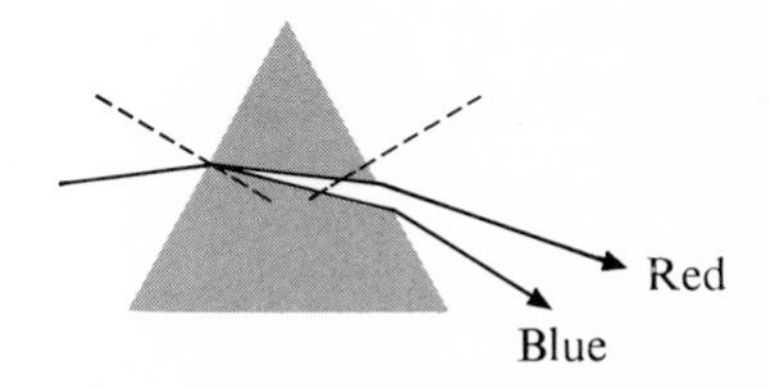

Figure 3.8. Dispersion of light in a prism. Blue light is refracted more than red light because the index of refraction is larger for shorter wavelengths.

Physicists in the middle of the last century had seen that observed spectra ordinarily were of three types: (1) continuous spectra from white-light sources showing a complete array of colors; (2) dark-line spectra similar to the continuous, but showing colorless gaps or spaces at certain positions; (3) bright-line spectra consisting of light of only a few particular colors and dark at all other points. Gustav Kirchhoff (1824–1887) stated the following three laws which specify the conditions under which each type of spectrum will be seen:

1. A *continuous spectrum* is seen in the light emitted from a hot solid or liquid or a gas under high pressure.
2. A *dark-line spectrum* is seen when the source of a continuous spectrum is viewed through a cool gas under low pressure.
3. A *bright-line spectrum* is seen in the light emitted from a hot gas under low pressure.

These three types of spectra and their sources are sketched in Figure 3.9, and examples of emission and absorption spectra are shown in Color Plate 1. An inexpensive hand spectroscope may be used to demonstrate these spectra in the classroom. The filament of an incandescent light bulb is a hot solid and therefore shows a continuous spectrum. The mercury vapor in most fluorescent lamps is a gas, and when excited by an electric current emits a bright-line spectrum. The scattered sunlight coming from any point in the sky

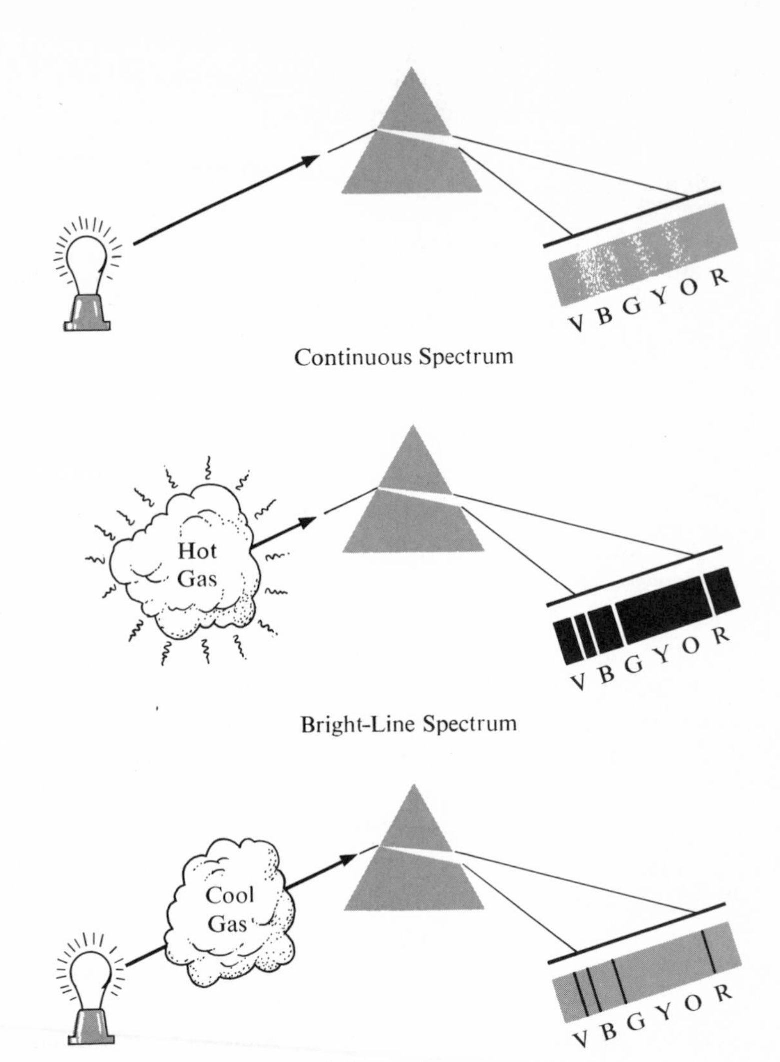

Figure 3.9. The circumstances under which the three types of spectra may be seen.

(even on a cloudy day) shows a dark-line spectrum since gases in the earth's atmosphere and in the sun's atmosphere lie between the sun's surface and the observer and are cooler than the sun's surface.

One of the most significant factors relating to bright-line and dark-line spectra is that they may be used to identify positively the gas (or the vapors of normally solid elements) which emitted the light. The pattern of lines in the spectrum of each element is unique and results from certain characteristics in the atoms themselves. These patterns never change whether the source is a few feet from the spectroscope in the laboratory or thousands of light-years away in space. The spectra of a few common chemical elements are shown in Color Plate 1. The pattern of lines will be the same whether the gas is isolated and emitting its own light or is in front of some hot solid and therefore causing dark lines to be superimposed on the continuous spectrum. A chemist can study the spectrum of light emitted when an unknown sample is vaporized and make positive identifications of the elements in the sample. Or the astronomer can study the spectrum of a star and tell from the patterns of lines what elements are present as gases in the atmosphere of the star. In later chapters it will be shown that the astronomer is able to tell much more about a star from its spectrum than chemical composition alone.

Returning to the concept of wavelength and the earlier mention that observed color is an indication of the wavelength, it should be clear that the wavelength of any spectrum line, either dark or bright, may be determined. The scale on the bottom of Color Plate 1 is calibrated in angstrom units, and from this scale the approximate wavelengths of the spectrum lines may be read off. In high-dispersion spectra the measured distance between two selected wavelengths is relatively large, and the wavelength of some specific spectrum line may be determined with great precision.

3.5 Doppler Effect in Optical Spectra

The pattern of lines and the wavelengths of the lines in optical spectra are dependent upon the nature of the atoms emitting or absorbing the light. Thus the wavelengths are fixed and unchangeable. There is, however, an important factor which can cause small changes in the positions of spectrum lines. This is the radial velocity, which may be defined as the velocity of the light source along the line of sight from the observer to the source. Or it may be said to be the rate of change of the distance between the observer and the source. The radial velocity is defined as positive when the distance is increasing and negative when the distance is decreasing.

If a light source is not moving relative to the observer, the number of waves per second reaching the observer will be given by

$$\frac{300{,}000 \text{ km/sec}}{\lambda} = \text{waves per second}$$

where λ is the wavelength. If the source is approaching the observer, however, a greater number of waves per second will reach his eye, and he will have the impression that the wavelength is shorter. Similarly, if the source is going away from the observer, fewer waves per second will reach his eye, and he will have the impression that the wavelength is longer. This change in wavelength resulting from the relative motion of the source and the observer is known as the **Doppler effect** after Christian Doppler, a Czech physicist of the last century who described these effects for sound. The Doppler effect results in a systematic displacement of the spectrum lines away from their normal positions. The lines will be shifted toward the blue if the source is approaching (or if the observer is approaching the source) and the lines will be shifted toward the red if the source is receding (or if the observer is receding from the source).

Doppler himself tried to extend the principle from sound to light and said, incorrectly, that stars which appear red are going away and stars which appear blue are coming toward us. He overestimated the effect by a considerable amount; in fact, the Doppler effect is not detected as a change in a star's color at all. If the velocity of some object is v and the speed of light is c, the displacement of some spectral line $\Delta\lambda$ is

$$\Delta\lambda = \lambda \frac{v}{c}$$

This is actually only an approximate formula, but it is accurate when the velocity v is only a small fraction of c. The amount by which the strong red line in the hydrogen spectrum ($H\alpha$) would be shifted by a rather large radial velocity of, say, 300 km/sec may be computed as an example. The wavelength of the $H\alpha$ line is 6563 Å and so the displacement is

$$\Delta\lambda = 6563 \frac{300}{300{,}000}$$

If the arithmetic is carried out, $\Delta\lambda$ is 6.6 Å. Referring again to Color Plate 1, it may be seen that even on the scale of this diagram 7 Å is measurable but very small. In spite of the very small shifts due to the Doppler effect, radial velocities of celestial bodies are routinely measured by astronomers. Furthermore, radial velocities are measured with surprising accuracy even when the actual radial velocity is considerably smaller than the 300 km/sec used in this example. It is absolutely necessary that the student understand this thoroughly, because the Doppler effect is of fundamental importance in astronomy.

3.6 Spectroscopic Proof of the Earth's Revolution

The Doppler effect gives us a measure of the radial velocity of a light source, and, as we have said, this radial velocity can result either from the motion of the observer or the motion of the light source. The orbital motion of the earth should, therefore, be reflected as a Doppler shift in the spectra of stars. With reference to Figure 3.10, the light from a distant star in the direction of the top of the page of this book reaches the earth in its orbital positions *A*, *B*, *C*, and *D*. When the earth is at position *A*, its orbital motion carries it in the direction of the arrow or toward the distant star. The result of this motion is that the lines of the star's spectrum are shifted slightly to the blue. Six months later, when the earth is at point *C*, the lines of the star's spectrum will show a displacement toward the red since the orbital motion is away from the star. At points *B* and *D* there is, of course, no part of the earth's motion directed toward the star, and the spectrum lines show no shift. The Doppler shift thus varies from zero at *B* and *D* to maximum positive and negative values at *C* and *A* and thus proves that the earth is actually in motion around the sun. Furthermore, the actual orbital speed in kilometers per second is proportional to the maximum Doppler shift observed at *A* and *C*. This orbital speed multiplied by the number of seconds in a sidereal year (31,558,150) gives the circumference of the earth's orbit in kilometers. The radius of the earth's orbit may be derived from the circumference, and thus the earth-sun distance could be found spectroscopically even if the earth were the only planet in orbit about the sun.

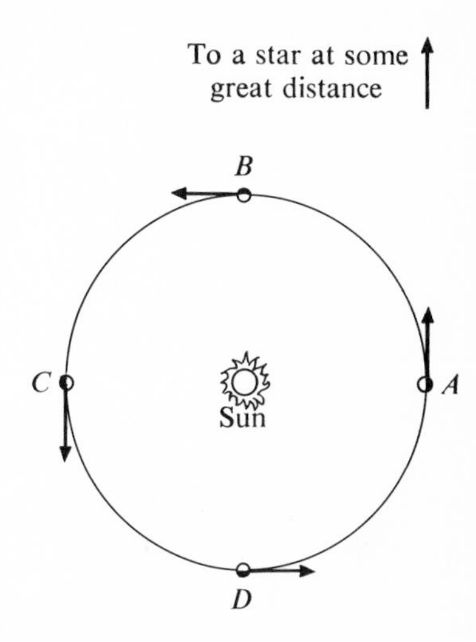

Figure 3.10. Because of the earth's orbital motion, the spectra of stars show small, seasonal Doppler shifts.

Astronomers attempting to measure the radial velocities of stars must obviously correct their observations for the radial component of the earth's orbital speed at the time of the observation. Another variable component which complicates the determination of stellar radial velocities arises from the daily rotation of the earth on its axis. The earth's rotational speed is, of course, well known and corrections for both rotation and revolution effects are easily applied.

3.7 Theoretical Structure of Atoms

We have spoken about the characteristics of light and the concepts of light waves and photons. We have mentioned also that the spectra of light emitted or absorbed by gases are unvarying characteristics of those gases. Since the gases are composed of atoms, it may be assumed that the differences in spectra of the chemical elements arise from basic differences in structure of the atoms themselves. An understanding of the structure of an atom might help us to understand how that atom absorbs and emits light and why the spectrum of each gas is different from all others. We cannot hope to obtain an actual microscopic view of a single atom, and so the best that we can do is try to devise a theoretical model of an atom. Such a model atom should have characteristics which will permit us to predict the appearance of its spectrum, and it should be consistent with other physical and chemical knowledge of that atom.

A successful theoretical model for the hydrogen atom was suggested in 1913 by the Danish physicist Niels Bohr (1885–1962). The hydrogen atom is the simplest of all atoms, consisting of just one proton and one electron. Astronomers know that hydrogen is also the most abundant element in the universe. It is appropriate that the simplest atom should have been the first for which a model was attempted. This model was then extended to other more complex atoms.

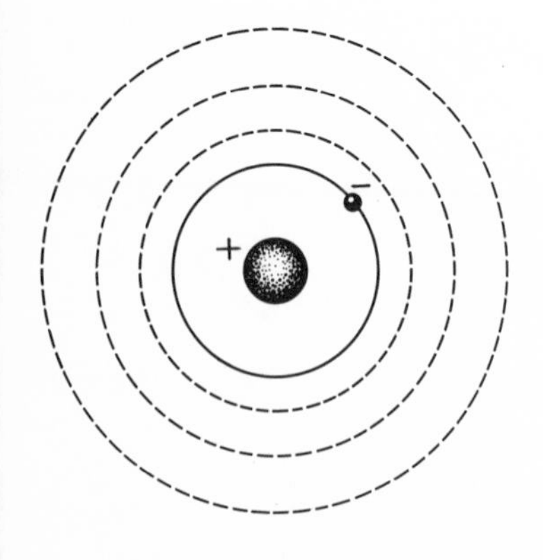

Figure 3.11. The Bohr model of the hydrogen atom.

In the Bohr model a single electron is said to be in orbit around the hydrogen atom's simple nucleus. This normal orbit is referred to as the ground state and is shown as the solid ring around the positive nucleus in Figure 3.11. A number of dashed rings are also shown to represent other possible orbits in which the electron may exist. If the electron should find itself in one of these larger orbits, it will return very quickly to the smallest orbit or ground state. In order that the electron may go from the ground state to any higher level, the atom must gain a certain amount of energy. When the atom has received energy and the electron has gone to a higher orbit, the atom is said to be **excited.** If the atom receives too much energy, the electron becomes completely detached and the atom is then **ionized.** One of the important points in Bohr's model is that only certain specific orbits are possible. This means, then, that the atom can receive energy only in certain discrete amounts. If an atom suffers a collision with an electron, for example, some of the kinetic energy of the colliding electron's rapid motion is transferred to the atom. Only the amount required to raise the atom to one of its particular energy levels can be absorbed. The colliding electron rebounds with its original kinetic energy reduced by the amount used to excite the atom. The possible states to which the atom may be excited are referred to as **energy levels,** and any atom undergoes an upward **transition** when it absorbs energy as described above.

The excited atom becomes de-excited again in a very short interval of time; during this process the orbital electron makes a downward transition and returns to the ground state. The atom now possesses less energy than when it was excited, and it loses this energy by emitting light. Since the atom could make only certain discrete upward transitions, it can likewise make only the same downward transitions. Light will be emitted, therefore, only in definite pulses, and these are the photons or particle-like bits of energy referred to earlier in this chapter. The amount of energy possessed by a particular photon will be determined by the size of the transition from which the photon resulted. A large transition gives rise to an energetic photon which may, for example, be detected as blue light. A shorter transition similarly gives rise to a less energetic photon which may be detectable as red light.

Bohr based his theory largely on the optical spectrum of hydrogen in which the pattern of lines is relatively simple. In Figure 3.12 a series of numbered horizontal lines represent the various possible energy levels in the hydrogen atom. A few upward transitions are indicated on the left. Upward transitions may actually originate at any energy level, but the atom must already be excited to some degree if the transition is to originate anywhere above the ground state. Downward transitions begin at the level to which the atom has been excited and may end in any lower level. Figure 3.12 shows two groups of downward transitions ending respectively in the ground level and in level 2. Consider now those transitions ending in level 2. Of these transitions, $6 \rightarrow 2$ will clearly result in the most energetic photon. If a large number of hydrogen atoms are making this transition, violet light will be seen. Large numbers of atoms making the other transitions, $5 \rightarrow 2$, $4 \rightarrow 2$, and $3 \rightarrow 2$, will give rise to light of other colors of longer and longer wavelengths. In the actual spectrum, therefore, color is seen only at certain specific places.

The lower half of Figure 3.12 shows the normal hydrogen spectrum. Hα is a strong line in the red at 6563 Å and the others are

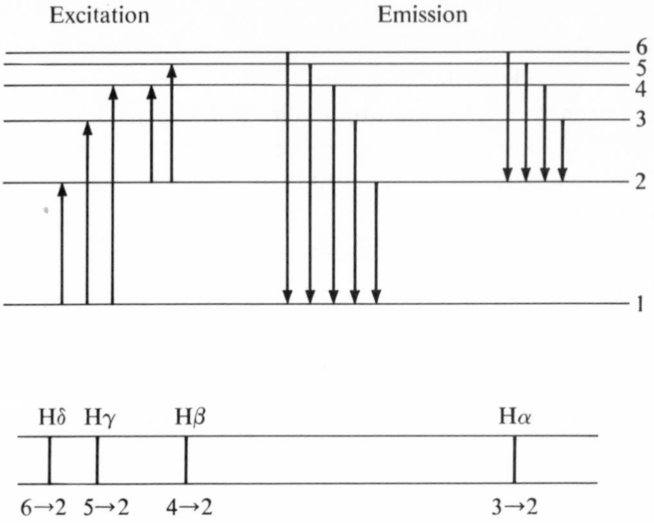

Excitation Emission

6
5
4
3
2
1

Hδ Hγ Hβ Hα

6→2 5→2 4→2 3→2

Figure 3.12. Upward and downward transitions between energy levels for the hydrogen atom. Since only a limited number of discrete downward transitions are possible, the hydrogen emission spectrum shows light only at certain particular wavelengths.

all in the blue and violet. The lines become progressively closer together in the violet, and many other possible lines are not shown here. This is, of course, the pattern of lines resulting from the transitions which end in level 2. This particular pattern of lines is known as the **Balmer series.** Transitions ending in level 1 give rise to more energetic photons and so the resulting series of lines lies entirely in the ultraviolet. This series is called the Lyman series. The hydrogen spectrum in the infrared shows two more series of lines which result from transitions ending in levels 3 and 4. Thus, each downward transition results in an emission line at some wavelength according to the size of the transition.

3.8 Excitation Processes

We have spoken only in general terms of the fact that atoms can become excited when they receive an increased amount of energy from some outside source. It was mentioned that a collision between an electron and an atom can excite the atom. More specifically, there are three important excitation mechanisms. The first of these may be referred to as **collisional excitation** and is usually accomplished by raising the temperature of the gas. Kirchhoff's laws state that a hot tenuous gas emits a bright-line spectrum. The kinetic theory of gases tells us that the atoms or molecules of a gas are continuously in motion and suffer frequent collisions with each other and with the walls of the container. The average velocity of the atoms will increase as the temperature of the gas increases. (It is perhaps more realistic, in a sense, to say that temperature, as we customarily measure it, is really an indication of the average velocity of the atoms or molecules.) In the course of many collisions, then, the kinetic energy of motion is transferred to the atoms which thereby become excited and subsequently emit photons.

A second excitation process, similar to the first one, involves the bombardment of the atoms of a gas with electrons. This is the method which causes gas in neon signs and fluorescent lamps to emit. Electrical energy is transferred to the atoms through collisions between atoms and electrons, and the atoms are thus excited.

The third important process may be referred to as **photo-excitation.** An atom is able to absorb any of the photons which it is able to emit. Photons of other energies, however, have no effect on the atom. If an atom absorbs an extremely energetic photon, it can become ionized. Sometimes an atom may be excited to a high energy level by the absorption of a single rather energetic photon. The atom may then return to the ground state in either a single long transition or by **cascading** through a series of smaller transitions. Cascading offers an important means by which the energy of high-energy radiation such as gamma rays or ultraviolet light is transformed into light energy in the visible part of the spectrum.

3.9 Absorption Spectra

The above theory, which seeks to explain the processes of excitation and emission in atoms, also makes possible an understanding of the dark-line spectrum. As described above, the dark-line or **absorption** spectrum is seen when a cool gas (or a mixture of cool gases) lies between the observer and some source of a continuous spectrum.

White light, being a mixture of light of all colors, contains photons of all wavelengths. Thus when the white light from some source reaches a cloud of gas, there will be present photons of the same wavelengths which the atoms of the gas could emit. These photons can excite any of the atoms with which they happen to collide. To put it another way, the atoms of the cool gas can absorb from the incident white light the same kinds of photons which they themselves are able to emit. (The word "cool" is used here in the sense that the gas is cool relative to the temperature of the source of the continuous spectrum.) Thus certain photons will have been removed from the original beam and will be missing from the beam as it is finally observed. We observe these missing wavelengths as colorless areas or dark lines in the normal continuous spectrum. This is the basic process, but it is in fact complicated somewhat by several other factors.

We have said that the excited atom emits a photon and returns to a lower energy level almost instantaneously. If the atom has been excited by a photon, it may re-emit a photon of the same wavelength as that which it absorbed. If this is the case, one might ask why we should then see an absorption line at all. The lines are seen for two reasons. First, all of the original light is coming in one direction, as indicated in Figure 3.13, but the photons emitted in the gas cloud may be travelling in any direction. Some of the new photons may even be emitted in the original direction, a process resulting in the observed condition of absorption lines not completely dark but only dark by comparison with the surrounding continuous spectrum. Second, if an energetic photon is absorbed, the atom may return to the ground state by cascading with the emission of photons of several wavelengths. The second process is especially important because it helps to provide atoms in various intermediate degrees of excitation which are in turn capable of absorbing other photons from the original beam. For example, in the case of hydrogen the Hα line in the red results from the transition $2 \rightarrow 3$ if it is an absorption line. This means then that unless some atoms are already excited to level 2 the Hα line will not be seen. The upward transition $1 \rightarrow 2$ can, of course, provide some atoms in level 2. Others can be provided also by any downward transitions which happen to end in level 2 such as $6 \rightarrow 2$ (Hδ), $4 \rightarrow 2$ (Hβ), etc.

The examination of any typical absorption-line spectrum shows that there are distinct differences among the absorption lines.

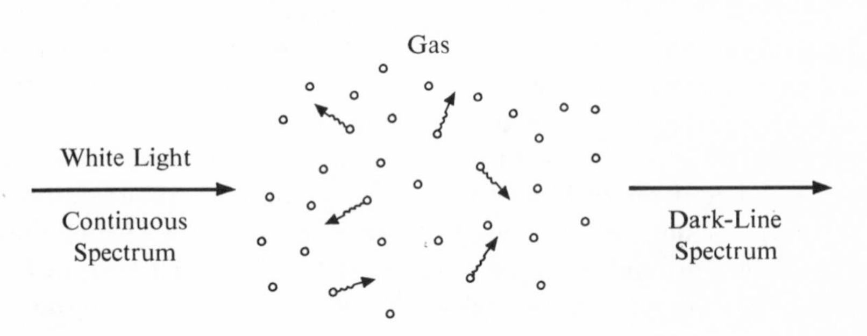

Figure 3.13. The atoms in a gas absorb energy from the beam containing photons of all wavelengths. The excited atoms quickly re-emit the energy which they have absorbed, but these new photons do not necessarily travel in the original direction. Energy at these wavelengths is therefore missing from the light beam to the right of the gas.

Some lines are much more prominent than others, and it is evident that the strength of an absorption line depends, first of all, on the number of atoms making a given transition. Physicists have learned to predict statistically the likelihood of any particular transition under given conditions of temperature and pressure in a gas. As a result, physicists are able to go far beyond a mere chemical analysis from spectra and deduce a great deal about physical conditions in stellar atmospheres and in clouds of interstellar gas simply from a detailed analysis of the relative strengths of various absorption lines.

3.10 Spectra of More Complex Atoms

The preceding discussion has been related mainly to the simple hydrogen atoms and the rather uncomplicated spectrum of hydrogen. All of the other atoms, of course, have heavier nuclei and greater numbers of electrons surrounding the nuclei. As a result, for atoms such as iron there are a great many energy levels and possible transitions. The spectrum is accordingly complex. Bohr's theory is meaningful only for a few of the lightest atoms. The complete analysis of more complex spectra requires the tools of quantum mechanics.

The transitions giving rise to the emission and absorption lines in the visual spectrum all involve the electrons in the outermost of the electron shells familiar to chemistry students. Considerably more energy must obviously be required in order to disturb the electrons in the inner shells. The resulting transitions involving these inner electrons must naturally result in more energetic (i.e., shorter wavelength) photons. These higher-energy photons are what we commonly refer to as x rays, notable for their ability to pass through many types of material and expose photographic films.

3.11 Molecular Spectra

In molecules several atoms of the same element or of different elements are held together by the electrical attractions between the individual atomic nuclei and the electrons. Again, largely through optical techniques, chemists and physicists have developed a workable theory of the structure of molecules and of the manner in which a molecular gas can absorb and emit energy.

The simplest molecules are composed of only two atoms. The molecule can rotate about an axis between the two atoms, and the two atoms can vibrate back and forth along the line joining them together. There may also be electron transitions similar to those described above for atoms. The result of all of these processes is that molecular spectra can be much more complex than atomic spectra, and they can show large numbers of closely spaced lines. These lines can give the overall appearance of broad absorption bands in the spectrum. In the case of O_2 the absorption due to molecules in our atmosphere appears as a converging series of lines. (See also Section 15.4 for a further discussion of emission from molecules in gas clouds between the stars.)

QUESTIONS

1. Describe situations in which light is best discussed in terms of waves, rays, and photons.
2. How would Römer's calculated velocity of light have been affected if the delay in satellite eclipses had been greater than the 1000 seconds he had noted?
3. Why did Römer's determination of the velocity of light help to support the concept of the earth's orbital motion?
4. A light ray makes an angle of 28° with the normal to the surface of a pan of water. What will be the angle of refraction of the light ray in the water?
5. Define wavelength and angstrom unit.
6. Why is it possible for a prism to disperse a beam of white light into a spectrum?
7. A glowing cloud of hot hydrogen gas is viewed through a cloud of cool sodium vapor. What will the spectrum look like? Explain.
8. Describe the appearance of the bright-line spectra of hydrogen, helium, and mercury.
9. Discuss fully the one known effect which can cause a change in the wavelength of light. How are such changes detected and what is the magnitude of such a change in wavelength?
10. In order to demonstrate the spectroscopic proof of the earth's orbital motion, what is the advantage of studying the spectrum of a star on or near the ecliptic?

11. In terms of the Bohr theory of the atom, account for the appearance of bright-line spectra of the elements.

12. If an excited atom quickly becomes neutral by emitting one or more photons, why is it that absorption lines show up at all in the spectra of gases?

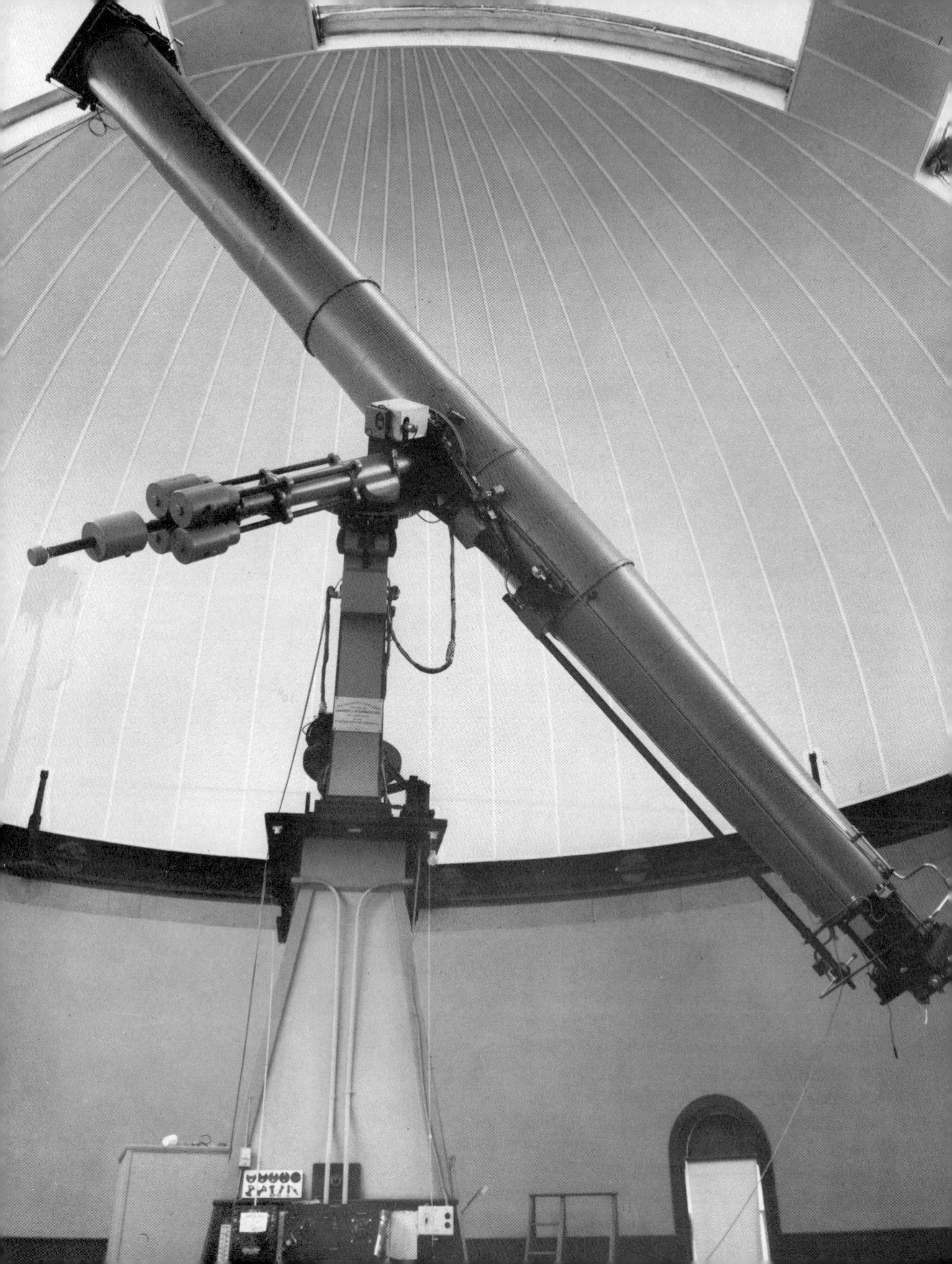

4

Telescopes and their Applications

There are several types of telescopes in use by astronomers, some of which do not fit the layman's concept of a telescope at all. More important than the optical and mechanical designs, perhaps, are the methods in which the telescopes may actually be used. Astronomers are not really much concerned with "looking through" their telescopes. Their valuable data are gained, for the most part, by means of certain accessories employing either standard photographic techniques or modern electronic equipment.

4.1 Refractors

There are two important types of telescopes; one type, the refractor, makes use of conventional glass lenses which—as we have seen in the previous chapter—refract light rays. A good example of the classical astronomical refractor is seen in Figure 4.1. Rays traveling parallel to each other in a beam of light will be refracted in such a way that they will come together at a point behind the lens. If the incoming beam is at right angles to the lens, it is said to lie along the **optical axis** of the lens. In such a situation (Figure 4.2a), the parallel light rays converge at a point on the optical axis known as the **focal point** of the lens. The distance between this focal point and the center of the lens itself is the **focal length,** and

◄ **Figure 4.1.** The 26-inch McCormick refractor at the University of Virginia. (McCormick Observatory photograph.)

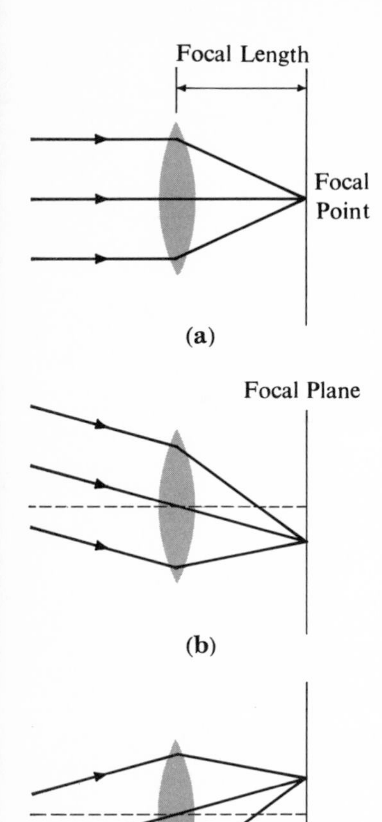

Figure 4.2. Parallel beams entering a lens from many directions are focused in the focal plane to form an image.

the diameter of the lens is usually referred to as the **aperture.** We shall see that certain important properties of lenses depend upon the combination of aperture and focal length in a particular lens. Parallel beams or bundles of rays which do not lie along the optical axis also converge behind the lens, as shown in Figures 4.2b and 4.2c. Thus, light coming to a lens from distant points in a variety of directions is focused at a number of corresponding points behind the lens, and an inverted, reversed image is formed. This image is not usually apparent when a simple lens is held in the hand, but it may easily be focused onto a white card or piece of paper held behind the lens.

In telescopes intended for visual use a second lens called the **eyepiece** is placed behind the focal plane. The eyepiece is used to examine the image in the focal plane and magnify details in this image. An eyepiece functions in exactly the same manner as a hand magnifier used to read fine printing or to look for detail in small objects. The simplest eyepiece is a single lens, and the focal length of this eyepiece lens may be specified in the same manner as it was for the objective lens above. In more complete discussions of geometrical optics it is customary to show the manner in which the magnification of a two-lens system depends upon the focal lengths of the objective and the eyepiece. This simple relationship is

$$\text{magnification} = \frac{\text{focal length of the objective}}{\text{focal length of the eyepiece}}$$

Thus, a small portable telescope might have a focal length of sixty inches and an aperture of five inches. If an eyepiece with a one-inch focal length were to be used with such a lens, the magnification (or power) of the system would be sixty times. An object seen through the eyepiece would appear sixty times larger than it would to the unaided eye. It should also be noted that the magnification may be changed simply by changing to another eyepiece. In order to increase the magnification, an eyepiece with a shorter focal length would be used. Likewise, to obtain a low power a longer focal length eyepiece would be used. Most astronomical telescopes are equipped with several eyepieces so that the magnification may be varied by simply changing from one to another (Figure 4.3). Typical eyepieces usually have two or more lenses, depending upon the applications intended. The effective focal length of the combination is marked on the eyepiece. Since the eyepiece magnifies the

Figure 4.3. A series of interchangeable telescope eyepieces. (Courtesy of Vernonscope and Co.)

image exactly as it appears in the focal plane, the observer using an astronomical telescope sees the original object inverted and reversed. The inverted image should be no problem for the user, however, in view of the nature of the objects usually under study. Binoculars and telescopes intended for non-astronomical uses have only one fixed eyepiece but are equipped with an extra lens to reinvert the image for the convenience of the observer.

In experimenting with simple lenses held in the hands as objective and eyepiece, one will usually notice that the image as formed by the objective and magnified by the eyepiece is not always a well-defined representation of the original scene. Usually only the central part of the field looks the way it should. The outer portions are distorted by a number of **aberrations** which are inherent in the nature of the lenses. Many of these aberrations can be corrected by the proper choice of curvatures and types of glass in multiple-lens systems. Such corrective efforts result in a larger undistorted image in the center of the field of view, and have reached a high degree of perfection in the wide-angle lenses available for expensive cameras. Some three- and four-lens systems are in use in astronomical refractors designed specifically for photography of large areas of the sky. A very common problem in refracting telescopes arises from the fact that the lens acts like a prism and disperses the colors in an incoming beam of white light. As indicated in Chapter 3, the blue light is refracted most. The result is that the simple lens does not focus all colors at the same point. Figure 4.4 depicts the blue rays coming to a focus closer to the lens than the red rays. In attempting to focus the eyepiece on the image, then, the observer finds that he cannot obtain a very clear picture. Red

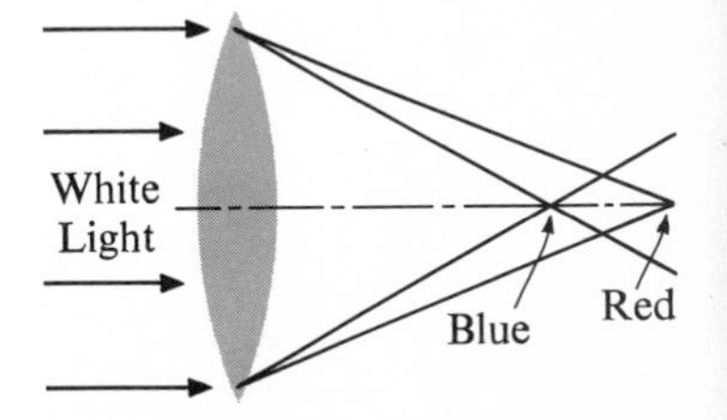

Figure 4.4. Chromatic aberration in a lens results from the fact that the blue light is focused closer to the lens than is the red light.

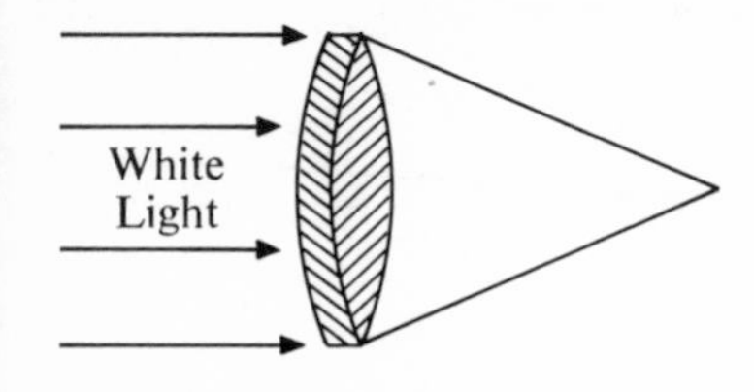

Figure 4.5. The use of a second lens made of a different type of glass makes possible the design of optical systems in which chromatic aberration is almost completely eliminated.

light is out of focus when blue light is in focus and vice versa. When viewed, the image tends to have a colored fringe around it. This troublesome effect is known as **chromatic aberration** and became very serious to telescope users in the early seventeenth century.

Modern refracting telescopes were made possible by the correction of chromatic aberration through the use of a second lens as part of a compound objective. It was discovered that a concave lens made of glass of a different index of refraction and used in conjunction with the objective could refract almost all of the light to a new focal point, as shown in Figure 4.5. This relatively simple solution works rather well, but it is still impossible to design a lens system which eliminates chromatic aberration completely. As a result, one will usually see a halo of color around the edge of an object, such as the moon, viewed even with a well-made modern refractor.

4.2 Reflectors

Newton had been troubled in his optical experiments by chromatic aberration and had despaired of any simple solution. He recognized, however, that light could be effectively focused by a concave mirror as well as by a lens. If the cross section of such a mirror is a parabola rather than part of a circle, all rays in a parallel beam will be focused at a single point as indicated in Figure 4.6. By constructions similar to those in Figure 4.2, off-axis beams from many directions combine to form an image in front of the mirror at F. An eyepiece placed beyond F could be used to view and magnify the image just as in the case of the refractors. The focal length of the mirror is the distance from the mirror to the focal point F.

Figure 4.6. A paraboloidal reflector forms an image in front of itself as beams from several directions are brought to focus. Compare this figure with Figure 4.2.

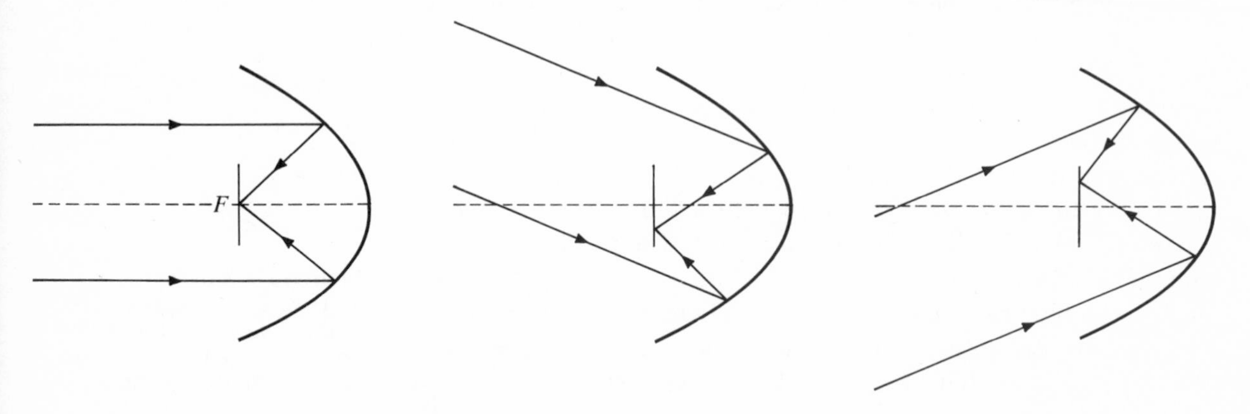

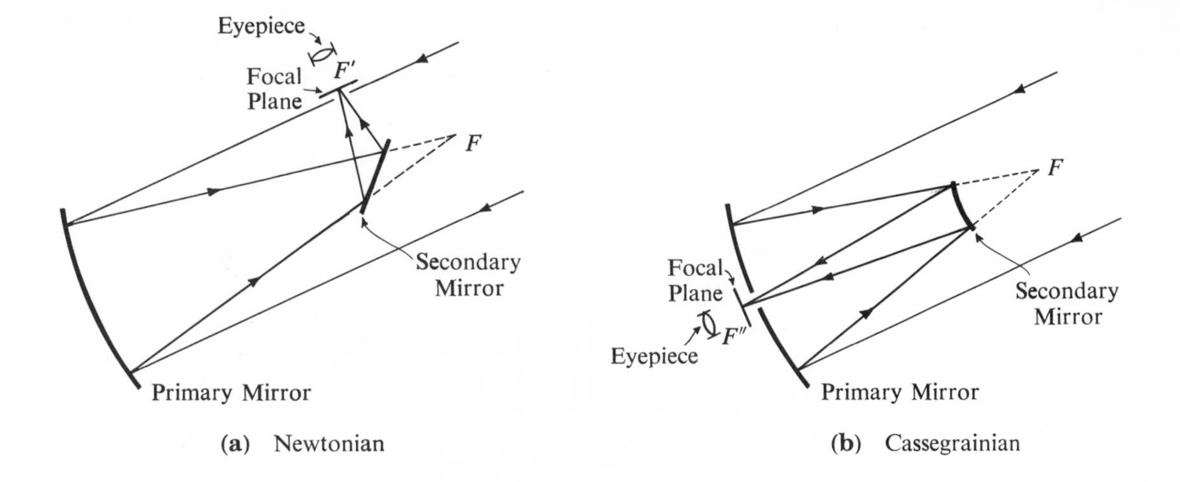

Figure 4.7. The Newtonian and Cassegrainian forms of the reflecting telescope.

In Newton's original design and in most small reflecting telescopes, a secondary mirror is used to reflect the converging beam out to one side of the tube where the image may be conveniently viewed. This is necessary, of course, since the observer's head would block the entering beam if the eyepiece were used on the optical axis. Two common arrangements for the secondary mirror are shown in Figures 4.7a and 4.7b. In the first of these a flat secondary mirror (or a right-angle prism) reflects the converging beam to the Newtonian focus at F'. The observer then looks into the side of the telescope. This is the simplest type of reflecting telescope and is most commonly made by amateurs.

The second arrangement employs a convex secondary mirror which reflects light back along the optical axis through a hole in the primary mirror. The **Cassegrainian focus** is then located behind the main mirror and the observer looks in the direction in which the telescope is pointed. In both the Newtonian and Cassegrainian telescopes the secondary mirror and its support structure will block some of the incoming light. The amount blocked is usually only a small percentage of the total, however. Two large reflectors, the 200-inch (508 cm)* at Mt. Palomar Observatory and the 120-inch (305 cm) at Lick Observatory, may actually be used at the prime focus, F. In these two cases the observer rides inside the telescope in a small cage (Figure 4.8).

*Although the apertures of many large telescopes are best known in inches, their measurements will usually be converted to the metric system, which is used throughout this book.

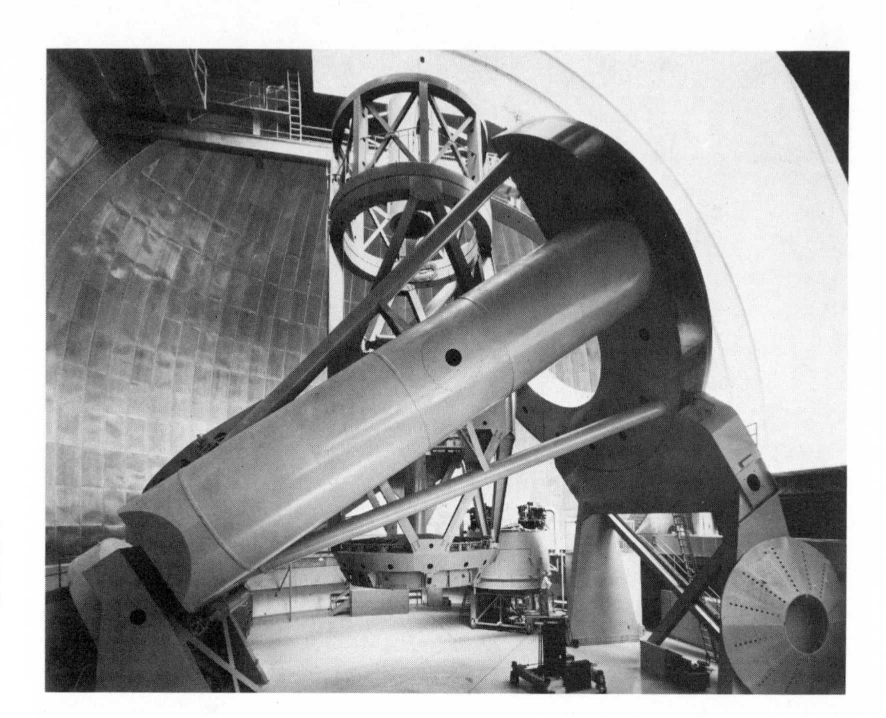

Figure 4.8. The 200-inch Hale telescope at the Mount Palomar Observatory. (Hale Observatories photograph.)

The great advantage of reflecting telescopes is that all colors are reflected in the same manner and so there is no chromatic aberration. Furthermore, since the objective mirror has only one optical surface to be ground and polished, the reflector is usually less expensive than the refractor, in which the two-element objective has a total of four optical surfaces. Telescope mirrors are usually made of Pyrex glass since it is cheap and has a low coefficient of thermal expansion. Changes in temperature can cause changes in the shape of the mirror and therefore affect the focus, so that materials which are not much affected by temperature changes are desirable. Fused quartz, though rather expensive in large blanks, is also used as a material for telescope mirrors since its thermal characteristics are even better than those of Pyrex. One side of the mirror is ground to the proper paraboloidal shape and is then coated with aluminum or silver. It is this thin metallic film which forms the actual reflecting surface. No light actually passes through the glass.

The size of a telescope is usually specified by the diameter of the objective, whether the telescope is a refractor or a reflector. Among refractors, the largest in the world is the 40-inch (102-cm) telescope of the Yerkes Observatory of the University of Chicago. This telescope has been in operation since 1897, and few large refractors have been built since that time. A 40-inch reflecting telescope, on the other hand, is only moderately large. The largest telescope in the world was a 200-inch reflector at Mt. Palomar in California, but this has been surpassed by the 6-meter (236-inch) reflector recently built in the USSR. It is likely that by 1975 there will be more than a dozen reflectors with apertures larger than 2.5 meters.

There are good reasons for the great gap in size between the largest refractors and the largest reflectors. The greater cost has already been mentioned, but other factors enter in as well. Large, heavy lenses and mirrors tend to bend very slightly under their own weight as they are moved from one position to another. Any such bending affects the focus and must therefore be minimized. In a large reflector the mirror may be adequately supported on the back and thus maintained in its proper shape. The refractor's lens must rely on its own structural strength to avoid bending. This strength can be increased by making the lens thicker in its overall cross section, but the light must pass through the lens, and light is going to be absorbed in the glass. The thicker the glass, the greater will be such losses due to absorption. Thus two factors act together to place a practical limit on the size of refracting telescopes, while the ultimate size of large reflectors depends mainly on the willingness of men to divert large sums of money into such projects.

4.3 The Schmidt Telescope

A third general class of telescopes combines both lenses and mirrors in order to achieve well-defined images over a wide-angle field of view. The **Schmidt telescope** is one of the most widely used of these systems, and its chief parts are sketched in Figure 4.9. The principal element is a mirror with a spherical rather than a parab-

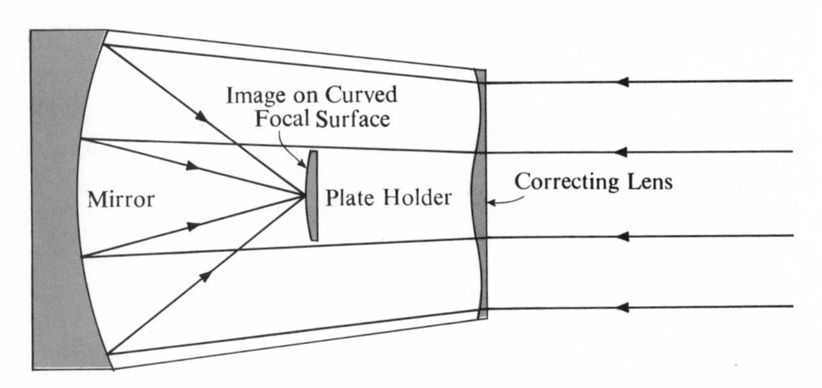

Figure 4.9. The optical system of the Schmidt telescope.

oloidal figure. At the front of the telescope is the refracting corrector plate which has the cross section shown in exaggerated form in the figure. The shape of the corrector is chosen so that the paths of the incoming rays are changed by just the amounts necessary to make the spherical mirror produce a good image. A minor problem is that the focal surface is curved rather than flat. However, since these telescopes are used exclusively as cameras, this is overcome by pressing the film against a surface of the proper shape inside of the telescope. The largest Schmidt telescope is at the Hale Observatories and has a mirror 182.8 cm (72 in.) in diameter and a corrector plate 121.9 cm (48 in.) in diameter.

4.4 Resolving Power and Seeing

In using a telescope to study a bright, extended object such as the moon, the observer is interested in the magnifying power of the telescope so that as much detail as possible may be discerned. It might seem that magnification could be increased almost without limit simply by decreasing the focal length of the eyepiece. There is, in fact, a practical limit known as the **resolving power** of a telescope. From considerations of the wave nature of light, it is shown in textbooks on optics that resolving power is given by

$$a = 2.1 \times 10^5 \times \frac{\lambda}{d}$$

where a is the angular size in seconds of arc of the smallest object which can be distinguished and d is the diameter of the objective lens. λ is the wavelength of the light being considered and must be expressed in the same units as d. For a wavelength of 5000 Å and telescope diameters measured in centimeters, the above equation becomes

$$a = \frac{10.5}{d}$$

Thus, a 20-cm telescope can resolve an object with an angular diameter of 0.52 second of arc. (See Figure 4.10.) Using this telescope an observer could, in theory at least, distinguish as separate points of light two stars separated from each other by this very small angle. At the distance of the moon, a crater 0.9 km in diameter would have an angular size of 0.52″ and so would be the smallest detect-

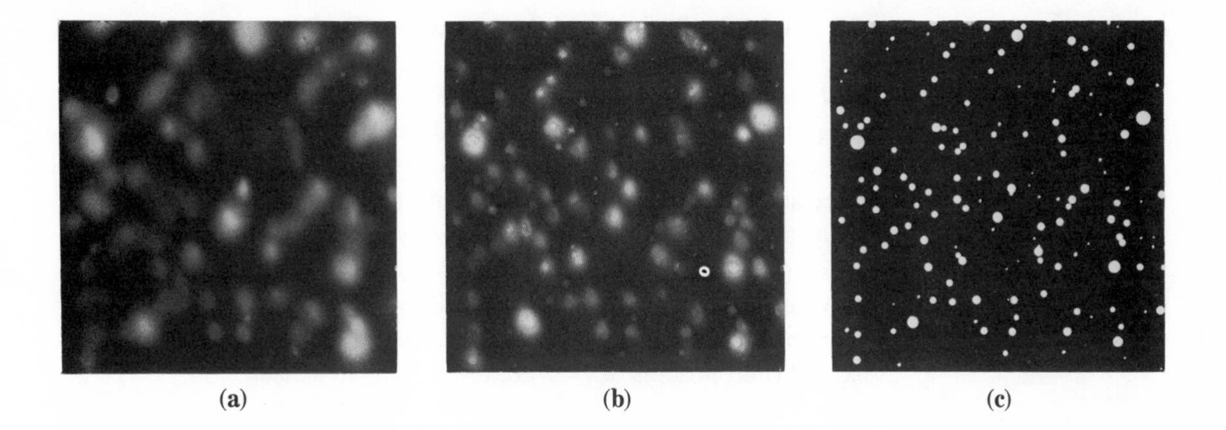

Figure 4.10. Hypothetical views of the sky as it might appear if it were seen under conditions of very low resolving power. In (a) the resolution is 3 minutes of arc; in (b) it is 1 minute; and in (c) the resolution is considerably higher.

able with the 20-cm telescope. Larger telescopes would, of course, be capable of greater resolution, but attempts to use very high magnifications with the telescope are ultimately limited by these inherent properties of lenses and light.

A second factor usually limits the resolving power long before the theoretical limit for a given telescope has been reached. This factor is the earth's atmosphere, which affects what astronomers have come to call **"seeing."** With only a little experience at the telescope, one notes that on certain nights the objects under study appear to be quite steady, while on other nights the objects appear to be in continuous motion and fine detail is completely blurred. These seeing effects arise in the lower layers of the atmosphere within a few hundred feet of the ground. Turbulence in the air results in continuously changing regions of cool and warm air. Since the index of refraction of air depends upon its temperature, the light rays are thus refracted in many directions as they pass through the atmosphere. In a small telescope the object appears to be in more or less continuous motion. In a large telescope, poor seeing causes the image of a star to appear unduly large and to fluctuate in diameter. Weather patterns such as the approach of a cold front may bring about poor seeing temporarily. Local conditions such as the height of the hill or mountain above the surrounding terrain and the nature of the ground cover have also been found to have a close relation to seeing. Astronomers planning large new observatories, therefore, make extensive surveys to try to find sites in which the seeing is exceptionally good throughout most of the year.

4.5 Light Gathering and Optical Speed

Stars seen through a telescope appear only as dimensionless points of light. Regardless of the magnification used, they never appear as anything else. As Galileo noted, however, the telescope reveals more stars than the unaided eye, and it is for this reason that over the years astronomers have continually strived for larger and larger telescopes. The larger the telescope, the fainter the stars which may be seen. The faint stars are likely to be the distant ones and therefore of increasing interest to astronomers as they attempt to extend their studies to ever greater depths in space.

By enabling astronomers to see faint stars, the telescope functions as an instrument for **light gathering.** Since the pupil of the eye is only about 1 cm in diameter, the retina can respond only to the energy in a beam of that size. If an object is very faint, a beam of light 1 cm in diameter may not contain enough energy to stimulate the retina, and the object remains invisible. The telescope, however, may be used as a sort of funnel to collect light from a large beam and concentrate it into a small beam which then enters the eye. The telescope functions thus whether used visually or with some of the auxiliary equipment to be described below. Light energy in a large beam is concentrated into some smaller area, and faint objects become detectable. It is for this reason that astronomers are interested in light gathering perhaps even more than in magnification. The giant telescopes of Kitt Peak, Lick, and the Hale Observatories are designed to collect light over a large area and thus make possible the study of extremely faint objects.

The objective of a telescope, whether it be a refractor or a reflector, always forms an image in the focal plane, and the size of this image depends upon the focal length of the objective. The longer the focal length of the objective, the larger is the image formed in the focal plane. Let us consider two telescopes of the same aperture of 20 cm but with focal lengths of 1 meter and 3 meters, respectively. Both telescopes concentrate the same amount of light into an image since their apertures are the same. The smaller image will therefore be brighter than the larger image, and could as a result be photographed in a shorter time if a film were to be placed in the focal plane. The ratio between the focal length and the aperture has come to be called the **optical speed** or **focal ratio** of the lens system. In hand cameras this ratio may be varied by means of an adjustable diaphragm. Astronomical refractors usually have focal ratios of about 15; and reflectors usually have focal ratios of 4 or 5. An advantage of the Cassegrainian optical system is that the curvature of the secondary may be chosen to change the convergence of the reflected beam and produce a desired focal ratio at the eyepiece in a relatively compact tube. Many modern Cassegrainian reflectors are thus equipped with interchangeable secondaries. The choice of the correct focal ratio for use depends on the application at hand. A fast system such as *f*:2 would be used to photograph faint extended objects such as luminous gas clouds, while an *f*:15 system would be used where greater separation among star images on a photograph is desired.

4.6 Spectroscopes

Since our chief contact with celestial bodies is by means of the light which they emit or reflect, it is important to analyze and measure this light as thoroughly as possible. The spectroscope permits such analysis and, when used on a telescope as in Figure 4.11, permits us to acquire a remarkable amount of information about the planets, stars, and galaxies.

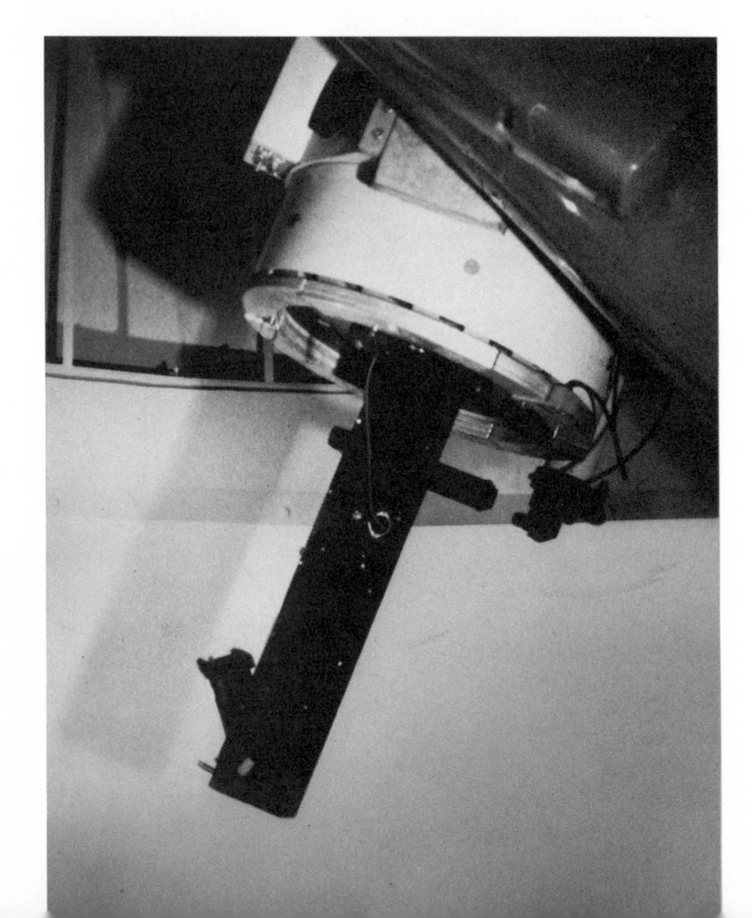

Figure 4.11. A grating spectrograph mounted on a 24-inch reflector. (Wellesley College photograph.)

The optical parts of a simple spectroscope are indicated in Figure 4.12. Light from some source falls on a narrow slit, and this entrance slit itself now acts as a narrow rectangular light source. A collimator lens is situated in such a way that the distance from slit to lens is equal to the focal length of the lens. The light from the slit then leaves the collimator as a beam of parallel rays. This parallel light enters a prism in which it is dispersed, and parallel beams in each color leave the prism. A second lens focuses all of the beams, producing a spectrum of the original light. An eyepiece may then be used to view and magnify the spectrum. When the eyepiece is omitted and the instrument is adapted mainly for use with photographic materials in the focal plane, it is usually referred to as a **spectrograph.** Photographs of spectra obtained with such equipment are called **spectrograms.**

Because of the rectangular shape of the entrance slit, the image of the slit in the focal plane will be a rectangle. As white light enters the slit, overlapping images in all colors or wavelengths are formed, and the shape of the slit is not obvious. When the original source of light is a hot gas, however, images of the slit are seen only at the few selected wavelengths characteristic of that gas. Since the slit is ordinarily so narrow that it looks like a line, the images of the slit look like lines also. For this reason it has become customary to speak of **spectrum lines.** The student, however, should understand and remember just what a spectrum line really is. If the entrance slit should be made in the shape of the letter *s*, the spectrum lines would also be shaped like the letter *s*. In an absorption spectrum, light of certain wavelengths is missing and so images of the slit are missing at corresponding points in the spectrum. We see, in this case, the dark lines. If the entrance slit is widened, more light can enter the spectrograph and faint sources may be studied. The resulting spectrum lines are also widened with an eventual loss of detail. The width of the slit in this type of spectrograph is usually in the neighborhood of 0.10 mm.

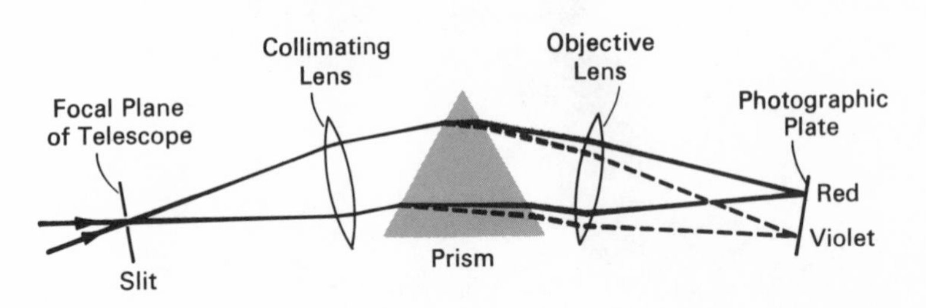

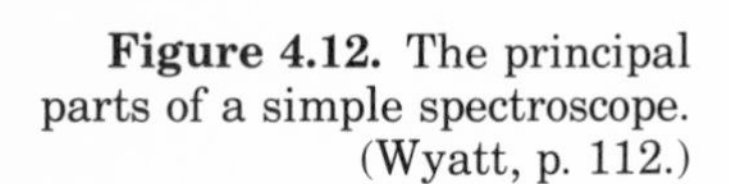
Figure 4.12. The principal parts of a simple spectroscope. (Wyatt, p. 112.)

In order to make accurate determinations of the wavelengths of spectrum lines, it is necessary to establish some wavelength standards on a spectrogram. This is usually done by putting a bright-line source in front of the slit. This source could be a neon lamp or an iron arc, for example. In practice, the ends of the slit are usually covered while the star's spectrum is being photographed. Then the central part of the slit is covered and the ends are uncovered while the comparison spectrum is photographed. The result is a spectrum of a star with a reference or comparison spectrum on either side. If the wavelengths of the comparison lines are known, then the wavelengths of lines in the star's spectrum may be determined.

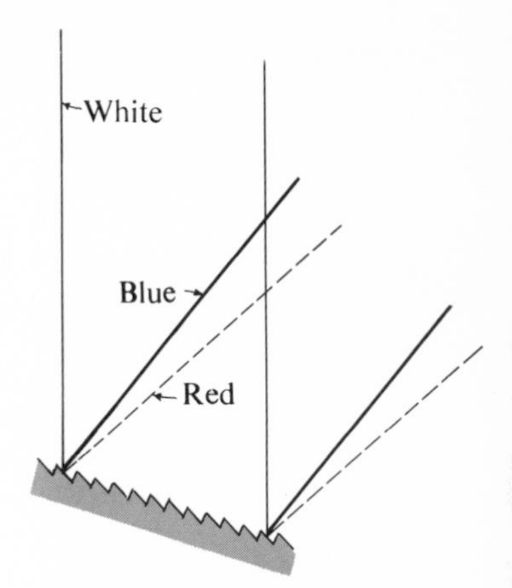

Figure 4.13. Schematic diagram of a diffraction grating. A parallel beam of white light strikes the grating. Parallel beams in each color are reflected from the grating. The greater the number of grooves per inch, the greater the dispersion will be.

A second type of spectrograph in common use employs a diffraction grating rather than a prism to disperse the light. The grating indicated in Figure 4.13 consists of a surface on which 300 or more lines per millimeter have been scribed. Interference phenomena associated with the reflection of light from such a surface result in dispersion of light. Transmission gratings may also be employed in a manner similar to that described for the prism spectrograph. In this case the lines are scribed on a transparent medium. It is very difficult to manufacture high-quality diffraction gratings and in the past they have been quite expensive. Since about 1950, however, techniques have been developed by which relatively inexpensive but effective replicas of an original grating may be made. The reflection-grating spectrograph has two advantages over the prism type. First, it can be used to study ultraviolet light which could not pass through ordinary glass lenses and prisms. Second, wavelength is directly proportional to position in the spectrum, which is not the case in a prism spectrograph. A disadvantage is that the available light is actually dispersed into a number of overlapping spectra or orders, but proper design can produce spectra in which up to 80% of the incident light is concentrated into one desired order.

As described above, the collimator lens functions to produce parallel rays of light which fall on the prism or grating. All stars are so far away that their light rays are essentially parallel as they reach the telescope, and astronomers are able to eliminate the slit and the collimator in certain types of spectroscopic studies. If a large prism is placed in front of the telescope objective, the entire telescope is essentially converted into a spectrograph. Such an arrangement, sketched in Figure 4.14, is called an **objective-prism spectrograph** and has the great advantage that spectra of many stars may be recorded on one photograph. The light of each star is

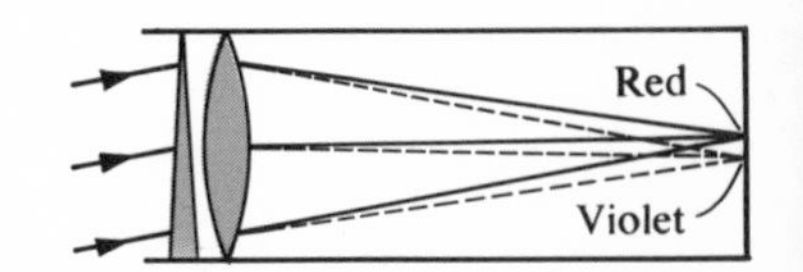

Figure 4.14. Schematic diagram of an objective-prism spectrograph.

Figure 4.15. Objective-prism spectra showing the variety of absorption-line patterns which are commonly seen. (Warner and Swasey Observatory photograph.)

spread out into a small spectrum. These spectra are very narrow since the stars are essentially point sources, and few spectrum lines are ordinarily detectable. When the spectra are widened, however, a considerable amount of detail shows up. Figure 4.15 shows a portion of an objective-prism spectrogram in which the Balmer lines of hydrogen are visible in several spectra. The spectrum "lines" seen here are not images of a slit but are wavelength regions in which color or light has been absorbed in the star's atmosphere and is therefore missing. The appearance of lines comes from the widening of the spectra at right angles to the direction of dispersion. Objective-prism spectra have very low dispersion and show only the more conspicuous features. Nevertheless, stars can be effectively classified from these spectra into one of several groups to be described in Chapter 10. Detailed studies of particular types of stars may then be made one star at a time with the higher-dispersion slit spectrograph.

4.7 Photography at the Telescope

In a great many applications the astronomer uses the telescope as a camera. The photograph gives astronomers a permanent record which may be studied at leisure or compared with previous or subsequent photographs to determine changes in relative positions or brightnesses of stars. For use in photography the telescope functions in a very simple manner. Just as in a hand camera, the light-sensitive photographic materials are placed in the focal plane of the lens, and the image formed by the lens is recorded. The astronomer usually uses photographic emulsions prepared on glass plates rather than on rolls of flexible film, and his exposure times may be many minutes or even hours rather than the fraction of a second required for an ordinary snapshot.

The same factors regarding light gathering which were previously described for telescopes used visually apply also, of course, to telescopes used photographically, but with one important difference. The photographic plate can accumulate light during long exposures and thus is able to record light sources which are too faint to be seen visually. For example, tenuous clouds of luminous gas may be seen even with a small telescope in the Orion Nebula (Figure 15.1). Long-exposure photographs reveal that these clouds extend over tremendous areas of the sky. The faint light of their outer portions affects the photographic emulsion cumulatively and after a sufficient time produces a detectable image on the plate.

Most of the important astronomical data acquired photographically are on black and white plates. Only in a few cases have astronomers ventured to use color films to record celestial objects. One important reason is that accurate color representation is difficult to obtain in long exposures when the emulsions are designed for short exposures. On the other hand, however, astronomers have learned to make particularly effective use of certain properties of common black and white emulsions. Even though the final picture may be recorded only in shades of black and gray, the most common photographic materials are more sensitive to blue light than to other colors. As a result, if a red star and a blue star close to each other in the sky appear to be about equal brightness to the eye, the blue star will look brighter on a photograph. This property, which could result in confusion, has turned out to be a great advantage. Photographic engineers have learned to make special emulsions sensitive to light of colors other than blue, and some of these, when combined with certain filters, actually reproduce the spectral sensitivity of the human eye. In Figure 4.16 two photographs of the same star field are reproduced. The first one was made using an ordinary blue-sensitive emulsion. The second was made with an emulsion-filter combination which is most sensitive to red light. The second photograph shows the stars about as they would appear if one were to look at this area with a telescope. We have essentially isolated in each photograph the light from a different region of the spectrum, and we are able to obtain from such pairs of photographs a tremendous amount of information. Obviously, a blue star is brighter on the blue photograph than on the red one since most of its light is in the blue part of the spectrum. A red star will likewise be brighter on the red photograph. An accurate quantitative measure of the difference in brightness on two such photographs is known as a **color index** and will be fully discussed in Chapter 10. Other emulsion-filter combinations are also in use to isolate spectrum regions in the infrared and in the near ultraviolet. The number of possible color systems which could be devised is limitless, but astronomers have standardized a few

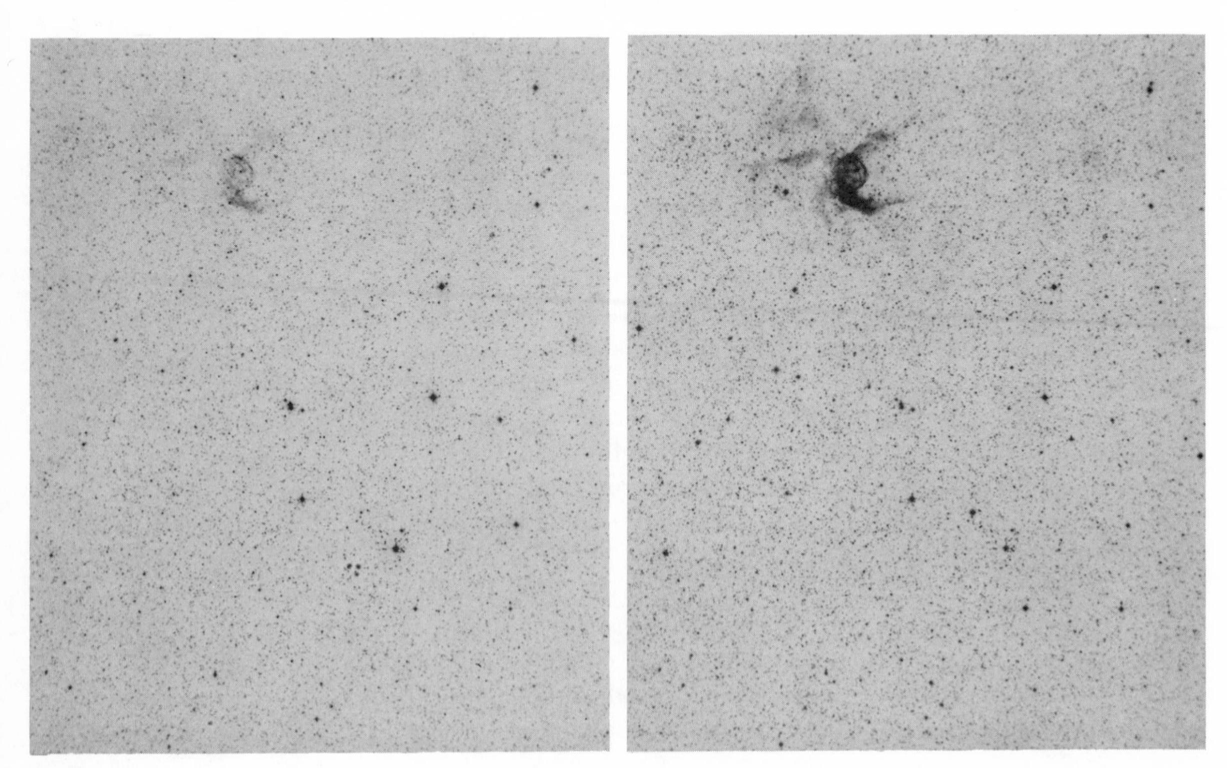

Figure 4.16. A star field photographed in blue light (left) and in red light (right). Note the difference in appearance of the stars which are basically red and blue. (Hale Observatories photographs, © copyright by National Geographic Society.)

particularly useful systems, which are in general use throughout the world.

4.8 Photometers

The size and blackness (density) of the star images in Figure 4.16 depend, of course, on the brightnesses of the stars in the photographs. Astronomers have learned to measure these images in order to determine the relative brightnesses of stars. A degree of accuracy far surpassing visual estimates may be obtained, but astronomers have found their greatest precision in brightness measurements by the use of **photoelectric photometers** attached to their telescopes.

A somewhat simplified photoelectric system is shown in Figure 4.17. The light from a single star to be studied is allowed to pass through a small hole or diaphragm in the focal plane of the objective. The light beam then passes through a filter so that some desired wavelength band may be isolated and eventually fall on a photocell. In the cell, light striking the photocathode, or light-sensitive surface, causes the emission of electrons from that surface. An exterior voltage source applied to the photocell causes a potential difference between the cathode and an anode, and thus the emitted electrons are drawn to the anode. A current then flows in the external circuit—the amount of current depending upon the brightness of the incident light. Other refinements such as optical viewing systems are usually added for convenience.

In astronomical applications the light of a faint star causes very little current to flow in the circuit, and so it is necessary to amplify the weak signals in order to measure the light with accuracy. Part of the amplification is accomplished with conventional electronic amplifiers in the external circuit, but the important part of the amplification is accomplished inside a somewhat complex type of photocell called a **photomultiplier.** Within the photomultiplier the weak current from the photocathode is amplified about a million times. Just as in the photographic color systems, various combinations of photocathode type and filter may be selected to fit some desired color system. A three-color system (ultraviolet, blue, and visual or simply UBV) has been established and is the standard system in general modern use. The UBV system was standardized from photoelectric observations but can very closely be reproduced photographically as well.

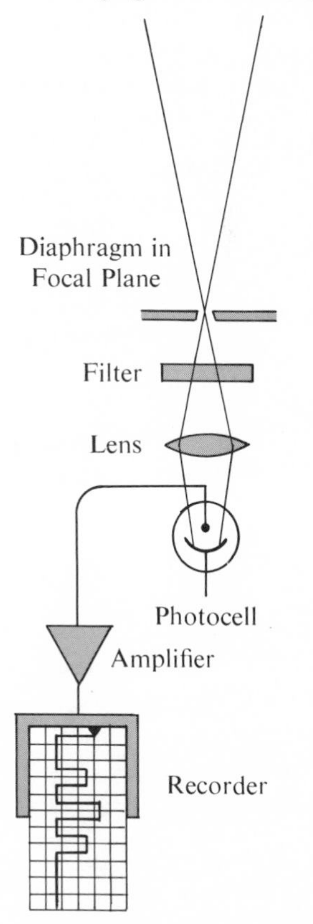

Figure 4.17. Schematic diagram of the principal parts of a photoelectric photometer.

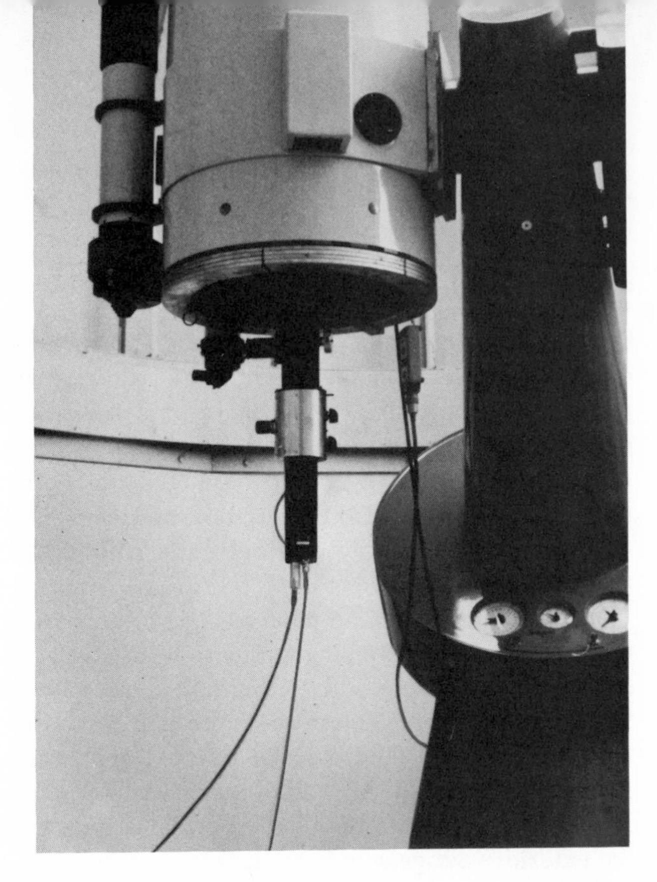

Figure 4.18. A modern photoelectric photometer at the Cassegrainian focus of a telescope. The photomultiplier tube is in the lowest section. The next section contains filters, diaphragms, and the centering eyepiece. (Wellesley College photograph.)

The complete photometer, except for the external electronics, is usually rather compact and may be easily interchanged with spectrographs and cameras in the focal plane of a telescope. Figure 4.18 shows a photoelectric photometer in position. Thus, most of the astronomer's important data today are collected from direct photographs, spectrograms, and multicolor photoelectric measurements. Time devoted to actual visual observations is usually required simply for the identification of the objects and star fields under study.

4.9 Observations beyond the Visible Wavelengths

Astronomers have traditionally studied the heavens only by means of the light which could be seen and photographed. The available range of wavelengths was originally limited by the sensitivity of the eye to wavelengths from about 3400 Å to 7000 Å. The introduction of photography brought with it the possibility of extending this range since photographic materials could be made to respond to light from below 2000 Å in the ultraviolet to about 12000 Å in the infrared. As a practical matter, however, the limitations on the wavelengths at which astronomers could work were imposed by the earth's atmosphere. At the short-wavelength end of the visible range, molecules of ozone (O_3) are very effective in blocking ultraviolet photons high up in the atmosphere. At the other end of the visible range, the absorption is due mainly to water-vapor molecules. The characteristics of the atmospheric gases are now known in considerable detail, and absorption by these gases over a very wide range of wavelengths can be predicted. Figure 4.19 shows the range of wavelengths covering most of the electromagnetic spectrum and the locations at which extraterrestrial radiation is blocked in the earth's atmosphere. In the years since 1950, some of astronomy's most exciting new developments have occurred in the search for ways to study the heavens in as many regions of the spectrum as possible.

This great new era in astronomy actually began in the 1930's with the positive discovery by Carl Jansky that radio waves came from beyond the earth's atmosphere. A few years later, Grote Reber confirmed the work of Jansky and showed that the Milky Way was a strong emitter of radio radiation. With a rapidly advancing technology in electronics after World War II, the period of truly remarkable discoveries began in radio astronomy. The successes and surprises at the relatively long wavelengths of the radio spectrum have encouraged astronomers to make observations at all possible wavelengths along the electromagnetic spectrum. Technology developed in many areas unrelated to astronomy has been adapted to astronomical problems, and astronomers have been successful in having their new and specialized detectors carried

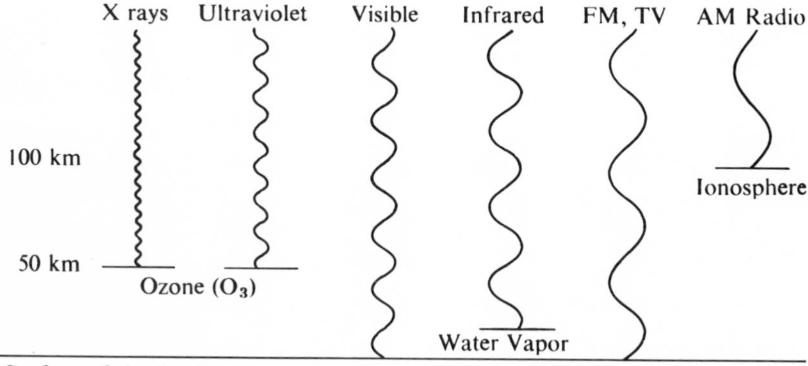

Figure 4.19. Principal factors limiting passage of radiation to the earth's surface. Visible radiation is blocked by clouds only a few miles above the surface.

aboard balloons, rockets, and satellites to record the celestial radiation which does not reach the earth's surface. We are now accumulating data in a routine way at the wavelengths of gamma rays, x rays, ultraviolet, infrared, and the radio radiation (see Figure 3.7). Detailed consideration of the instruments used in these studies is beyond the scope of this book, and we shall describe only the basic principles of some of the new techniques.

4.10 Radio Telescopes

There are actually a number of different types of radio telescopes, but one widely used style functions in a manner quite similar to that of optical reflectors. A paraboloidal surface collects radio waves over a large area just as an optical reflector collects light. A radio receiver then acts as the detector at the focal point. The wavelength of radio waves is considerably larger than that of light waves and may be said to range from a few millimeters to many meters. For the longer waves in this range, a reflecting "surface" may actually consist of a wire mesh. Figure 4.20 shows the important parts of a radio telescope system. The receiver or detector placed at the focal point of the paraboloid is one of the most crucial units in radio telescopes. Radio astronomers have pioneered the development of sensitive receivers which are free from unwanted signals or "noise" originating in the equipment itself. In operation the radio telescope sweeps across the sky, and the intensity of the detected signal is continuously recorded. Maps showing areas of equal intensity may then be constructed.

Certain types of radio data from space are effectively collected on extremely large antenna arrays. One type of array, the Mills Cross in Australia, has two intersecting arms 500 meters long. Arrays such as this make use of intereference phenomena to obtain rather

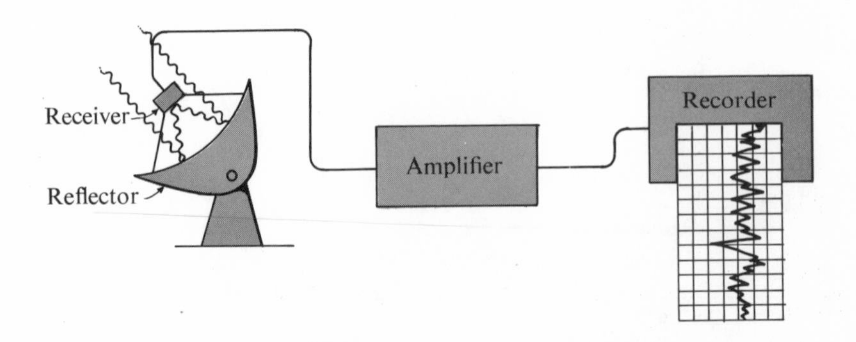

Figure 4.20. The radio telescope measures and records the intensity of the signals from particular directions in the sky. In practice the receiver is rapidly tuned through a variety of wavelengths.

good resolving power with a relatively small actual collecting area. By tilting the individual antenna elements, the array becomes somewhat directional in the north-south plane. The rotation of the earth, of course, sweeps the antenna beam around the entire sky in any 24-hour period.

Within the solar system, **radar** techniques have been employed successfully in many investigations, and their method of operation should not be confused with that of normal radio astronomy studies. The radio astronomer merely "listens" and records the incoming signals from distant self-emitting sources. The radar astronomer, on the other hand, actually transmits a signal and records the weak echo when the signal is reflected (from the moon, for example) and returned. Since radio waves travel at the speed of light, the elapsed time between transmission of the signal and return of the echo gives a measure of the distance to the reflecting object.

In the earlier discussion of resolving power an expression was given on page 72 which related resolving power to telescope diameter. The general expression

$$a = 2.1 \times 10^5 \times \frac{\lambda}{d}$$

applies just as well to radio waves as it does to light, and if a is to be made small, a radio telescope intended to detect wavelengths of a few centimeters must have a diameter of many meters. Thus to gain greater resolving power, the radio astronomers must use immense apertures. The largest antenna (300 ft) at the National Radio Astronomy Observatory (Figure 4.21) has a resolution of 8

Figure 4.21. The 300-foot-diameter paraboloidal reflector of the National Radio Astronomy Observatory. A transit-type instrument, it can be directed only toward points along the meridian. (National Radio Astronomy Observatory photograph.)

minutes of arc for wavelengths of 21 cm. For many years, radio astronomers have struggled to increase resolving power. In many cases the optical identification of strong radio sources has been seriously hampered by the indefinite locations of the radio sources.

4.11 Radio Interferometers

The desire to obtain better and better resolving power has led radio astronomers to apply the techniques of interferometry to their observations. The basic method uses a pair of antennae separated by a known distance as in Figure 4.22. As seen in the figure, the signal from some distant source will reach antenna A slightly earlier than it reaches antenna B. This time difference becomes smaller as the earth's rotation brings the source closer to the meridian, i.e., as angle q becomes smaller. At the same time, the distance, a, becomes progressively smaller. If the length of a is an even number of wavelengths, then a wave maximum will reach each of the two antennae at the same time. The two signals will reinforce each other when they are combined at C. When a has become smaller by one-half of the working wavelength, a wave minimum from B and a wave maximum from A will cancel each other at C. Thus, as a radio source approaches the meridian, the output at C will fluctuate as shown in Figure 4.22b. Because the two antennae are most sensitive in the direction of the meridian, the output signal will become progressively stronger until the source has crossed the meridian. The time at which the highest peak occurs is related to

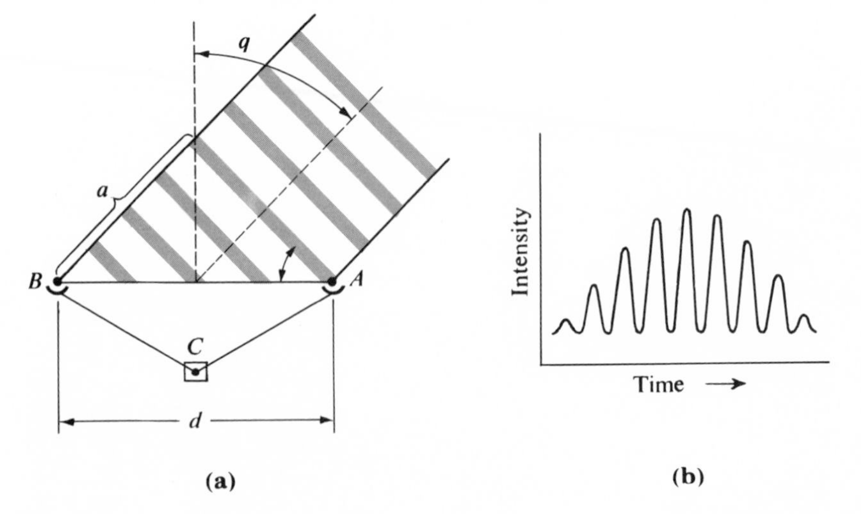

Figure 4.22. Principles of the radio interferometer. Part (b) represents the change in intensity of the signal received at C as the radio source crosses the meridian.

the direction in space of the radio source. This basic technique of combining signals from two or more antennae has been developed to remarkable levels, and radio astronomers have actually been able to achieve resolution better than that of optical telescopes. Since the angular resolution of this method depends upon the distance between the two antennae, radio astronomers have found ways of combining signals received at stations separated by nearly the diameter of the earth. This is done by tape recording the signals and carefully combining them electronically later on.

4.12 Infrared Astronomy

Between the visible and the radio spectral regions lies the infrared. The wavelengths of the infrared spectrum may be thought of as lying in the range from roughly 8000 Å, or 0.0008 mm, up to 0.01 mm. At the longer wavelengths within this range, the infrared begins to overlap the radio region. At the shorter wavelengths, infrared radiation may readily be detected as radiant heat. Thus, if one holds a hand above the burner of an electric stove, one can feel the radiated heat long before the burner begins to glow visibly. Not only is the hand a good infrared detector, but it is an infrared emitter as well. All warm or hot objects radiate energy in a manner to be described more fully in Chapter 5, and some of this energy is in the infrared. This can present serious problems for astronomers because their telescopes radiate the same sort of energy they are trying to detect. For such reasons the actual infrared detector must be placed inside a container maintained at a very low temperature. The cold container is a poor infrared emitter and so it very effectively limits the detector to the desired radiation from celestial sources. At some wavelengths the gases of the earth's atmosphere also emit infrared radiation; to avoid this, telescope have been carried aloft by huge balloons and by airplanes. Once above a large fraction of the atmosphere, the telescope then measures only the radiation from celestial sources such as stars.

Modern technology has produced several semiconductor devices which are excellent infrared detectors. Such devices are often small crystals of lead sulfide which change their electrical conductivity when infrared radiation falls on them. The lead sulfide crystal is then placed at the focus of a telescope and the infrared photons from the sky reach it through a small window in the cooled container.

In spite of the problems involved, several infrared surveys have already been made from the earth's surface. These studies include photographic observations using special infrared films and filters. Maps of the sky in the infrared indicate that many of the stars which are faint at visible wavelengths are very bright at the longer wavelengths. This distribution of energy with wavelength can be indicative of an object with a temperature less than 1000°K. It can also indicate that the star's light has passed through an extensive cloud of interstellar dust on its way to our telescopes. These possibilities will be discussed more fully in Chapters 10, 15, and 16.

4.13 Ultraviolet and X-Ray Observations

Theoretical and laboratory studies some years ago suggested that the ultraviolet spectra of hot stars should show some very interesting and significant features. The study of these features had to wait, however, until the U.S. space program could provide orbiting observatories to carry telescopes beyond the atmosphere. It is the atmosphere, of course, that protects us from severe sunburn incurred by exposure to the ultraviolet light from the sun, so we must go above the atmosphere in order to study the ultraviolet light from the stars. Today, unmanned satellite telescopes routinely measure ultraviolet radiation from the stars and transmit the data back to ground stations. Astronomers are at last able to study ultraviolet phenomena in the atmospheres of hot stars.

To most of us, x rays are tools of the medical profession and we understand only vaguely that x rays can penetrate the body to record such things as broken bones, ulcers, and tooth decay. To astronomers in today's world, however, x rays in the wavelength range from about 400 Å to 1 Å have given unexpected surprises and many new insights. The penetrating power of x rays has made it difficult to design telescopes which can focus the radiation to form an image. As a result, the earliest x-ray detectors were designed only to count the x-ray photons coming from wide-angled areas of the sky. These detectors were carried aboard rockets to heights above most of the atmosphere. Then, as the rocket rolled on its long axis, the detector scanned a limited area of the sky. Just from the meager data of a few early rocket flights, astronomers were able to know that strong celestial x-ray sources existed. This knowledge stimulated the development of another group of satellite observatories designed to survey the "x-ray sky" and to provide more detailed observations of specific sources of x rays. The early successes have also led to the design of true telescopes capable of producing and recording an x-ray image. Such equipment has already shown, for example, that on the sun the areas that are bright in x rays are usually the areas in which we see sunspots by conventional techniques.

Thus there has been a rapid expansion in the range of wavelengths accessible to astronomers, and our observational capabilities now extend from the radio region at one end to the x rays at the other. The traditional visible wavelengths still provide the basic observational data, but the invisible radiation has shown that the universe of stars, gas, and galaxies is considerably more complex than formerly imagined.

QUESTIONS

1. What serious problem inherent in refracting telescopes stimulated the development of the reflecting telescope?
2. Sketch the arrangement of the parts in three types of reflecting telescopes.
3. How do resolving power and seeing limit the amount of fine detail which may be seen with a telescope?
4. Why are light gathering and optical speed usually more important than magnification in astronomical telescopes?
5. Compare the radio telescope with the optical reflecting telescope.
6. Why is resolving power a more serious problem to the radio astronomer than to the optical astronomer?
7. List the principal parts of a spectrograph and describe the function of each.
8. What will be the effect on the appearance of spectrum lines if the width of the entrance slit of the spectrograph is widened? Why is it desirable that the entrance slit be very narrow?
9. What are some advantages and disadvantages of a grating compared with a prism in a spectrograph?
10. How can the color of a star be estimated from a pair of black and white photographs made with different filters?
11. Sketch the principal parts of a photoelectric photometer.
12. Why has it been impossible until recent times for astronomers to study the ultraviolet and x-ray emissions from the sun and the stars?

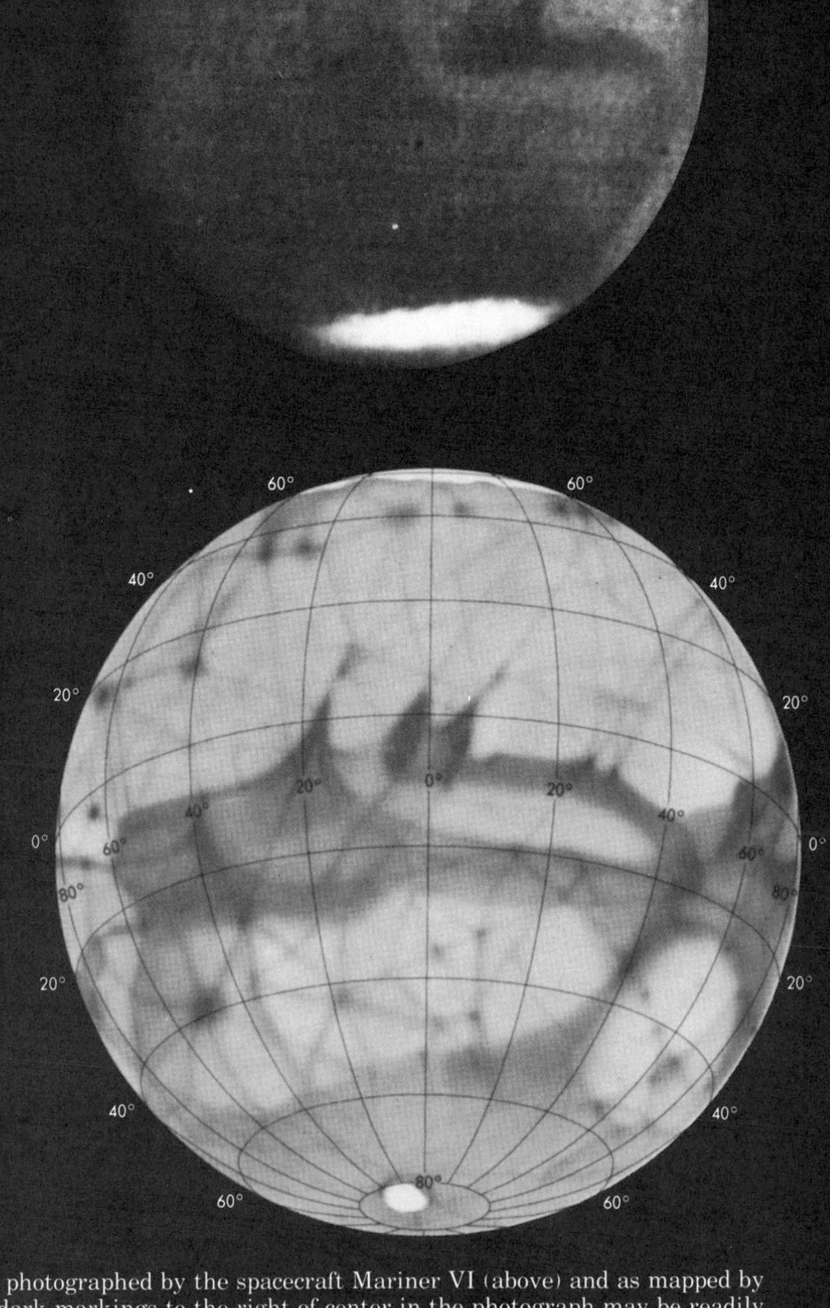

Figure 5.1. The planet Mars as photographed by the spacecraft Mariner VI (above) and as mapped by astronomers (below). The dark markings to the right of center in the photograph may be readily identified with permanent features shown in the drawing. (Photograph from NASA; drawing from Aeronautical Chart and Information Center, U.S. Air Force.)

5

The Physical Nature of the Planets

Few college students begin a study of astronomy without some basic knowledge of the distinguishing characteristics of each of the nine planets of our solar system. Lengthy tables of planetary dimensions and periods have often been memorized in elementary and secondary school. This information is important and interesting, but our greatest concern is the actual means by which the popular general picture has been derived from original observations. Though often ignored in elementary discussions, the methods of astronomy themselves are interesting and instructive. In some cases, many steps lie between the first observations and the final picture. Interpretation of the data gathered by these methods is also important, and the student should realize that astronomers living in two different eras may not interpret the same basic observation the same way. We shall, therefore, consider the various physical and orbital characteristics of the planets and note particularly the manner in which each of these charcteristics is ascertained.

The student should try to recognize the interdependence of many of the points to be discussed in the next few pages. Certain quantities must be known before other quantities may be determined, and this fact dictates the order in which the properties of the planets are discussed in this chapter.

5.1 Revolution Periods

The first quantity in the chain of interrelated properties of the planets is the revolution period. As mentioned in Section 2.3, the observable quantity from which the revolution period is calculated is the planet's synodic period: the interval between successive conjunctions or oppositions of the planet. For a planet in an orbit larger than that of the earth, the following formula may be used to compute the revolution period from the observed synodic period:

$$\frac{S \times P_e}{S - P_e} = P_p$$

P_e is the earth's revolution period, P_p is the desired revolution period of some planet, and S is the synodic period. If the planet being considered is in an orbit smaller than that of the earth, the negative sign should be changed to positive. As an example, we may apply the data of Table 5.1 (pages 98–99) and find the revolution period of Jupiter:

$$\frac{S \times P_e}{S - P_e} = \frac{398.88 \times 365.256}{398.88 - 365.256} = 4333.015 \text{ days} = 11.863 \text{ years}$$

The period determined in this manner may then be used in Kepler's third law to find the mean distance of the planet from the sun.

5.2 Angular Size and True Diameter

To the naked eye, planets are distinguished by their continuing motion among the stars. With the telescope, however, the brighter planets show a disclike appearance in contrast to the stars, which are only dimensionless points of light even in the largest telescopes. The observed size of the planet's disc depends partly upon the telescope and the magnification in use. You will recall from Chapter 4 that the size of an image in the focal plane depends upon the focal length of the telescope. Small distances between images in the focal plane represent small differences in direction between objects in the field of view, as indicated in Figure 4.2. Thus the scale of a telescope is usually specified in terms of some number of seconds of arc per millimeter in the focal plane. The scale of the McCormick refractor (see Figure 4.1) is 21 seconds of arc per millimeter. A distance of 35 mm between two stars on a plate made with that telescope represents an angle of 12 minutes of arc be-

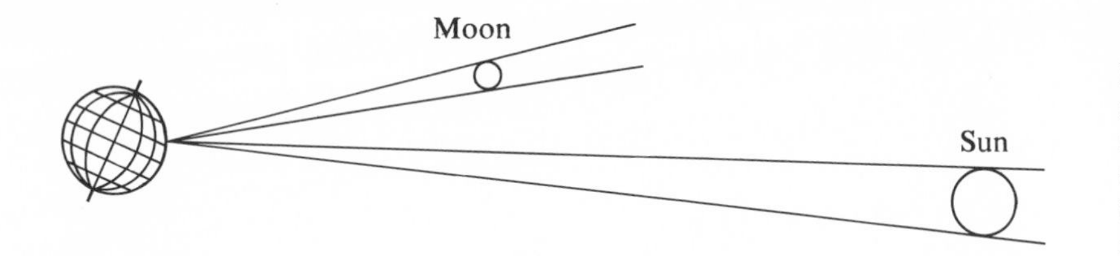

Figure 5.2. The angular diameters of the sun and moon are very nearly the same even though the sun is much larger than the moon. The figure is not drawn to scale.

tween the directions to the two stars. When we see or photograph a planet, then, the planet's image is a disc in the focal plane. From the measured size of this image we may compute the **angular diameter** (that is, the angle between the directions to the two ends of any diameter) of the planet.

Clearly, the observed angular diameter of a body must depend upon the true linear diameter of the body and its distance from the observer. We have, for example, the interesting case of the sun and the moon, which are quite different in both actual size and in distance from us. Nevertheless, by a fortuitous circumstance the moon is at such a distance that its smaller diameter subtends the same angle as the much larger diameter of the more distant sun (Figure 5.2). The angular diameters of both of these bodies are approximately one-half degree. For the planets, the angular diameters, which can only be determined telescopically, can be converted to linear diameters only when the earth-planet distance at the time of the observation is known. These distances in turn are obtained from the calibrated model solar system described in Chapter 2. (See Figures 5.3, 5.4, and 5.5.)

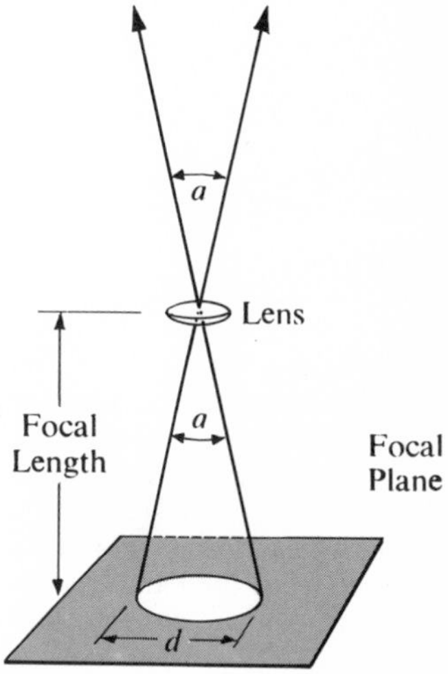

a
Lens
Focal Length
a
Focal Plane
d

Figure 5.3. The angular diameter of a planet may be determined from the measured size of the planet's image in the focal plane of the telescope.

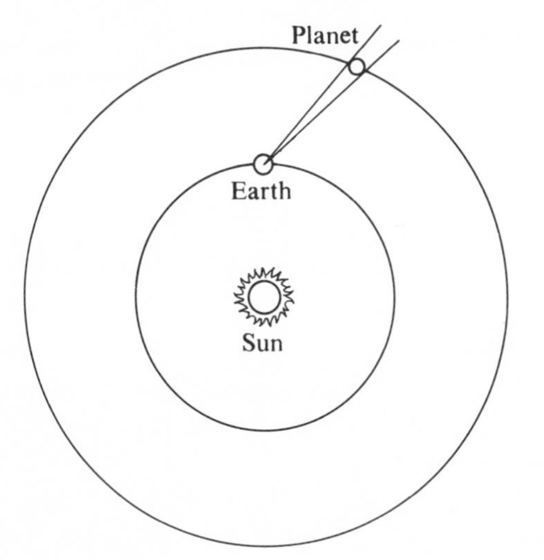

Figure 5.4. The diameter of a planet may be calculated after the angular diameter of the planet has been observed. The distance from the earth to the planet is computed from the geometry of earth, sun, and planet at the time of observation.

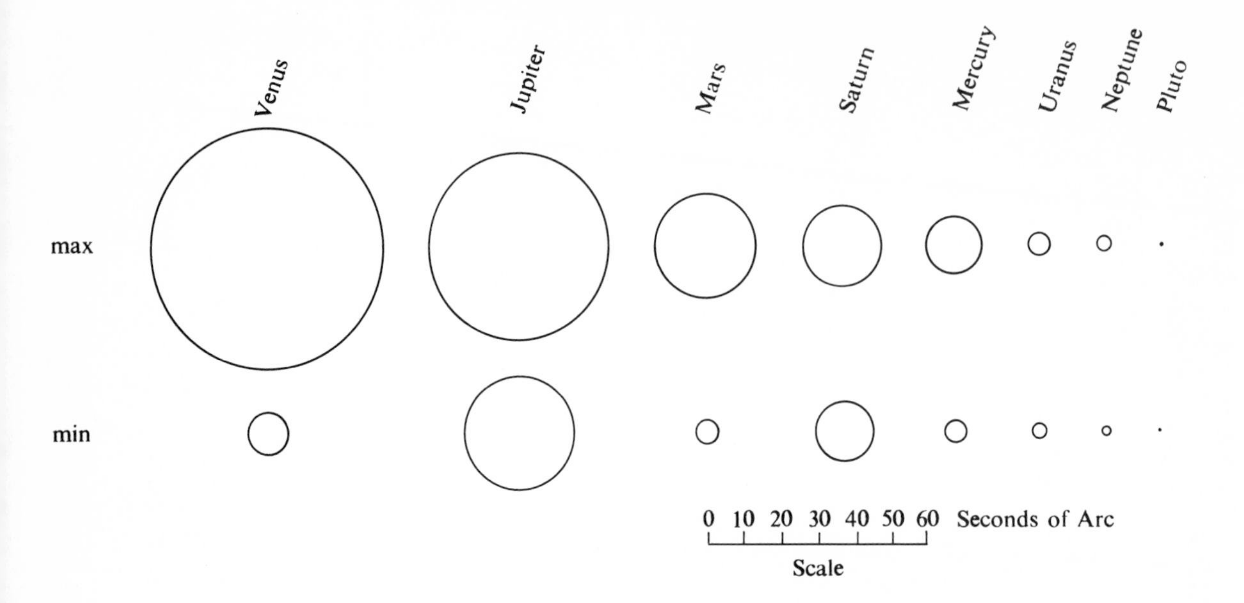

Figure 5.5. Largest and smallest angular diameters of the planets drawn to scale. The observed angular diameter depends upon the planet's true diameter and its distance from the earth at the time of the observation.

5.3 Rotation Periods

Long before the rotation of the earth had been proved, observers had seen quite clearly that some of the other planets were rotating. The actual surface of Mars is plainly visible through the thin Martian atmosphere, and conspicuous permanent markings are numerous. An observer can note the changes in position of these surface features over a period of time and measure the rotation period with relative ease (see Figure 5.6). In the case of Jupiter, the surface markings are not visible. We see only the tops of the clouds. There are in the clouds, however, semipermanent irregularities which persist long enough for an observer to note the rotation period of just under ten hours.

In the case of Venus, Saturn, Uranus, and Neptune, however, features suitable for rotation measurements are seldom present, and the astronomer must fall back on other methods. If a planet is rotating with its axis at right angles to the line of sight from the earth, one side of its equatorial plane should be approaching the earth as the other side moves away from it. Thus the rotation imparts a radial velocity to each edge of the equator. If the rotation of the planet is rapid enough, these radial velocities should be detectable spectroscopically. The spectrum of light reflected from one edge should show a Doppler shift to the blue, and the spectrum of light from the other edge should show a Doppler shift to the red.

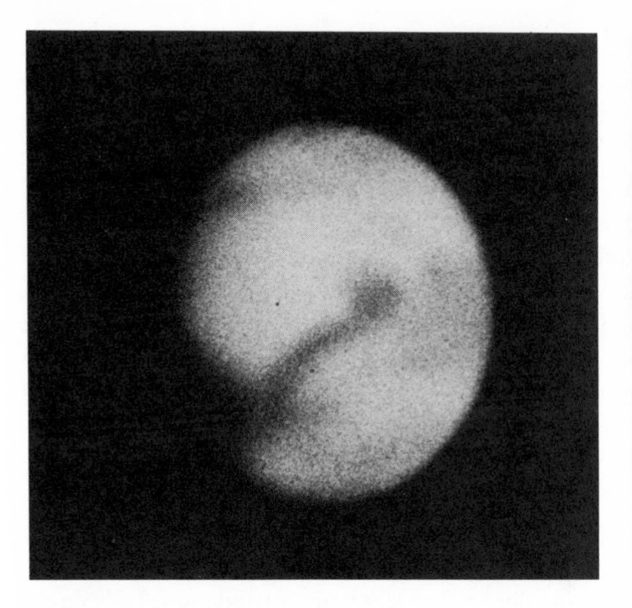

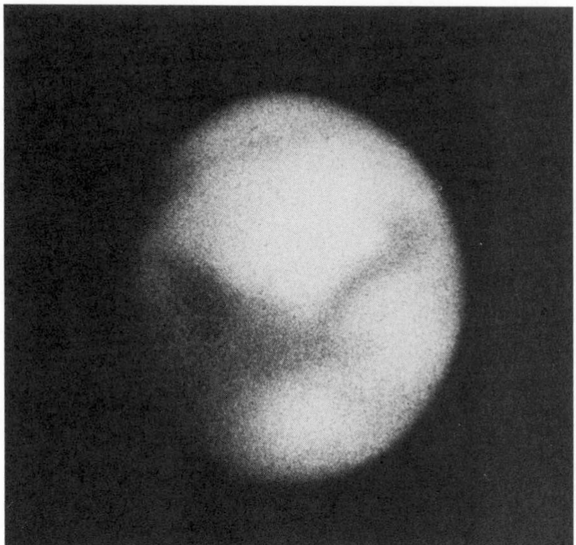

Figure 5.6 Two views of Mars taken several hours apart. The change in position of the surface markings clearly indicates the planet's rotation. The clarity of the surface features is due to the fact that these photographs were made in red light. (Hale Observatories photographs.)

Figure 5.7 shows this effect rather strikingly for the planet Saturn. Here the slit of the spectrograph has been placed along Saturn's equator. The resulting spectrum lines show quite a definite tilt due to the fact that the radial velocities at opposite sides of Saturn's disc are equal but opposite.

In the case of Venus, clouds have prevented visual measurement of the rotation period; in the case of Mercury, observations are

Figure 5.7. Spectrum of Saturn showing a conspicious tilt due to the rapid rotation of the planet. (Lick Observatory photograph.)

difficult to make and surface markings are not conspicous. As a result, an incorrect rotation period was accepted for Mercury for many years. In recent years, however, radar has been used to measure the rotation periods of both Mercury and Venus. Under ordinary circumstances, radar pulses are transmitted to some distant object, reflected back to the earth, and received again at the original source. The travel time of the pulse then becomes a measure of the distance since it is known that the pulses travel at the speed of light. This is the customary application of radar for measuring simple distances, and the pulse is normally transmitted at some desired wavelength in the microwave region. In rotation studies those who analyze the returned pulses are interested in the wavelengths of the returned signals. Reflection from the rotating planet causes Doppler shifts, which, as in the case of Saturn above, depend upon the portion of the disc from which the signal is reflected. The result is a measurable spread, in wavelength, of the returned signals around the wavelengths of the transmitted signal. From an analysis of this spread in wavelength and the changes in this spread, the rotation periods of Venus and Mercury have been derived.

Pluto is so small and so distant that no very reliable rotation period has yet been determined. Variations in Pluto's brightness have been used to give a tentative rotation period on the assumption that the reflecting surface of Pluto includes limited areas of high reflectivity.

5.4 Mass, Density, and Escape Velocity

In Chapter 2 it was shown that Kepler's third law was modified by Newton to include the masses of the sun and the planets. If period and distance are expressed in years and astronomical units, respectively, and if the masses are expressed in terms of the sun's mass as a unit, this equation may be written

$$P^2(m_1 + m_2) = D^3$$

This expression will hold for any pair of bodies revolving one about the other (or actually revolving about their common center of gravity). It applies just as well to a satellite orbiting a planet as it does to a planet orbiting the sun provided, of course, that the quantities are always expressed in the proper units. It follows, then, that for any planet with a satellite the quantity $(m_1 + m_2)$ becomes known as soon as the satellite's revolution period and distance from the planet are known. The quantity $(m_1 + m_2)$ is the sum of the masses of the planet and the satellite. Again, in parallel to the sun-planet case, this sum is essentially the mass of the planet if the mass of the satellite is relatively small.

These relationships may be illustrated by comparing the earth-moon system with Neptune and its satellite Triton. Triton is approximately the same size as our moon and lies just about as far from Neptune as the moon does from the earth. Whereas our moon takes 27⅓ days to go around the earth once, Triton makes its revolution around Neptune in only 7 days. The difference in these two revolution periods is due to the much larger mass of Neptune.

The formula mentioned above permits us to know the mass of a planet only in units of the sun's mass (that is to say, as a fraction of the mass of the sun). To know a planet's mass in more familiar units, such as tons or kilograms, one must know the mass of the sun in these units. The sun's mass is calculated in terms of the earth's mass by knowing the gravitational attraction which the sun exerts to keep the earth in its orbit. And the earth's mass, the first quantity which must be known in this sequence, is determined in the laboratory. The experiment by which the earth is "weighed" is carried out as a demonstration in many introductory physics courses.

After the size and mass of a planet have been found, the density, or mass per unit volume, may be computed. For planets with satellites accurate masses are known, and for Mercury and Venus masses have been established from other effects. We have reliable density values for all the planets except Pluto. As may be noted from Table 5.1, these densities fall roughly into two groups. Some are comparable to the earth's density of 5 times the density of water, and some are only a little greater than 1.

The escape velocity is another important quantity which may be calculated when the mass and radius of a planet are known. Again, beginning with Newton's law of universal gravitation, it is possible to derive an expression for the actual velocity of a small body in orbit around a larger one. Kepler's second law implies that this velocity must be greatest when the two bodies are closest together and least when the bodies are at their greatest distance apart. The general expression for the orbital velocity is then

$$v^2 = G(m_1 + m_2)\left(\frac{2}{r} - \frac{1}{a}\right)$$

Table 5.1. Data on the planets.

Planet	Synodic Period	Rev. Period	Distance from Sun	Greatest Ang. Diam.	Diam.	Rotation Period
	(days)	(years)	(A.U.)	(sec of arc)	(km)*	(hours or days)
Mercury	115.88	0.241	0.387	10.90	4840	59^d
Venus	583.92	0.615	0.723	61.00	12,200	243^d
Earth		1.000	1.000	...	12,756	$23^h56^m4.1^s$
Mars	779.94	1.881	1.524	17.88	6760	$24^h37^m22.6^s$
Jupiter	398.88	11.862	5.203	46.86	142,700	$9^h50.5^m$
Saturn	378.09	29.458	9.540	19.52	120,800	10^h14^m
Uranus	369.66	84.013	19.180	3.60	47,600	10^h49^m
Neptune	367.49	164.79	30.07	2.12	44,400	15^h
Pluto	366.74	248.4	39.44	0.22	6000	6.39^d

*See Appendix Table A.4 for planetary diameters in miles.

As in other expressions in earlier chapters, G is the gravitational constant, m_1 and m_2 are the masses of the two bodies, and r is the distance between them at a given moment. The semimajor axis of the orbital ellipse is a. Consider next the case when the orbiting body is at its greatest distance from the principal body, and note the effect on the shape of the orbit when the velocity is increased. From the above expression the value of a must increase with an increase in v, and the orbit must become larger. When a particular value of v is obtained, a and r will be equal, and the orbit will be circular. Any further increases in velocity will result in a new series of elliptical orbits of ever increasing size. When the velocity becomes very large, the value of a becomes infinite. The orbit is no longer an ellipse but is now a parabola. Since the parabola is an open figure, the orbiting body never returns. This velocity, just sufficient to put a body into a parabolic trajectory, is the **escape velocity** and may be computed from the expression

$$v^2 = G(m_1 + m_2)\frac{2}{r}$$

Note that the escape velocity is $\sqrt{2}$ times the circular velocity.

In the case of a planet, the mass is very much larger than the mass of a gas molecule or of a spacecraft which might be launched from it. Therefore, only the major mass need be considered. The escape velocities calculated from the accepted values of mass and radius are listed in Table 5.1.

No. of Satellites	Mass (earth's mass = 1)	Density (water = 1)	Escape Vel. (km/sec)	Temp. (°K)	Atm.	Albedo
0	0.054	5.4	4.2	611	. . .	0.06
0	0.815	5.1	10.3	700	CO_2,H_2O	0.76
1	1.000	5.52	11.2	295	N_2,O_2,H_2O,CO_2	0.40
2	0.108	3.97	5.0	270	CO_2,H_2O	0.15
12	317.800	1.33	61	135	CH_4,NH_3,H_2	0.58
10	95.2	0.68	37	125	CH_4,NH_3,H_2	0.57
5	14.5	1.60	22	103	CH_4,H_2	0.80
2	17.2	2.25	25	108	CH_4,H_2	0.71
0	?	?	. . .	. . .	. . .	0.15

5.5 Albedo

The light which we receive from the planets and the moon is reflected sunlight, and several rather obvious factors affect the amount of this reflected sunlight which reaches our eyes and instruments. Consider first the distance of a planet from the sun. The intensity of light varies inversely with the square of the distance from the source, so the amount of light reaching a planet will depend on the planet's distance from the sun. On Mars, which is 1.5 times as far from the sun as the earth, each square centimeter of surface facing the sun will receive only $1/1.5^2$ or 0.44 times the amount of light reaching a similar surface on the earth. Second, the same inverse-square law causes the brightness of a planet to vary throughout the year as the distance from the earth to the planet changes due to orbital motions.

Most important, perhaps, the brightness of a planet depends upon the planet's **albedo.** The albedo is the ratio of the amount of light reflected by the planet's surface to the amount of light incident upon the surface. Even though the full moon is extremely bright, its albedo is only 0.07. The albedo of Venus is 0.76, which means that a rather high percentage of the incident light is reflected from that planet. Comparing the albedos of the moon and planets in Table 5.1, we see that no planet is a very efficient reflector, but that the cloud-covered planets are better reflectors than those whose light is reflected by the solid surface.

If only a fraction of the solar radiation incident on a planet is reflected, the rest of the incident energy must be absorbed. It is this absorbed energy that ultimately determines the temperature of a planet.

5.6 Theoretical Temperatures

Many factors affect the temperature of a planet. As we know from our experience with the earth, we may expect a variety of temperatures at different latitudes on a planet's surface and at different heights in a planet's atmosphere. The temperature, in turn, has profound influence on many conditions in which we are interested. The temperature of a planet, however, is quite difficult to measure. A first approximation of planetary temperature may be based on careful measurements made here on the earth and extended to the other planets. For at least some of the planets, direct telescopic temperature measures have also been made, and in recent years radio techniques have proved to be quite successful.

Extensive studies have led to values for a quantity known as the **solar constant.** This unit represents the amount of energy which is received from the sun on each square centimeter of the earth's surface during each minute. The accepted value of the solar constant today is 1.388×10^6 ergs per square centimeter per minute and includes corrections for the amount by which the direct solar radiation has been attenuated by the atmosphere of the earth. If this is the amount of energy passing each square centimeter at the earth's distance from the sun, we should be able to calculate in a straightforward way the amount of energy passing each square centimeter at any other distance from the sun. As mentioned above in the discussion of albedo, the solar energy not reflected by a planet is absorbed and has an important relationship to the temperature. With knowledge of the solar constant, the size of the orbit, and the albedo, one can compute a theoretical temperature for each of the planets. This derived temperature should take into account the rate of rotation of the planet since it will be important in determining the rate at which the planet cools by radiation to maintain an equilibrium. A rapidly rotating planet will be more or less uniformly warmed by sunlight and will radiate heat into space from all parts of its surface. A slowly rotating planet, on the other hand, might be expected to have one hot side and one cool side.

Direct temperature measurements may be made in two types of planetary studies, and such temperatures are in reasonable agreement with the theoretical values. An understanding of the methods used in such studies requires a review of the basic radiation laws of physics.

5.7 Planck's Radiation Law

It is well known that when a bar of iron is sufficiently heated it becomes luminous. First it begins to glow a dull red, and as the temperature rises, the color changes to a straw-yellow and then to white. At the same time the bar becomes brighter. If one were to look at the glowing bar with a spectroscope, he would see a continuous spectrum (Chapter 2). At first the spectrum would seem brightest in the red region, but as the temperature rose, the spectrum would be brighter in the yellow and then in the green regions. In 1901 the German physicist Max Planck discovered the following mathematical relationship governing this behavior.

$$E_\lambda = \frac{C_1 \lambda^{-5}}{\exp(C_2/\lambda T) - 1}$$

E_λ is the energy in ergs per second radiated at wavelength λ cm, T is the temperature in degrees Kelvin, and C_1 and C_2 are constants. This formula, known as Planck's radiation law, permits one to compute the intensity of the radiant energy emitted at a variety of wavelengths for a theoretically perfectly radiating solid at a particular temperature. When such computations are performed for many wavelengths, the intensities may be plotted and joined by a smooth curve to obtain a figure like that seen in Figure 5.8a. Such a curve is referred to as a **spectral-energy curve** or **spectral-energy distribution** or a **Planck curve.**

The curves defined by Planck's law are never matched exactly by the observed intensities radiated from actual objects. A perfect Planck curve would be produced only by a theoretically perfect radiator and such objects simply do not exist. In the early radiation studies, physicists coined the term **black body** to describe the theoretical object which would radiate energy precisely according to Planck's law. Theoretical curves for bodies at temperatures of 4000° and 8000°K are plotted in Figure 5.8b. Two points should be noted in regard to these curves. First, as the temperature increases, the area under the curve increases markedly, representing a significant increase in the total energy radiated by the body. Second, the wavelength of maximum intensity, or the peak on the curve,

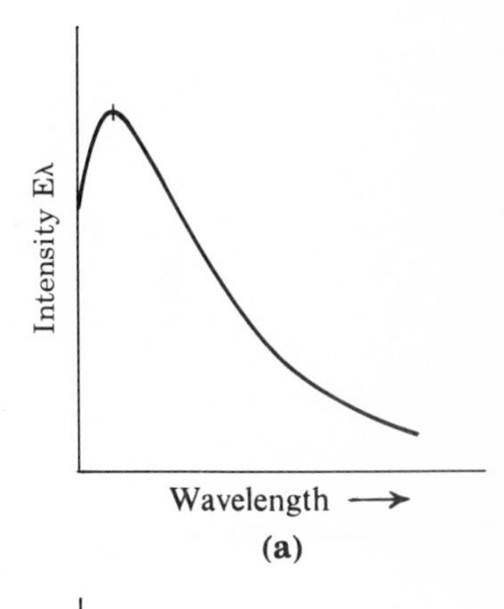

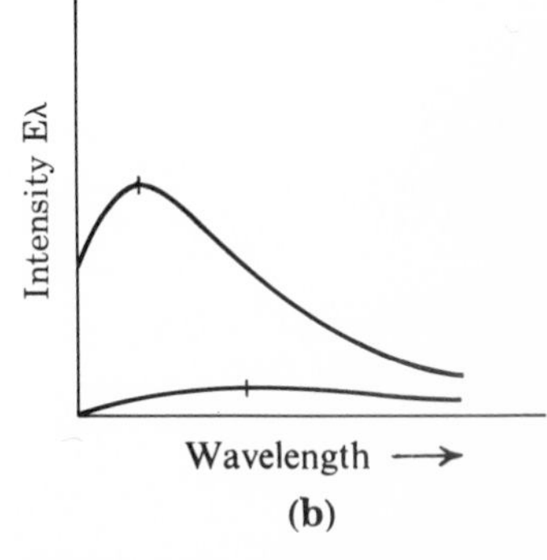

Figure 5.8. (a) The Planck radiation curve for a body at a temperature of 9000°K. (b) Similar curves for bodies at 8000°K and 4000°K.

shifts to shorter wavelengths as the temperature increases. This repesents the change from red to yellow to white in the color of the light emitted by the hot solid.

Well before Planck's derivation of this law, other workers had derived expressions to calculate the above two effects. The Stefan-Boltzmann law states

$$E=\sigma T^4$$

indicating that the total energy E is proportional to the fourth power of the temperature T; σ is a constant. Wien's law states

$$\lambda_{\text{m.i.}}=\frac{K}{T}$$

showing that the wavelength of maximum intensity is inversely proportional to the temperature. Both Wien's law and the Stefan-Boltzmann law are often derived from Planck's law in college calculus courses. Planck's law applies just as well at the temperatures of the planets as at the temperature of the sun and at higher temperatures as well.

If only a fraction of the solar radiation incident upon a planet is reflected, the rest of the incident energy must be absorbed. It is this absorbed energy which contributes significantly to the temperature of a planet's surface and atmosphere. The planet will be warmed as it absorbs energy, and the warmed planet itself then radiates. If the planet absorbs energy at a faster rate than it radiates, the planet's temperature will rise. As follows from the Stefan-Boltzmann law and from Planck's law, at a higher temperature the planet is able to cool by radiation more quickly. Eventually an equilibrium temperature is reached, when the planet radiates energy at the same rate as it absorbs energy, and its temperature becomes stable.

5.8 Observations of Planetary Temperatures

The spectral-energy curve of a planet, if it could be observed and plotted over a large enough range of wavelengths, should show two peaks, as in Figure 5.9. The larger peak represents reflected sunlight and is approximately the same shape as the spectral-energy curve for the sun. The second peak represents the energy radiated

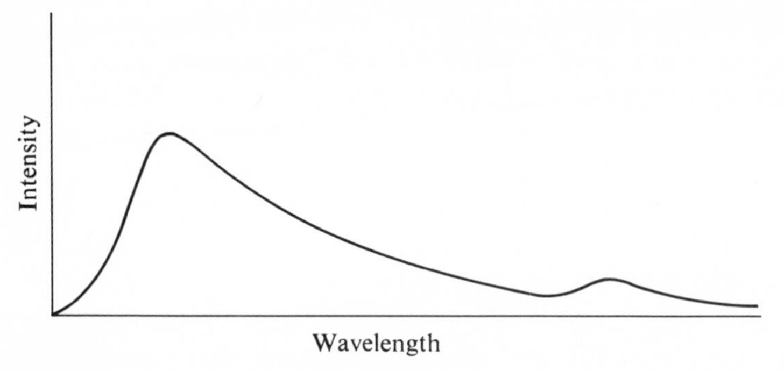

Figure 5.9. Idealized radiation curve for a planet. The higher peak represents reflected solar radiation; the lower peak represents radiation from the planet itself.

directly by the planet itself. The solar energy absorbed by the planet has raised its temperature, and the planet itself now radiates as a black body. Since the planets are quite cool in relation to the sun, the second peak is well over in the long-wavelength or radio region of the spectrum. From the surface of the earth we can not observe this entire double-peaked curve for a planet because the earth's atmosphere is opaque in much of the infrared and radio parts of the spectrum. Nevertheless, the black-body curves at various temperatures are different enough to permit fairly accurate determination of temperature from studies at the few available wavelengths.

The earliest attempts to measure lunar and planetary thermal radiation were made in the early years of this century by astronomers using a sensitive **thermocouple** at the focus of a large telescope. A thermocouple is simply a junction of two metals such as bismuth and tin. When the thermocouple is connected into a suitable electric measuring circuit and heated, a current will flow. When calibrated, this device can be used to make very sensitive measurements. Thermocouple measurements made in the infrared wavelengths indicated not only thermal radiation from the planet but reflected solar energy in the same wavelengths as well. One of the most difficult parts of this experiment was the isolation of the desired planetary radiation from the total energy received. This was accomplished essentially by observing the solar radiation alone and subtracting it from the combined solar plus planetary radiation.

Since the earth's atmosphere is transparent to radio waves of centimeter wavelengths, the first radio astronomers naturally looked for thermal radiation from the moon and the planets. (With reference to Figure 5.9, it may be noted that observations at just one wavelength in the radio region should give an intensity which can be related to the Planck curve for some specific temperature.)

When the first results of radio temperature studies became available, the values seemed to be in great disagreement with the older temperatures. More complete analysis, however, eliminated most of the disagreement for it was realized that the thermal radiation coming to us in the radio and optical wavelengths was being emitted from different depths in the bodies' surface layers. The longer wavelengths could escape from a deeper level than could the shorter wavelengths. In the case of the moon, radio astronomers recorded a lower and more stable temperature in the longer wavelengths than the infrared observations had given. Clearly the temperature of the lunar crust a few inches below the surface was nearly constant while the surface itself varied considerably in temperature throughout the lunar day.

For Venus the problem has become somewhat complex as data at a number of wavelengths have accumulated. The results of four studies are summarized in Figure 5.10, in which the temperatures of a number of possible atmospheric levels are indicated. For any planet with an atmosphere the picture must necessarily be somewhat similar. The recorded data must be analyzed in terms of some assumed structure for the atmosphere. If the assumptions are incorrect, then the computed temperatures do not relate to the assumed atmospheric levels.

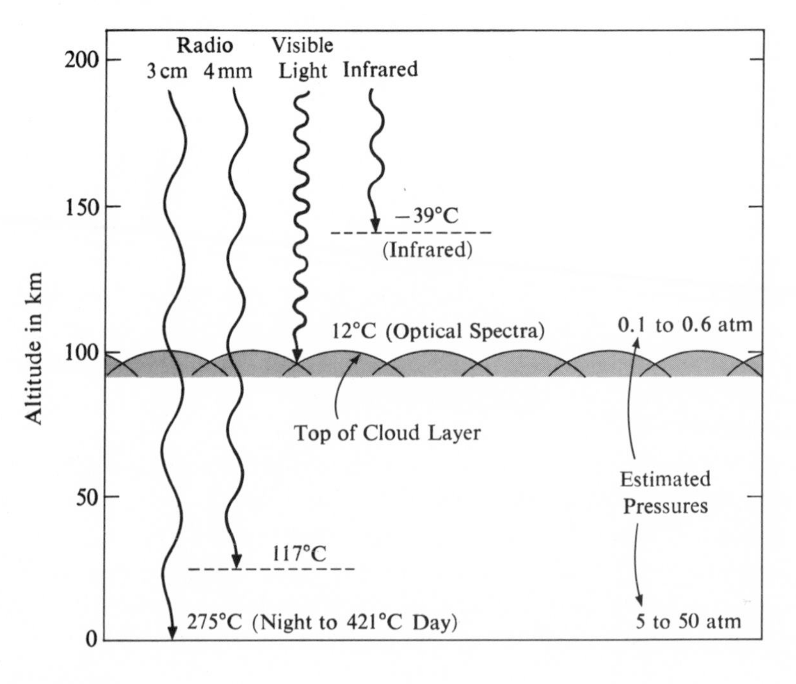

Figure 5.10. For Venus the measured temperature depends upon the wavelength of the observations. The differences in the observed values indicate the temperature at several heights above the surface. (After A. G. Smith and T. D. Carr, *Radio Exploration of the Planetary System*, Van Nostrand Reinhold Co., 1964.)

5.9 Atmospheres

The first evidence for the presence or lack of an atmosphere for a planet or the moon may be determined simply by telescopic inspection. If surface features are clearly seen, as on the moon, there is little or no atmosphere. If heavy clouds obscure the surface totally, then there must be a dense atmosphere. Mars lies somewhere in between these two examples, for a good bit of its surface detail is readily visible with large telescopes while, at times, clouds and haze obscure parts of the Martian surface.

We need, however, to look beyond mere observational data and attempt to identify gases present in the planetary atmosphere as well as measure their pressures and relative abundance. This problem, like many others, may be approached from both the theoretical and observational points of view. Let us consider first a theoretical approach based on the kinetic theory of gases. According to this theory, the molecules or atoms of a gas are in constant motion and the velocities of the individual particles are determined by the temperature of the gas. The higher the temperature, the faster the particles will be moving. It is out of the question to try to make direct measurements of the speeds of individual particles, but the effects of these velocities are easily detected. When the moving particles strike the walls of a container, some of their energy of motion is transferred to the container. We detect this energy as a rise in the temperature of the walls of the container. According to the theory, there is an average speed for particles of any particular gas at any specified temperature. Most of the particles will be moving at speeds somewhere near this average velocity—some faster and some slower. In the normal course of collisions among particles, a few particles will be moving considerably slower and a few considerably faster than the average. A **frequency distribution** may be drawn as in Figure 5.11 to show this effect. In a mixture of gases a separate curve could be drawn for each gas, and the average velocity indicated by the peak of the curve would be slower for the heavier gases than for the lighter gases.

Earlier in this chapter the escape velocities of the planets were discussed. Now we may see that escape velocity and temperature combine to determine the kind of an atmosphere that a planet may have. Suppose that the moon had once had an atmosphere the same as earth's. The moon's average temperature would have been similar to earth's, and so the atoms and molecules in the lunar

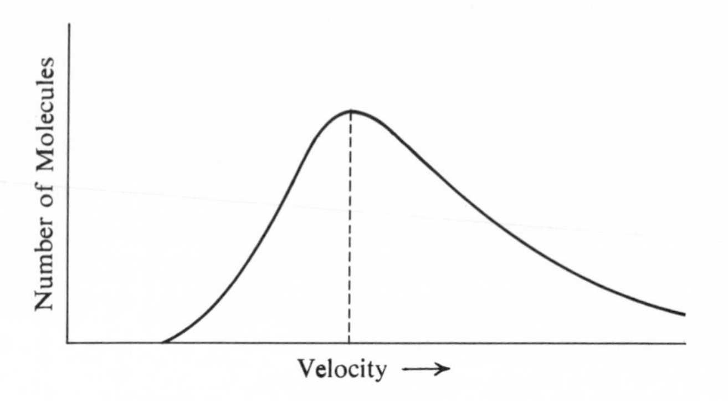

Figure 5.11. Frequency distribution of molecular velocities. As the temperature is raised, the whole curve moves toward the right, indicating that the average velocity of the molecules becomes greater.

atmosphere would have been moving at predictable speeds. These velocities are well below the escape velocity here on earth, but they are above or at least close to the escape velocity on the moon. If a particle is moving faster than the escape velocity and does not collide with another particle, it will be lost to the planet. Thus, atom by atom and molecule by molecule, the moon would have lost its atmosphere. Many years would have been required, since only in the uppermost levels of the atmosphere where the density is very low would collisions be unlikely. In the lower levels the paths of the particles are constantly changed by collisions and escape is impossible. It should be noted that planets will lose their atmospheres even if the temperature is such that the average velocity seems to be safely below the escape velocity. In the upper atmospheric levels there will almost always be a few fast-moving particles which will be lost. In Table 5.2, critical temperatures for a number of gases are listed for each planet.

Table 5.2. Planetary temperatures, in °K, required for escape from planetary atmospheres.

	H_2	H_2O	N_2	O_2	CO_2	Xe
Mercury	56	500	779	889	1,220	3,650
Venus	339	3,030	4,720	5,390	7,410	22,100
Earth	403	3,600	5,610	6,400	8,810	26,300
Mars	83	739	1,150	1,310	1,800	5,390
Jupiter	10,700	95,500	148,000	170,000	233,000	696,000
Saturn	3,470	31,000	48,200	55,000	75,700	226,000
Uranus	1,480	13,200	20,500	23,400	32,300	96,200
Neptune	2,000	17,900	27,900	31,800	43,800	131,000

To summarize, our laboratory knowledge of the molecular weights of the gases combined with our knowledge of planetary temperatures permit us to calculate which gases could persist in the atmosphere of each individual planet or satellite. The moon simply has too low an escape velocity to retain an atmosphere. An interesting parallel is seen, however, in comparing one of Saturn's satellites, Titan, with the moon. Titan is the only satellite known to have an atmosphere, and it retains its gases even though its mass is comparable to the mass of the moon. The big factor is, of course, temperature. At Saturn's great distance from the sun, the molecules simply do not move fast enough to exceed the escape velocity. This helps us to explain also why the large, cold planets (Jupiter, Saturn, Uranus, and Neptune) have such extensive atmospheres. Carrying this analysis to completion one may calculate a theoretical atmosphere for each planet. The task then remains to see if the observed makeup of the atmospheres agrees with the theoretical composition.

5.10 Planetary Spectra

The critical data needed for the identification of gases in planetary atmospheres come from studies of the spectra of the planets. Since the light from the moon and the planets is all reflected sunlight, the spectra of those bodies should primarily look like the spectrum of sunlight. Wherever atmospheres are present, the reflected sunlight should have superimposed on it the absorption lines of the atmospheric gases. It should be remembered, also, that the light ultimately has to traverse the earth's atmosphere to reach the spectrograph. The planetary spectrum is, then, a composite of three spectra: (a) that of the sun, (b) that of planet's atmosphere, and (c) that of the earth's atmosphere. The normal spectrum of sunlight is a combination of two of these, so we may compare the planetary spectrum with the spectrum of normal sunlight. Extra absorption features in the planetary spectrum not found in the solar spectrum are the clues to the composition of planetary atmospheres.

The sun's absorption features arise in the hot atmosphere of the sun and will be discussed at length later on. Because of the high temperatures in the sun's atmosphere, the gases are nearly all composed of atoms: it is too hot for complex molecules to remain. In the planetary atmospheres, on the other hand, the gases are cooler and fairly complicated molecules do exist. Most molecular spectra show complicated absorption structures in bands rather than the sharp, well-defined lines seen in atomic spectra. Thus, absorption features of certain molecules arising in planetary spectra are easily distinguishable from solar absorption features.

From laboratory studies we know quite well the composition of the earth's atmosphere, so we can ascribe certain absorption features to gases in the earth's atmosphere. These features are often referred to as **telluric lines** and are due to the presence of such gases as water vapor, carbon dioxide, and molecular oxygen. Some of these life-sustaining gases might well be present on other planets. However, their presence will be difficult to confirm because for a particular gas the absorption lines originating in the planet's atmosphere will have the same wavelengths as those originating in the earth's atmosphere. The Doppler effect offers us a way to solve this problem. If the planet's spectrum is photographed at a time when the distance between the earth and the planet is changing rapidly, the planet's spectrum lines will be shifted away from their normal positions. Observations to confirm the presence of carbon dioxide on Venus and Mars were made at such times.

In the spectra of the major planets, absorption bands of methane and ammonia are prominent, and it is interesting to trace the relative strengths of these features at progressively lower temperatures. As the temperature is decreased, ammonia liquefies sooner than methane and so would be removed from an atmospheric mixture of the two gases. The temperatures of Jupiter, Saturn, Uranus, and Neptune are related to their distances from the sun, Jupiter being the warmest. It is not surprising, then, to note that the ammonia bands become weaker as one progresses from Jupiter to even colder planets.

From their spectra, positive identification of many gases in planetary atmospheres has been made. These gases are listed in Table 5.1. Additional indirect evidence has suggested the presence of large amounts of molecular hydrogen on the major planets. This would be expected since the escape velocities are high and the temperatures are low. Even the light gases could be held. The most abundant gas in our atmosphere is molecular nitrogen, and it seems likely that nitrogen is an important constituent of other atmospheres also. Unfortunately, the confirmation of the presence of nitrogen must wait for planetary spectra obtained from above the earth's atmosphere.

5.11 Surface Mapping By Radar

Another of the spectacular triumphs of radar in astronomy has been the production of a preliminary contour map of Venus' surface. The map was produced by Doppler-delay techniques which are, in a way, extensions of the techniques used to find the rotation periods. Very slight changes in the wavelength of the reflected radar pulse are analyzed along with very small differences in the time of arrival of the returned signal. After removal of the effects due to rotation and the spherical shape of the planet, remaining irregularities are attributed to surface features.

5.12 Unique Features of Individual Planets

The basic physical data on the planets have been summarized in Table 5.1, and the methods by which these data have been acquired have been described. Beyond the tabulated data, however, there are interesting details which set each planet apart from all the others.

Mercury

We know from Chapter 1 that, as a morning and evening star, Mercury is never more than 28° from the sun in the sky. By the time the sky is dark in the evening, Mercury is very low, and the observer must look through a long path in the earth's atmosphere. As a result, Mercury is difficult to observe at best. Theory tells us that Mercury should not be expected to have an atmosphere. It has too low a surface gravity and too high a temperature. Telescopic observations reveal nothing to suggest that the theory is wrong in this case, and there are no traces of clouds or haze. However, a few indistinct surface markings have been noted, and it is from these that astronomers have tried to determine a rotation period for Mercury. Because of the geometry of the situation, the period during which the surface may be studied is a fairly small fraction of the revolution period.

Early observations of Mercury indicated that its rotation period was long compared with that of the earth and was, perhaps, the same as its own revolution period. Later, it was shown that for a planet of Mercury's mass and position near the sun, equal rotation and revolution periods might be expected. Gradually a rotation period of 88 days was established in astronomical literature. Other observers confirmed this period. The first serious questioning of this 88-day period arose only in 1964 with the publication of results of radar studies of this small planet. These data showed surprisingly that Mercury's rotation period was 57.8 days, rather than the traditional 88 days. Reexamination of some of the better visual observations shows that they are compatible with the 57-day period, and the new period is now generally accepted. Studies of tidal interaction of Mercury and the sun show that the observed rotation rate represents synchronization at nearly two-thirds of the orbital period and is certainly dynamically possible. (See also page 123.)

Venus

Through even a small telescope Venus shows interesting changes of phase as it circles the sun. At greatest elongation it appears as a semicircle. Then, as it comes closer to the earth, it becomes a progressively larger but narrower crescent. Venus' decreasing distance from the earth as it moves from elongation to inferior conjunction causes the planet to appear brightest during its crescent phase. The changing phases were seen by Galileo, and their recognition was helpful in efforts to convince men of the truth of the Copernican system. (See Figure 5.12.)

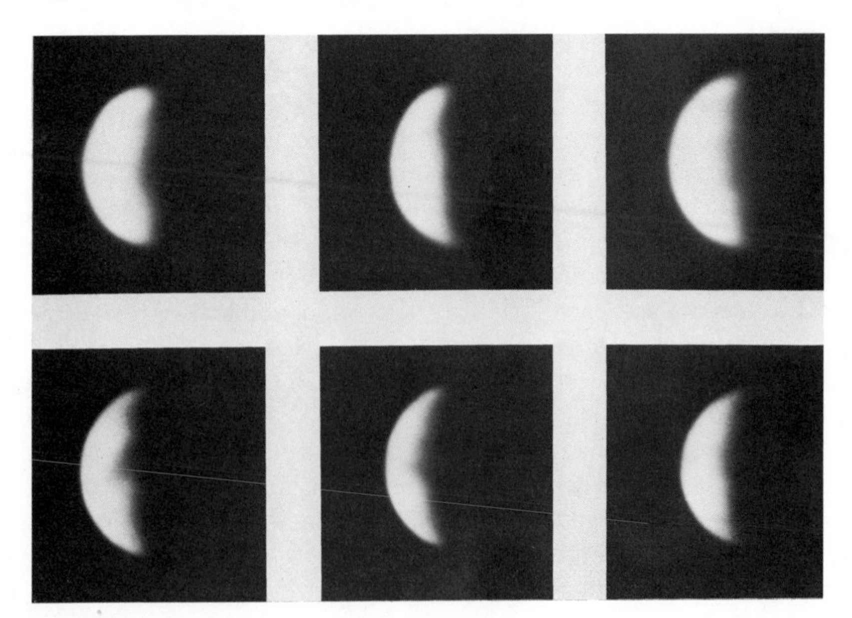

Figure 5.12. Venus showing rapid change of phase over a period of 25 days. (Hale Observatories photographs.)

The problem of Venus' rotation period was mentioned on page 96 in our discussion of radar techniques used to determine rotation periods. The application of these techniques in 1965 revealed a rotation period of 243 days in the retrograde direction. In the years since then, radar astronomers have been able to add discrimination in time to their analysis of radar returns from Venus. The signal reflected from the center of the disc is received slightly sooner than the signal reflected from other points on the surface. As the result of a difficult analysis in terms of both time and wavelength, the radar astronomers have been able to produce remarkable maps which show that there are shallow craters up to 150 km in diameter on some parts of Venus' surface.

The space program of the Soviet Union has successfully parachuted instrument packages to the surface of Venus to send back direct measurements of the surface temperature. These probes indicate temperatures in the neighborhood of 700°K (800°F).

Mars

Among the many wonderful feats of the space age, one must certainly include the close-up views of the Martian surface returned to earth from the spacecraft of the Mariner series. As has so often been the case in other situations, these new data provided immediate answers for some of the old questions and raised many new ones. For many years astronomers, using conventional telescopes, have studied the dark surface markings and white polar caps of Mars. Seasonal changes in color have been followed through the Martian year, and certain visible features were described as "canals." (See Figure 5.1.) The possibility of life on Mars has intrigued men to the point that when we think of creatures from other worlds, we usually think first of "Martians." Against these expectations, and perhaps hopes, concerning Mars, the pictures sent back by Mariner IV in 1965 proved to be extremely significant. As may be seen in Figure 5.13, some areas of Mars show a surface marked by numerous craters. The initial implication of these craters was that Mars had never been subjected to the weathering processes which have obliterated all but a few traces of craters on the earth. The craters suggested that the history of Mars may have been more like that of the moon than that of the earth.

Our ideas about Mars' surface began to change again in 1969. At that time pictures from Mariners VI and VII showed some areas

Figure 5.13. Surface features on Mars photographed from NASA space probe, Mariner IV. North is at the top. The area covered here is 270 km in the east-west direction and 240 km in the north-south direction (NASA photograph.)

that were essentially featureless and other areas near the south pole that seemed to be ridged or folded. Then in late 1971, Mariner IX went into orbit around Mars and began a mapping program which lasted for more than three months. This spacecraft has definitely proven that Mars is an "active" planet with slow processes which cause continuous change in the surface features. A major dust storm obscured much of the surface during the first weeks of the mission, but as the dust settled or cleared, the interesting surface features were again revealed. The results of the Mariner IX mapping have been so extensive that we have completely changed our major concepts regarding the red planet. The seemingly endless arguments over the canals will soon be forgotten. Some of the interesting types of terrain are pictured in Figures 5.14 through 5.17.

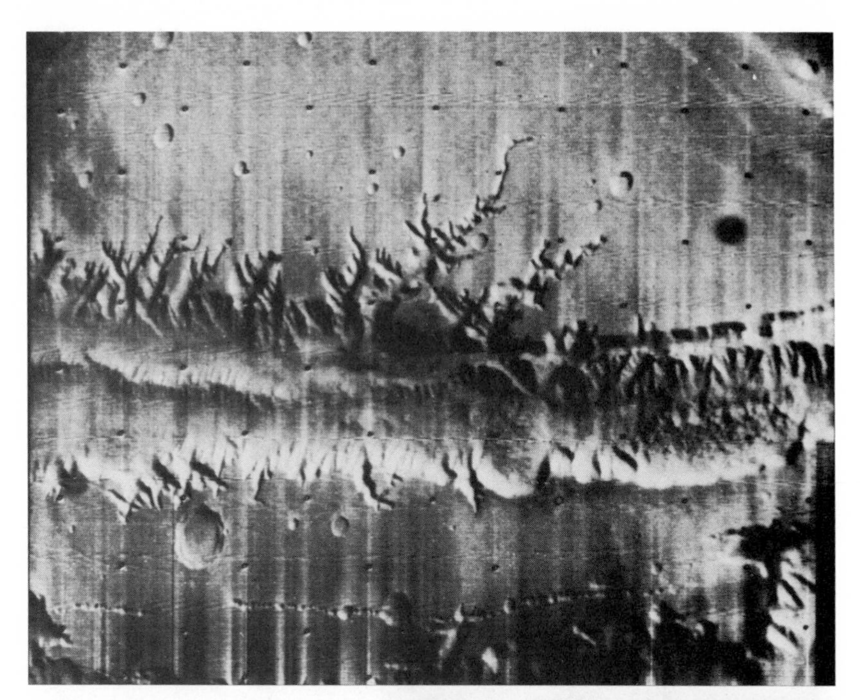

Figure 5.14. A vast Martian canyon which shows branching suggestive of erosion by water. (NASA photograph from Mariner IX.)

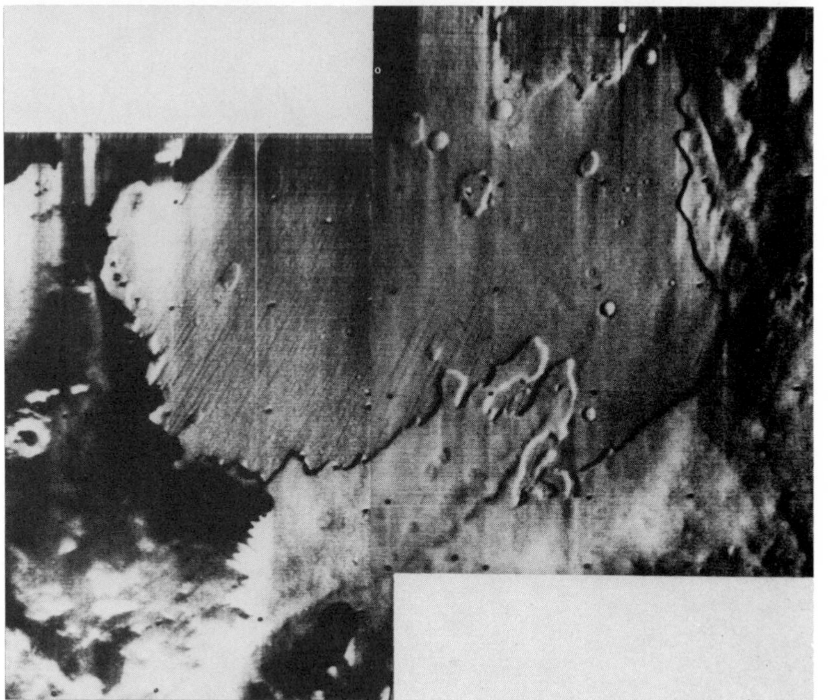

Figure 5.15. An irregular low plateau in Mars' southern hemisphere. This feature is about 70 km wide. Note especially the striations at the center and the numerous small craters. (NASA photographs from Mariner IX.)

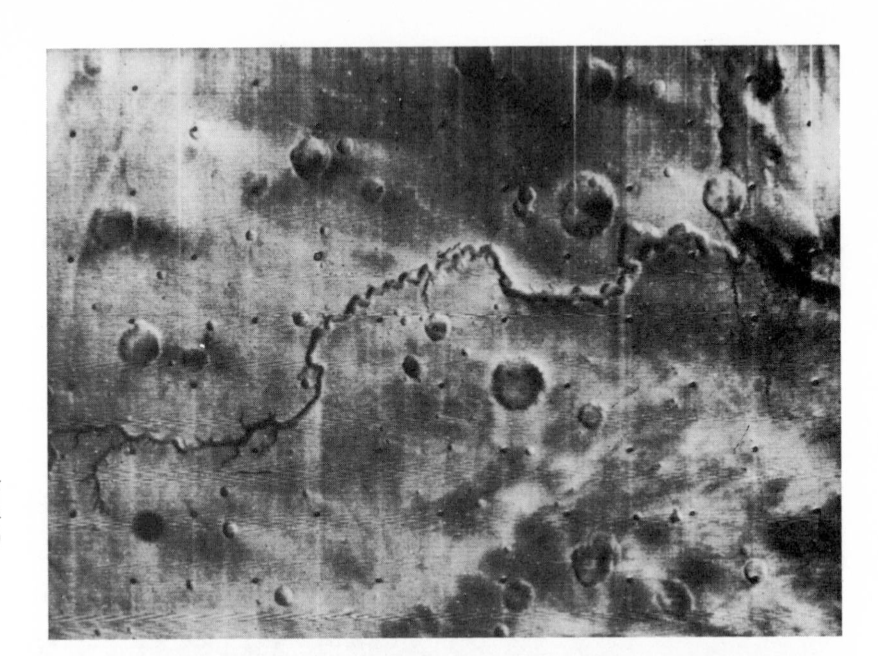

Figure 5.16. A curving valley some 400 km long. (NASA photograph from Mariner IX.)

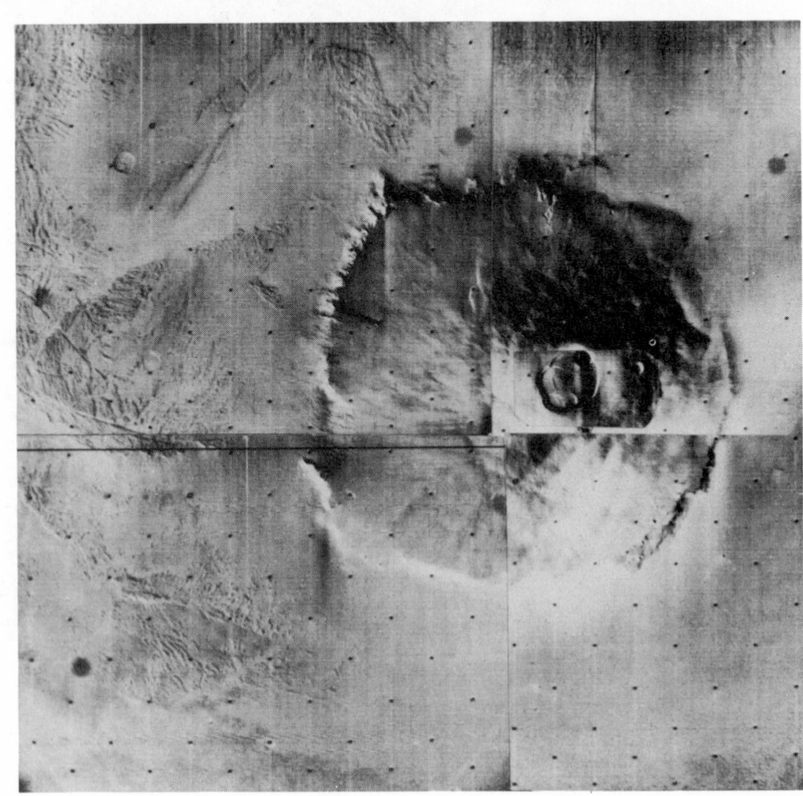

Figure 5.17. The feature Nix Olympica on older maps of Mars has been identified as this large volcano viewed from the top. (NASA photographs from Mariner IX.)

Jupiter

One of the most interesting objects for the observer with even a small telescope is Jupiter. Because of its rather slow motion against the background stars and because of its great brightness, Jupiter is conspicuous in the sky for fairly long periods each year. The large diameter and high albedo combine to make this planet brighter than any of the stars in spite of the fact that it is some five astronomical units from the sun. Through his telescope the observer may see the effect of rapid rotation on a body of such dimensions in the noticeable flattening through Jupiter's polar axis (see Figure 5.18). Careful study reveals shades of pink and brown in the belts which run parallel to the equator. The four bright "Galilean" satellites change their pattern from hour to hour and night to night, and occasionally the transit of a satellite or its shadow may be seen across the disc. Another prominent Jovian feature is the great red spot, an elliptical area some 48,000 km long. This pink or red area has varied considerably in size, color, and longitude since 1831, when systematic observations began. No one is yet sure just what the red spot is chemically or how it can remain more or less fixed in latitude and yet drift forward and backward in longitude.

Figure 5.18. Jupiter photographed in blue light. The red spot is visible as a dark oval at the upper left. The satellite Ganymede is visible at the upper right and its shadow appears as the dark dot near the red spot. (Hale Observatories photograph.)

Jupiter's great mass, 318 times the earth's mass, results in an escape velocity of 61 km/sec. Considering this and the low temperature, 135°K (−220°F), one would expect Jupiter to retain even the lightest gases in its atmosphere. Direct evidence of hydrogen and helium is difficult to obtain, since at this low temperature these gases do not absorb and emit in the visible spectrum. The best we can do is see spectrum features of hydrogen compounds which can absorb energy at Jupiter's low temperature. Thus the presence of methane, CH_4, and ammonia, NH_3, which are minor constituents, suggests large amounts of hydrogen in the atmosphere as a whole. Molecular hydrogen, H_2, and helium have been identified in infrared spectrograms of Jupiter.

A few hundred kilometers below the top of Jupiter's opaque clouds the pressure is probably great enough to liquefy the atmospheric gases. This means that the cold, dense, gaseous layer is really quite thin when compared with the radius of the planet. Early models indicated a solid mass below the first few hundred kilometers of gas and liquid. It was suggested also that Jupiter may have a small, high-density core comparable in size to the terrestrial planets. Recent calculations show, however, that it is unlikely that *either* Jupiter or Saturn has any truly solid portions.

It was pointed out earlier that all warm bodies should emit radio radiations, and Jupiter is no exception. Measurements at wavelengths near 3 cm were made by a number of observers beginning in 1956, and their results indicated a temperature of 135°K (−220°F). This temperature was in close agreement with the temperature derived from infrared studies some years earlier. In addition to this thermal radiation, however, Jupiter emits radiation at longer wavelengths in a very interesting and peculiar manner. The radiation at wavelengths from about 10 to 21 cm and longer is too intense to come from a simple thermal radiator. Variations in the intensity of this radiation have been related to the kinds of solar activity which cause auroras and magnetic disturbances on the earth. Most of the radiation in this microwave region very likely comes from solar particles trapped in Jupiter's magnetic field. These trapped electrons form a zone quite similar to the earth's Van Allen belts.

Even though this explanation cannot be called positive proof of Jovian radiation belts, the theory does seem to fit the observations rather well. Theorists have been less successful in explaining other radiations at wavelengths of 10 meters and more. These were first found in 1955 and early analysis showed that the signals were emitted in three well-localized longitudes, probably on Jupiter's actual surface below the clouds. There is a certain degree of correlation between these emissions and activity on the sun, and theories ranging from volcanoes and thunderstorms to cyclotron radiation have been proposed. A striking correlation with the inner satellite Io has also been found. So far, however, no theory has withstood all tests. To the radio astronomer, Jupiter will no doubt be an extremely interesting object for a number of years to come.

Saturn

Saturn's dramatic system of rings makes it one of the most interesting objects for telescopic study, and to many people the familiar ringed planet has become a symbol of astronomy. Without the rings, however, Saturn would have much less to offer to those interested in planetary studies. Saturn has the lowest density of all the planets, and in structure and composition it may be quite similar to Jupiter. Unlike Jupiter, though, Saturn emits radio radiations only as a thermal radiator. Belts and bands are to be seen in Saturn's cloud layers, but it is rare that any features persist long enough to permit direct rotation studies. The rotation period was found from studies of spectra, as in Figure 5.7, described earlier.

From the time of their discovery by Galileo, the rings of Saturn have offered a considerable challenge to those who have sought to understand their nature. As the study of mechanics developed following the work of Newton, it became possible for Maxwell to show in 1857 that the rings could not possibly be of the nature of a solid or liquid disc, but must be composed of individual particles in separate orbits around the planet. Convincing support for this was seen in spectra such as that in Figure 5.7. The upper and lower portions of this spectrogram represent light reflected from the ring system on either side of the main body of the planet. The effects of Saturn's rotation are clearly seen in the slanted absorption lines in the center section. The lines from the rings would show the same slant if the rings rotated as a solid system attached to Saturn. What is actually seen is that the absorption lines from the rings slant in the opposite direction from those in the disc. For the rings the Doppler shift is greatest for the portions nearest to the body of the planet. This is the way in which individual particles moving according to Kepler's laws would behave. The particles in the smaller orbits would have the greatest velocities. There seems to be no doubt that the rings are composed of particles.

The nature of the actual particles themselves has also stimulated much study. The most that may be said to date, however, is that the particles must be small, perhaps only a few centimeters in diameter or less, and the particles must be good reflectors of light. The best speculation as to origin suggests that the ring material may be left from a period during which the satellites were forming. The rings are at such a distance from Saturn that a satellite-sized object would be broken up by the strong tidal forces set up by the planet or could not have formed in the first place.

Figure 5.19 shows the ring system in four different orientations with respect to our line of sight from the earth. Since the plane of the rings is more or less fixed in space, there are occasionally times when we view the rings edge on. At these times the rings practically disappear. Photometric observations and dynamical theory indicate that a given particle will only be about a *meter* above or below the mean plane. The actual thickness depends on the size of the particles as well as their velocities perpendicular to the plane. During 1966, the earth went through the plane of the rings, but they remained visible in very large telescopes. During this same time, the tenth satellite, Janus, was discovered very near the outside edge of the rings.

Figure 5.19. Four views of Saturn showing the apparent change in orientation of the ring system as the relative positions of earth and Saturn change. (Lowell Observatory photographs by E. C. Sliper.)

An interesting dynamical demonstration of the particle nature of Saturn's rings is found in a study, done by Kirkwood about 1860, of the gaps in the ring system. Kirkwood showed that orbiting particles at certain distances from Saturn would be perturbed by the attraction of Saturn's satellites, particularly Mimas, the innermost one. At the distance of the principal gap, particles would have orbital periods just equal to half of the period of Mimas. At regular intervals, then, Mimas would exert a gravitational attraction on such particles and gradually alter their orbits. In other words, if the orbital period of the particles is one-half of the orbital period of Mimas, the orbit of a small particle is unstable. Orbital periods which are other simple fractions (1/3, 1/4) of the period of Mimas would also be unstable, and other gaps in the ring system are seen at the appropriate distances from Saturn. (See also page 152.)

Uranus

It was mentioned in Chapter 1 that Mercury, Venus, Mars, Jupiter, and Saturn were the five planets known in antiquity. The confirmation of the earth as a sixth planet has also been described. The seventh planet, Uranus, was discovered in 1781 by William Herschel in the course of his attempts to detect stellar parallax. Herschel was a musician of German birth living in England. Astronomy and telescope making were originally only hobbies with him, but these pursuits eventually occupied most of his time. His son, John, and his sister, Caroline, came to share William Herschel's interest in astronomy and they did much important work on their own.

In the course of his systematic searches of the sky, Herschel found a strange object which showed a definite disc in his 7-inch reflector. He thought that he might have found a comet, but continued observation showed none of the changes in appearance which are characteristic of comets. The peculiar object remained unclassified but attracted wide attention. When another astronomer at last showed that this faint object must be an additional planet beyond Saturn, credit for the discovery was naturally given to Herschel. King George III of England added fortune to Herschel's fame by providing him with a comfortable pension. Herschel responded by naming the planet "Georgium Sidus" or George's Star. For some years the planet was actually called "Herschel," and it was many years later that the name "Uranus" was internationally adopted. (See Figure 5.20.)

The study of the records of earlier observers showed that Uranus had actually been seen on a number of previous occasions. Not having Herschel's curiosity and open-mindedness, however, the others who saw the planet had ignored it.

For most of the other planets the axes of rotation are more or less at right angles to the planes of their orbits. For Uranus, however, the axis of rotation is only 8° from the plane of the orbit. The rotation is in the retrograde direction. The orbits of the five satellites lie right in the equatorial plane. Even in a very large telescope, Uranus shows virtually no features.

Neptune

Kepler's laws permit a very precise determination of the positions of the planets once the characteristics of their orbits have been determined. As Uranus was followed from year to year, however, astronomers noted some irregularities. Observed positions did not agree with predicted positions even when the gravitational perturbations of Jupiter and Saturn were considered. A young and relatively unknown English astronomer, John Couch Adams, began to consider the effect that a more distant and unknown planet might have on Uranus. He concluded that Uranus' peculiar behavior could in fact be attributed to the attraction of such a body, and he predicted the position at which such a body might be found with a telescope. Adams sent his predictions to the Greenwich Observatory and asked that the astronomers there look for the planet. How-

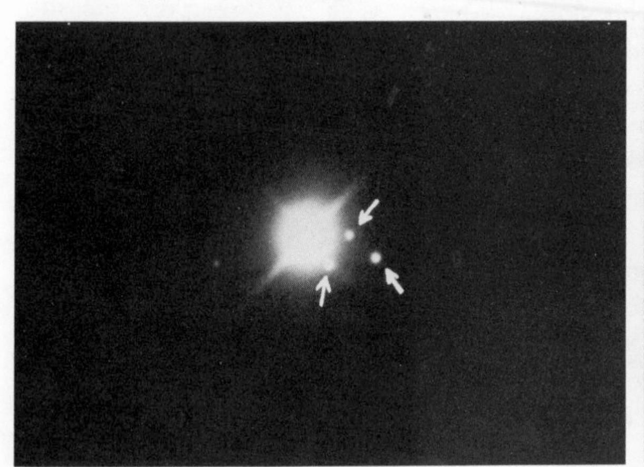

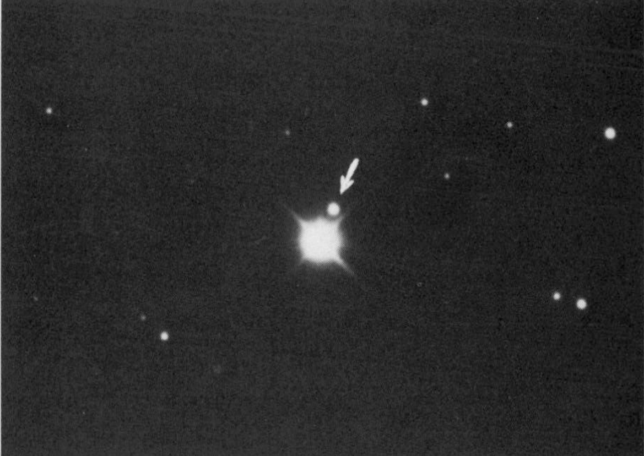

Figure 5.20. Uranus (left) and Neptune (right). Arrows indicate some of the satellites. (Lick Observatory photographs.)

ever, the Greenwich astronomers were too busy to take time for a careful search for Adams' planet.

Shortly afterward, a French astronomer, Urbain Leverrier, independently made his own predictions of the position of the undiscovered planet, reasoning along the same lines as had Adams. The request made by Leverrier was given immediate attention by observers in a German observatory and the planet was quickly found in 1846. Today, credit for the discovery is generally given to both Adams and Leverrier.

As is the case with Uranus, Neptune has very few features visible through even large telescopes.

Pluto

Convinced that the gravitational influence of Neptune could not account for all of the peculiarities of Uranus' motion, the American astronomer Percival Lowell began in 1906 a systematic search for a planet beyond Neptune. The search continued after Lowell's death and lasted until 1930. On photographs made in January of that year, Clyde Tombaugh found a star-like image which moved slowly against the background of stars. Continued observations confirmed that this faint object was the long-sought planet, Pluto (Figure 5.21). The orbit, now well known, is quite elliptical and brings Pluto inside the orbit of Neptune during part of each revolution.

Using the 200-inch telescope, Kuiper has obtained an angular diameter for Pluto and computed a diameter of about 6000 km. This was a difficult measurement and it is to be hoped that confir-

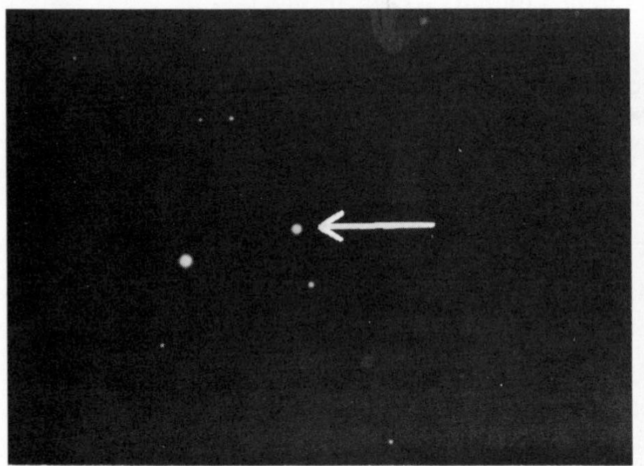

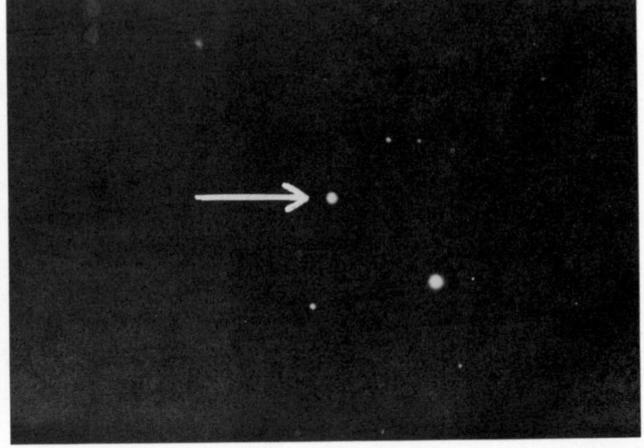

Figure 5.21. Pluto seen as a small body changing its position against the background stars. The two photographs were made 24 hours apart. (Hale Observatories photographs.)

mation may be made if Pluto eclipses a star at some future time. The duration of such an eclipse seen from several places on earth will give a good value for the diameter. Periodic variations in brightness have suggested a rotation period of about 6.4 days. It seems unlikely that Pluto has enough mass to hold the lighter gases, and it is too cold for other elements to remain gaseous. Thus the presence of an atmosphere seems to be unlikely. The gaps in our knowledge of Pluto are still wide, and much of what we know is uncertain or difficult to interpret.

5.13 The Terrestrial and the Major Planets

Study of Table 5.1 shows clearly that the planets fall naturally into two well-defined groups. Simply on the basis of size the terrestrial planets, Mercury, Venus, Earth, and Mars, are distinctly different from the major or Jovian planets, Jupiter, Saturn, Uranus, and Neptune. In other aspects the two groups are also well defined. The terrestrial planets rotate slowly, have relatively thin atmospheres if any, have high densities, and are characterized by few satellites. The opposite is true for the major planets.

QUESTIONS

1. What observations must be made before the true diameter of a planet may be known?
2. Describe two methods by which the rotation period of a planet might be determined.
3. How can the mass of a planet be determined?
4. Account for the fact that there is such a wide range in the measured albedos of the planets.
5. Discuss three factors which together influence the temperatures of the planets.
6. What is a *Planck curve* and how would such a curve be affected by an increase in temperature?
7. In terms of the radiation laws, what happens as the temperature of a planet comes to equilibrium?
8. Why should there be two peaks on the spectral-energy curve of a planet? Why is it impossible to observe the entire curve from the earth's surface?

9. Discuss the roles of temperature and mass in the theoretical study of the gases which might exist in the atmosphere of a planet.
10. How is the existence of a planet's atmosphere confirmed, and how can the gases in a planet's atmosphere be identified?
11. Describe the circumstances under which the planets Uranus and Neptune were discovered.
12. What are the principal differences between the terrestrial and the major planets as groups?

The Motion and Surface of the Moon

The methods used to study the moon are quite similar to the methods described for the study of planets. Because of the nearness of the moon, there are, of course, many details concerning the lunar surface which have become available sooner than similar data for the planets. The first great advance in lunar study came with Galileo's telescopic studies. The fascinating and beautiful pattern of craters could at last be seen (Figure 6.1). Efforts to record these features occupied mapmakers for a hundred years or more. The second great period of lunar exploration began in 1964 with the Ranger series of moon rockets and continued through the exploration of the moon's actual surface by the astronauts of the Apollo program. As in the case of Mars, years of argument over the possible nature of the moon's surface have been forgotten as a great wave of very positive information has been acquired. As the analysis of these data proceeds, we shall come ever closer to an understanding of the forces and processes which shaped the lunar topography. We shall come closer, also, to eventual knowledge of the historical relationship between the earth and the moon.

6.1 Distance from the Earth

The study of the moon may logically begin like a study of the planets, with the determination of distance from the earth. Unlike the

◄**Figure 6.1.** The moon near first quarter. North is at the bottom as seen through the telescope. (Lick Observatory photograph.)

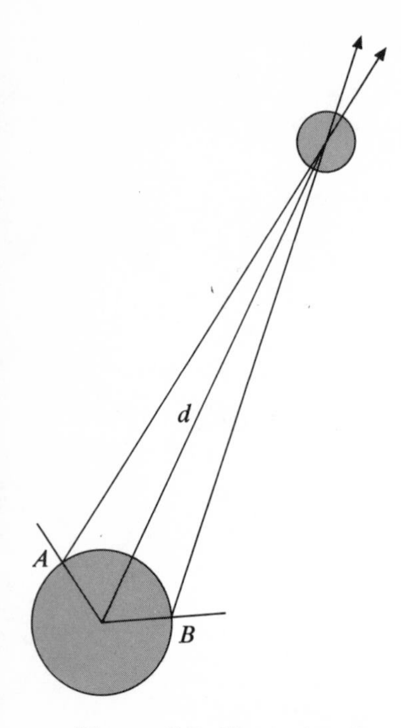

Figure 6.2. Measuring the distance from the earth to the moon.

planets, the moon is close enough to the earth so that its distance may be determined by direct parallax measurements.

Parallax measurements for the determination of the distance from earth to Mars or to the asteroid Eros are described in Chapter 5, and the situation for the moon is quite similar. The geometry is shown in Figure 6.2, where observers at two positions on the earth make simultaneous observations of the moon's position with respect to their horizons. Knowing the distance between them on the earth, the observers are able to compute the distance to the moon. From the actual observations, another quantity usually computed is the angle that would be subtended by the equatorial radius of the earth as seen by an imaginary observer at the center of the moon. This angle is called the moon's **horizontal equatorial parallax** (Figure 6.3). Astronomers take this extra, seemingly unnecessary, step in order to be sure that regardless of where and how the observations are made, everyone will be talking about the same quantity. The best value for the parallax of the moon is 57′2.62″, and this permits the computation of the moon's distance based on the radius of the earth at the equator. The resulting average distance from the earth to the moon is 384,404 km.

Kepler's laws also mean that the moon's orbit around the earth should be an ellipse. Because of this the earth-moon distance should vary in a regular way. Rather simple observations show that this is, in fact, the case. The angular diameter may be measured throughout a month and will be seen to vary by a few percent as the moon moves closer and then farther away in its orbit. The distance actually ranges from 356,400 km to 406,700 km. Sometimes when the moon is just rising above the eastern horizon or is still low in the sky, it seems to be conspicuously larger than when it is higher in the sky. Simple measurements of the moon's angular diameter during the course of an evening demonstrate that this peculiar effect is, however, only an illusion.

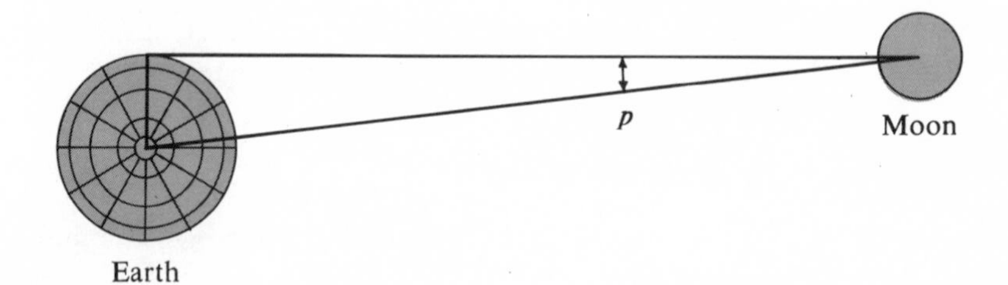

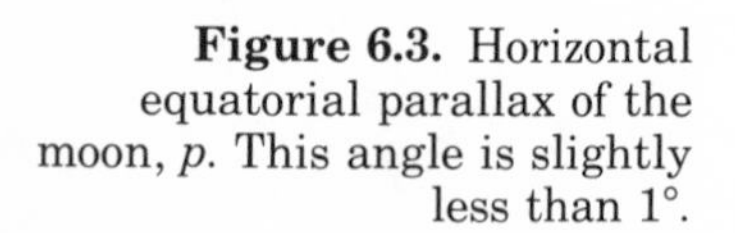

Figure 6.3. Horizontal equatorial parallax of the moon, p. This angle is slightly less than 1°.

6.2 Diameter

The angular diameter of the moon, as it moves in its orbit around the earth, is on the average 31 minutes of arc. If this figure is combined with the average distance of the moon from the earth, the moon's diameter can be calculated to be 3476 km. The moon is more nearly spherical than the earth. Flattening due to its slow rotation is negligible, and tidal distortions due to the attraction of the earth are very small. While the earth approximates an ellipsoid, the moon's shape is approximated by a triaxial ellipsoid (different polar, equatorial, and earthward radii) with a number of irregularities.

The earth-moon system as depicted in newspapers and many elementary books leaves the reader with a badly distorted concept. A simple and approximate scale model helps to put things in their proper perspective. If we consider that the earth's radius is roughly 6400 km, then the moon's distance of roughly 384,000 km is equivalent to 60 times the earth's radius. Imagine now a model earth 5 cm in diameter. On this scale the moon would be only 1.25 cm in diameter, and it would be located 150 cm from the model earth. Drawn to scale in Figure 6.4, the moon's great distance from the earth is more impressive. A feeling for the moon's great distance helps one to appreciate the observational precision needed to find its distance and diameter as well as the technical skill required for landing space probes at specific locations.

6.3 Motion

As the moon moves around the earth, the side facing the sun is illuminated by the sun, and the other side is dark. The relative positions of earth, moon, and sun determine the amount of the illuminated side which will be visible to us on earth at any time, that is, the **phase** of the moon. Thus when the earth, moon, and sun are in a straight line, the lighted side of the moon is turned away from the earth, and we see nothing. This is the **new moon**, as indicated in Figure 6.5. (In this case the moon would be hard to see even if it were self-luminous because of the intense glare of the sun's light.) At **first quarter** the moon has moved 90° with respect to the earth-sun line and we now see a full half of the illuminated side. Between new moon and first quarter the moon appears as a **crescent,** growing larger every day. After first quarter the moon continues to

Figure 6.4. The earth-moon system drawn to scale.

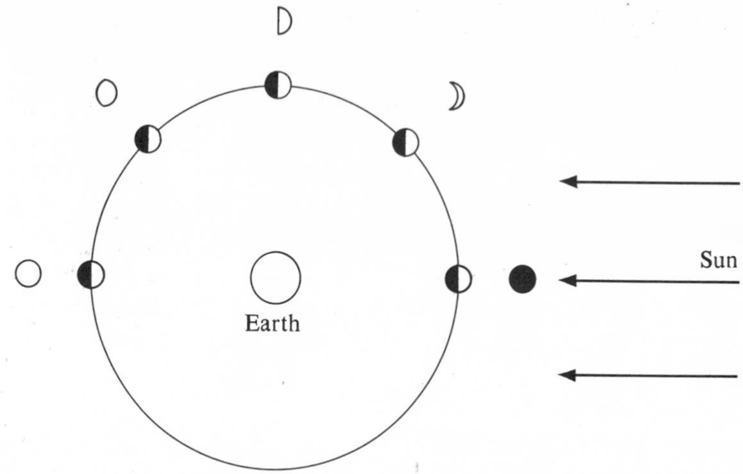

Sun

Earth

Figure 6.5. The phases of the moon. Half of the moon is always illuminated by the sun. The outer images show the moon as it appears from the earth.

grow and it is now said to be **gibbous.** As seen from the earth, the angle between the sun and the moon increases until it becomes 180°, and the hemisphere facing the earth is the same as the hemisphere facing the sun. The moon is then **full**. After full moon, the moon goes through the gibbous phase again to reach **third quarter.** Then the moon again appears as a crescent, and finally it returns to the new phase. From full moon to third quarter and back to new, the angle between the sun and the moon, as seen from earth, decreases. This complete cycle of **synodic period** lasts about $29\frac{1}{2}$ days. The moon really completes a full 360° revolution around the earth in less than the synodic period. This **sidereal** period is only $27\frac{1}{3}$ days and can be noted by recording the position of the moon against the background stars. The difference between the synodic and sidereal months arises because of the earth's own orbital motion during the period of the moon's revolution (Figure 6.6).

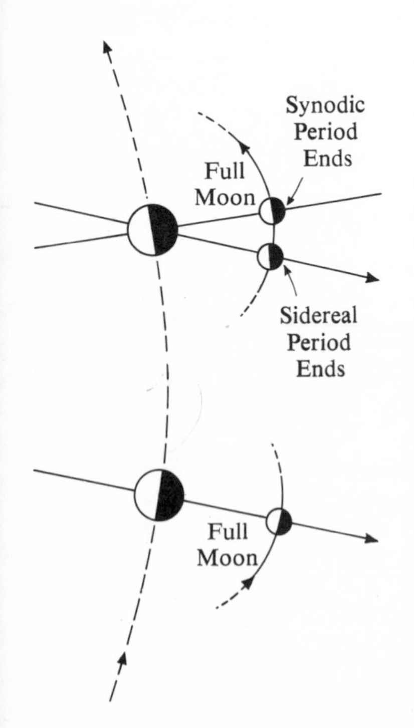

Figure 6.6. Synodic and sidereal periods of the moon.

As the earth rotates daily on its axis, the moon rises in the eastern sky, crosses the meridian, and sets in the west. Since the moon moves toward the east in its orbit and since the earth rotates toward the east, the moon will rise later each night than the night before. The earth has to rotate farther each day to catch up with the moon. The sidereal period is $27\frac{1}{3}$ days, and so the moon must move in its orbit 360°/27.3 days or 13°11′ per day. This is the extra angle through which the earth must turn from one moonrise to the next. Since the earth rotates 15°/hour on its axis, it will turn through the extra 13° in about 50 minutes. Therefore the moon rises about 50 minutes later each night than it did the night before. Clearly, the time of moonrise will be related to the phase of the moon. At new moon, the moon will rise for the observer on earth at sunrise. At first quarter, the earth must continue turning after sunrise before the moon will rise. Since the moon has moved along a quarter of its orbit, it will rise at noon and cross the meridian at sunset. By the same reasoning it can be seen that the full moon will rise at sunset, and that the third-quarter moon will rise at midnight (Figure 6.7).

In the course of its orbital motion the moon also rotates on its own axis. The period of this rotation is exactly the same as the period of the moon's orbital revolution. The same face is therefore

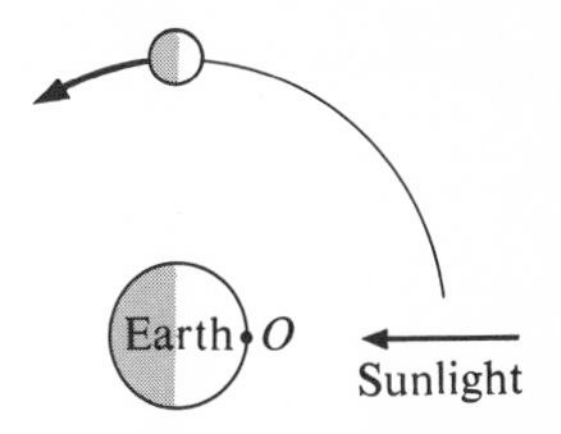

Earth and moon at first quarter. Moon is rising for an observer at *O*.

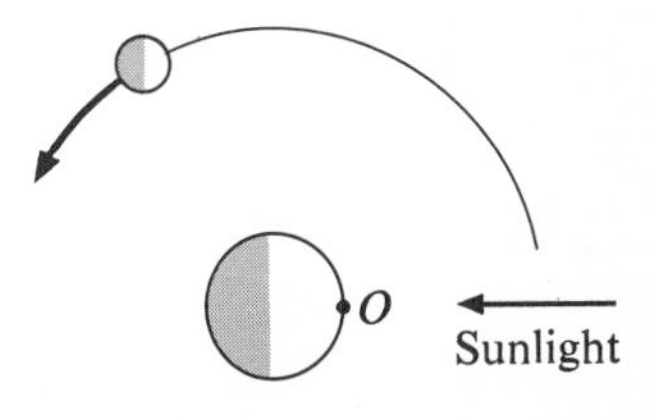

Earth and moon 24 hours later. Observer at *O* must wait 50 minutes for moonrise.

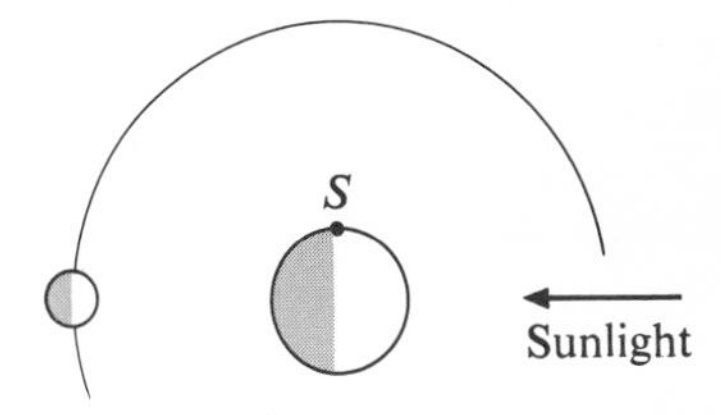

For an observer at the sunset point, *S*, the full moon is just rising.

Figure 6.7. Time of moonrise depends upon the moon's phase.

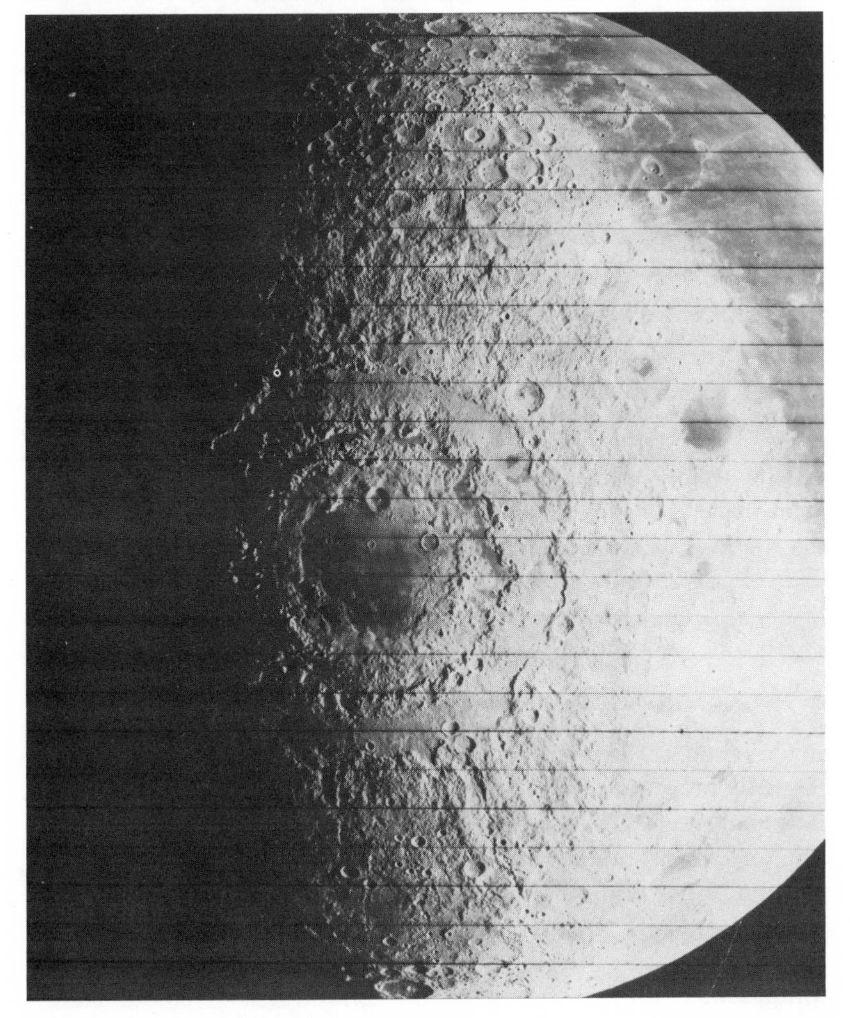

Figure 6.8. Orbiter IV photograph of the moon's far side. The spacecraft was 2720 km above the moon when this photograph was made. (NASA photograph from Langley Research Center.)

always turned toward the earth. Only since 1959, when the Soviet Union photographed the moon from a satellite, have we had any concept of the appearance of the unseen side. One extraordinary feature was photographed by Lunar Orbiter IV (Figure 6.8). Near the center of the photograph is what appears to be a triple-walled feature some 1130 km across. In general, the far side appears to be more severely broken up with crater features, and there are no large seas.

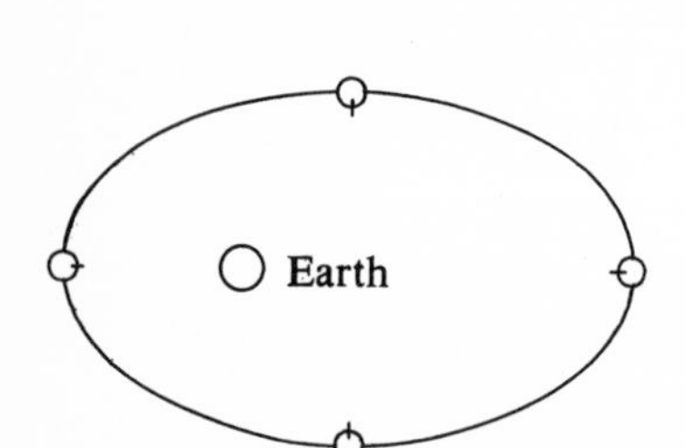

Figure 6.9. Librations of the moon in longitude and latitude permit us to see more than half of the surface.

Because of a uniform rate of rotation and a nonuniform rate of revolution, the moon appears to rock back and forth as it moves along its orbit. This is referred to as a **libration** in longitude and permits us to see a bit past the normal edge or **limb** of the moon on both the east and west sides (Figure 6.9). A tilt of the moon's rotational axis with respect to the plane of its orbit also permits us at certain times to look slightly past both the north and south poles and gives the moon a libration in latitude. These two geometrical librations combine with some lesser effects to reveal a total of 59% of the moon's surface area to the earthbound observer.

Both the earth and the moon have long shadows extending far out in space in the directions away from the sun. Because of the sun's great diameter, these shadows actually consist of two cone-shaped parts. As indicated in Figure 6.10, there is a dark inner cone called the **umbra** and a partially dark outer cone called the **penumbra.** If the plane of the moon's orbit coincided with the plane of the earth's orbit, then at new moon each month the moon's shadow would pass across the earth. Likewise, at full moon, each month the moon would pass through the earth's large shadow. On the earth we would observe eclipses of the sun at every new moon and eclipses of the moon at every full moon. Eclipses are actually rather rare and occur only a few times a year. The reason is that the moon's orbital plane is not coplanar with the earth's but is inclined about 5°. Most of the time, then, the moon at new and full phases is well above or below the earth's orbital plane, and eclipses do not occur. If the moon's position against the background stars is carefully noted from night to night through several seasons, a considerable range in the moon's position with respect to the celestial equator can be noticed. It is from a careful analysis of the range in position of the moon against the celestial sphere that the inclination of the moon's orbital plane has been determined.

The moon passes through the plane of the earth's orbit at two points called **nodes.** The line between these two points passes

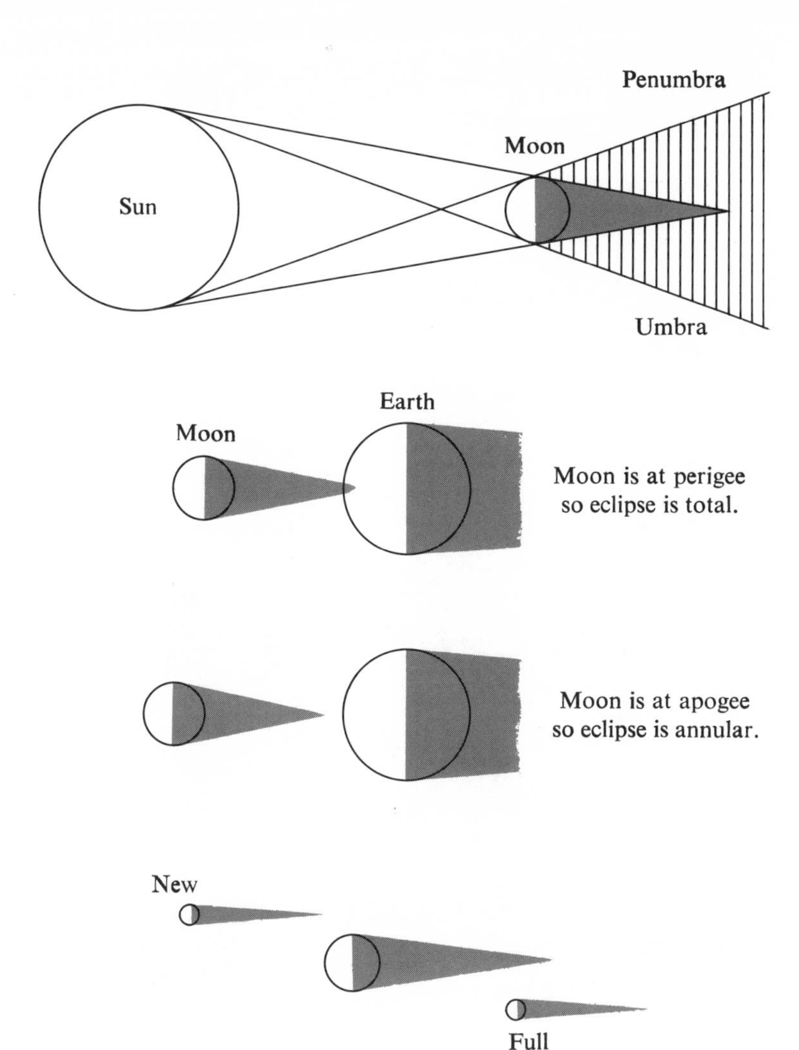

Figure 6.10. Shadows of earth and moon and eclipse situations.

through the center of the earth, and twice a year this line of nodes will point toward the sun (Figure 6.11). If a new moon occurs at one of these two times, there will be an eclipse of the sun and if a full moon occurs at such a time, there will be an eclipse of the moon. Eclipses of the sun are actually possible during a period from about 15 days before to about 15 days after the time when the line of nodes points toward the sun. This is due to the earth's size, which permits eclipses to occur even though the moon is a little bit above or below the plane of the earth's orbit. Two **eclipse seasons** six months apart each year are the result. Since each of these seasons is longer than a synodic month, there must be a new moon

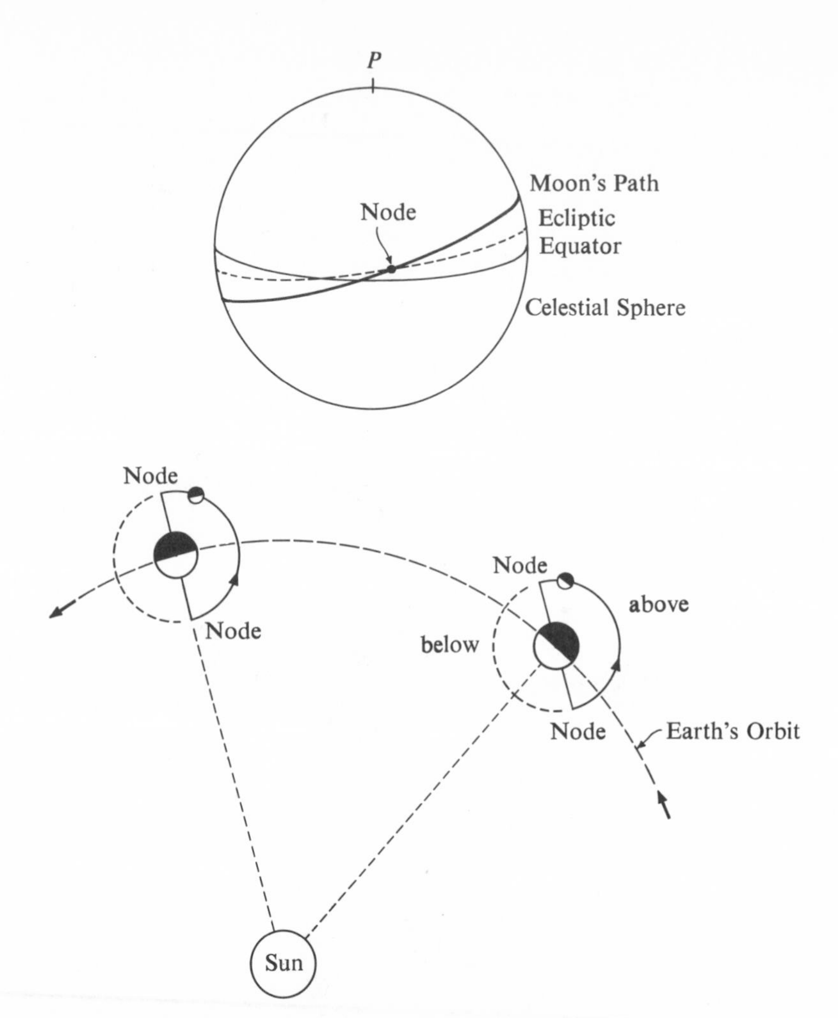

Figure 6.11. As the earth moves around the sun, there will be times when the line of nodes points toward the sun (left). At such times eclipses will occur.

during each of these periods. Thus, there are always two eclipses of the sun each year. When an eclipse occurs at the beginning of an eclipse season, a second one can occur before the end of the same season, and there can be four in the two seasons. (Because the line of nodes rotates slowly eastward, a fifth eclipse may occur in a calendar year.)

An eclipse of the moon may last several hours and is visible to everyone on the dark side of the earth. Aside from a reddening and dimming of the moon, however, there is little in a lunar eclipse to arouse much interest. Solar eclipses, on the other hand, last only a few minutes at best and are visible to observers in only a narrow band on the earth's surface. A solar eclipse is an impressive and beautiful sight, however, and a great deal of important scientific information regarding the sun's atmosphere has been derived from solar eclipse studies. Solar eclipses are so spectacular, in fact, that people throughout history have been awed by them and have recorded the events in their histories and folklore. Working the prediction process backward, one can establish with fair accuracy the date on which an eclipse occurred in a specific geographical location. Several important archeological dates have been established in this manner. Long-term changes in the earth's rate of rotation have been established by comparing known dates and eclipse records.

For most of us the rotation of the earth is an accurate time standard. We divide the period of the rotation into 24 hours, and so forth. The earth's rotation is slowing down by minute amounts, and the days are increasing in length by about one-thousandth of a second every hundred years. This hardly seems worth any consideration, but for many scientific problems the time must be known with such great precision that the earth's rotation is not a good enough clock. It is possible to measure time from observations of the moon in a most reliable and accurate manner by noting the moon's position among the stars from time to time. In order to do this, however, one must be able to understand completely just how and why the moon moves as it does. This is an extremely complex problem, because not only is the line of nodes rotating, but also the major axis of the orbit moves around the earth as well.

6.4 Mass

With an accurate knowledge of the moon's distance from the earth and the moon's revolution period around the earth, one should be able to use Kepler's third law to find the sum of the masses of the earth and the moon. If we write this as

$$P^2(m_e + m_m) = D^3$$

then the moon's period must be given as a fraction of a year, and its distance as a fraction of an astronomical unit. The sum of the masses will be determined as a fraction of the mass of the sun. The known mass of the earth could then be subtracted to give the mass of the moon as a fraction of the mass of the sun. This procedure is not very accurate, in this case mainly because it requires very precise knowledge of the earth-sun distance in kilometers.

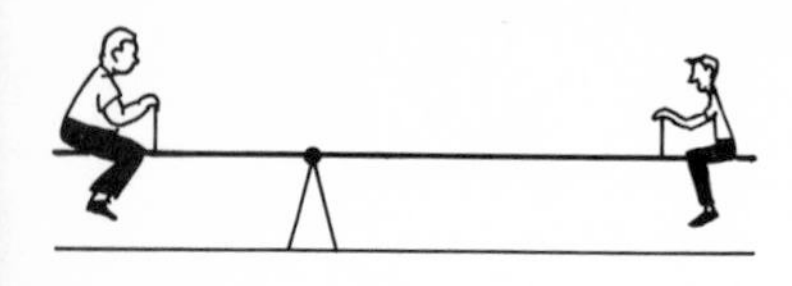

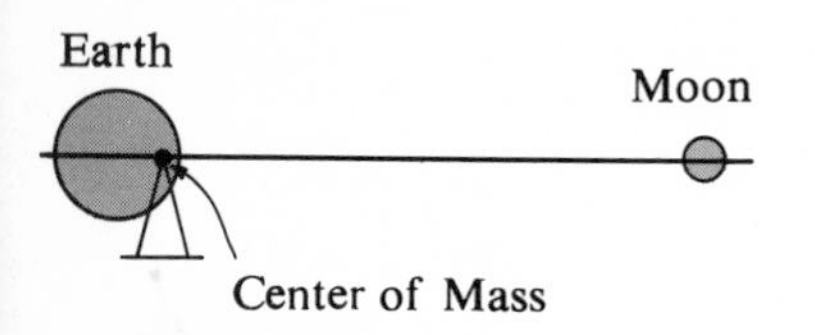

Figure 6.12. Location of the center of mass.

A better method of finding the moon's mass makes use of the fact that both earth and moon are really moving about a common center of mass between them. To balance any pair of masses (such as two boys on a seesaw) a point between them must be found such that

$$m_1 r_1 = m_2 r_2$$

where r_1 and r_2 are respectively the distances of the two masses from the center of mass (Figure 6.12). If we rewrite this relation as

$$\frac{m_1}{m_2} = \frac{r_2}{r_1}$$

we see that the ratio of the two masses is the inverse of the ratio of the two distances. In our problem here we already have found $(r_1 + r_2)$, the total distance from the center of the earth to the center of the moon. If we could locate the center of mass somewhere between the earth and moon, we could establish the mass ratio. Then knowing the mass of the earth in grams or tons from laboratory experimentation, we may find quite easily the mass of the moon. The center of mass is difficult to locate, but it can actually be located from careful observations of the sun's apparent motion along the ecliptic. The center of mass of the earth-moon system moves around the sun, so the center of the earth will be ahead of the center of mass at first-quarter moon and behind the center of mass at third quarter each lunar month. We detect this motion of the earth as small irregularities in the apparent motion of the sun. It turns out that the center of mass is about 4700 km from the center of the earth, well inside the earth's surface. The ratio r_2/r_1 is then about 81.3 and the moon's mass is $1/81$ times the mass of the earth.

6.5 Evidence Indicating Lack of Atmosphere

In Chapter 5 the effects of temperature and escape velocity on a planet's ability to hold an atmosphere were discussed. The same reasoning applies to the moon, and with knowledge of the escape velocity (2.38 km/sec), it becomes clear that the moon would have been unable to maintain an atmosphere for very long. Even before the lunar landings of the Surveyor and the Apollo programs, it was clear from other evidence that the moon has no permanent atmosphere. Temporary gaseous outbreaks have been observed from time to time, but the gases must necessarily dissipate.

We have already mentioned that the presence of planetary atmospheres is often manifest in observable clouds and lack of distinct and permanent surface detail. On the moon we are able to see small objects which approach the size limit dictated by the resolving power of the telescope. The actual limitations are imposed by the earth's atmosphere. Bright stars are often seen to pass behind the moon; at such times their light is cut off quite abruptly and not dimmed and reddened, as it would be by an atmosphere. Close-up studies from the actual surface of the moon show no evidence of wind or any meteorological erosion processes, and reveal a sharp well-defined horizon 16 km or more from the point of observation. On the Apollo XV mission, astronaut David Scott performed a classic demonstration of objects falling in a vacuum. While a television audience watched from the earth, he dropped a hammer and a feather at the same moment. The two objects struck the lunar surface together, since there was no atmosphere to impede the fall of the feather. Some radio observations have been cited as evidence for a trace of atmosphere and some of the heavy inert gases such as argon and krypton may actually be present. For practical purposes, however, the surface of the moon is in a very good vacuum.

Aside from the fact that there is no air to breathe, astronauts working on the moon find some perplexing problems. Voice communications are impossible except perhaps if two men are touching helmets to provide a path for the sound vibrations. One can not expect to hear noise such as the clatter of a dropped tool or the roar of an engine. Radio communications have to be on a line-of-sight basis because there is no ionosphere to reflect signals over long distances and behind hills.

6.6 Large-Scale Surface Features

Even to the naked eye, but especially on photographs or through a telescope, there seem to be two distinct types of surface regions on the moon. Long ago the dark smooth areas were given the name **maria,** or seas, and the rough lighter areas were called **terrae,** or lands. The Latin names are shown for a number of the seas on the maps in Figure 6.13. In spite of the fact that these are not really seas at all and in spite of their rather fanciful names, we find it convenient to continue to use the old, conventional designations. A number of observations suggest that the seas are really vast areas covered by lava flows which must have come through cracks lead-

Figure 6.13. The moon at first quarter (right) and at third quarter (left) showing the names of a few prominent features. (Yerkes Observatory photographs)

ing toward the interior. For example, there are easily visible in areas such as the Mare Imbrium circular features which are almost certainly buried craters. In many places, also, there are gentle ridges which suggest a flow of material of some sort. The dark flat bottom of the crater Plato resembles the nearby areas in the sea. Experts who have studied the close-up photographs and the rock samples from the moon now feel that the outflow of molten material has been confirmed.

Not quite visible to the naked eye but in great abundance on the moon are the **craters.** On the visible hemisphere the craters range in size from Clavius near the north pole – 230 km in diameter – to pits only a few centimeters in diameter as revealed on pictures made on the surface. The presence of overlapping craters and craters with lesser craters in their walls and floors indicates a range in age of rather indefinite degree. A few craters, Tycho, Kepler, and Copernicus, show patterns of **rays** which in some cases extend for thousands of kilometers. The fact that the light-colored ray material seems to overlay craters and other surface features has been taken to mean that the ray craters are the youngest of all the craters. Perhaps in time solar radiation will darken the rays to the extent that they will become invisible.

By studying the shadows cast by the crater walls both inside and outside the craters, it is possible to measure both the wall heights and the depths to the crater floors. Appearances are apt to be a bit deceiving because the craters are really quite deep. The crater Bullialdus, for example, is a well-formed one in the Mare Nubium but not a crater of spectacular size on photographs. Its diameter is 60 km and its walls rise 2700 meters above the crater floor. To the poor explorer who found himself in the center it would be a formidable climb to the top of the rim even in the moon's low gravity.

The origin of the craters has naturally evoked many opinions from astronomers through the years. Early opinions gave the craters a volcanic origin since volcanic craters were the ones best known on the earth at that time. It does not take much study to show that this is not a good general hypothesis, however. In the first place, terrestrial volcanic craters are never as large as the lunar craters and the profiles of volcanoes do not resemble the profiles of lunar craters. The best evidence indicates that these craters are the result of large explosions, and the explosions are attributed to large meteorites which crashed on the moon with great velocities. The energy of a large mass traveling at high speed would have to be released almost instantaneously and the result would be an explosion capable of blasting out a large crater. Comparison of bomb craters and known meteorite craters on the earth with lunar craters strongly supports this theory.

Although it seems likely that meteoric impact plays the most important role in the formation of lunar craters and in the pulverization of the lunar surface, this does not mean that internal geological processes are absent. In fact, such processes may still be going on. Even the impact of large bodies on the lunar surface might set off a series of secondary phenomena. Without an appreciable atmosphere to slow down the projectile, even small bodies will expend considerable energy when they hit. Any surface changes are then preserved until other processes or further impacts destroy them. In the final analysis, both impacts and internal processes have played a role in shaping the lunar surface features. The relative proportions of involvement of each, however, can be determined only after extensive lunar exploration.

Valid arguments have also been made that if the moon's craters were caused by meteorite impacts, there should be similar craters on the earth. The earth and moon are close enough to each other to have swept up meteorites in pretty much the same manner. Until recently there have been only a few well-studied craters of definite meteoritic origin on the earth. One of these is the famous meteor crater near Winslow, Arizona, in the southwestern United States (Figure 7.17). This one is estimated to be only about 30,000 years old, and it already shows obvious erosion effects. Older craters in less arid regions would be expected to be rather difficult to pick out after the passage of thousands of years. With the increase in high-altitude flying and extensive aerial mapping programs, a surprising number of large, probable meteor craters have been discovered. Several of these are in arid regions in Australia and Africa, which by their nature are largely unexplored from the ground. Others are in populated areas and are covered with farms, houses, trees, and villages. There is little doubt that craters were here on earth in the past just as they still are on the moon.

6.7 Detailed Nature of the Lunar Surface

Since about 1900, astronomers have been aware that the brightness of the moon's surface changes in a peculiar manner. At full moon all similar surfaces are equally bright and the same types of areas regardless of location are equally bright. Following an area in Mare Crisium, for instance, one would note that the surface becomes brighter as the moon grows from crescent to full. At full phase this area at which the sun's rays strike obliquely is just as bright as similar areas where the sun strikes the surface almost vertically. This curious phenomenon was attributed to a probable lunar surface covered with small cuplike depressions. Ordinarily we would see part of the cup in shadow. But at full moon we would see no shadows since we would be looking in the same direction as the sunlight which falls upon the surface. Thus the percentage of

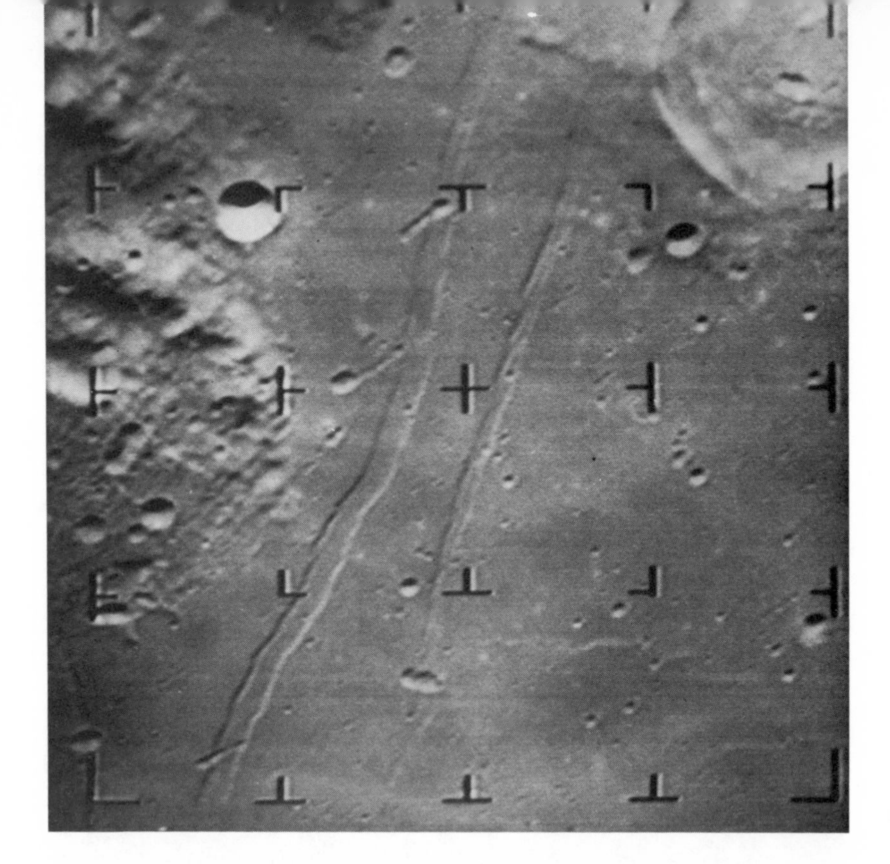

Figure 6.14. The lunar surface photographed from Ranger VIII in 1965, at a height of 432 km above the moon. The area here is Mare Tranquilitatis. (NASA photograph.)

the area that was composed of these small shadows would depend upon the angle between the incident sunlight and the direction to the observer. Careful theoretical studies showed that a simple surface covered with cuplike depressions could not completely account for the observed behavior of the light, and other more complex surface forms were considered. The close-up pictures returned from the Ranger cameras in 1964 and 1965 and the Surveyor camera in 1966 have in fact confirmed that the theory did give a reasonable idea of the small-scale character of the lunar surface.

The three Ranger vehicles launched in 1964 and 1965 by the United States were designed to record and transmit pictures of the moon as they fell toward the moon and crashed. The pictures showed progressively smaller craters as the cameras approached the surface. The final pictures were obtained from heights of about 10 km and showed details as small as one meter. One of the final Ranger pictures may be seen in Figure 6.14. Also in 1965, the Soviet Union landed a camera on the moon and was able to receive from it several pictures of surface details. These pictures from Luna IX showed an irregular surface with individual rocks a few centimeters long.

In June, 1966 the United States space probe Surveyor I landed on the moon and transmitted to the earth thousands of pictures made under varying conditions of sunlight and through several different colored filters. Pictures of the vehicle's own feet revealed grains only about one millimeter across and gave vital information on the ability of the surface layers to support a man or the larger vehicles which have followed. A great many puzzling questions were answered in the first cursory examination of the pictures. For example, the thermal properties of the moon's surface observed during lunar eclipses had led scientists to believe that the surface was covered by a layer of dust. For years astronomers discussed this dust and pondered its probable thickness. The Surveyor pictures and those from subsequent missions (see Figures 6.15 and 6.16) settled the question forever, and showed that the lunar surface material is more like sand than dust.

Figure 6.15. The astronaut Edward Aldrin standing on the lunar surface. The details of the surface may be seen here, and the footprints in the foreground give an idea of the powdery nature of this material. (NASA photograph from the Apollo XI mission.)

Figure 6.16. Another Apollo XI photograph of the lunar surface. (NASA photograph.)

In July, 1969 one of mankind's oldest dreams became reality as the American astronauts Neil Armstrong and Edward Aldrin reached the surface of the moon and opened a new period of exploration and study. The whole world has been able to watch the remarkable journeys of these men and of the Apollo astronauts who have followed them. After each Apollo crew has returned from the moon, a much less dramatic series of events has begun. World-wide teams of geologists and chemists have analyzed the valuable cargo of lunar rocks and soil samples brought back from each of the landing sites. In hundreds of laboratories, scientists have independently tested the samples using the most advanced analytical techniques. They have been able to compare the samples from the lunar seas with the samples from the highlands as they have tried to look for clues to the history of the lunar surface features and to develop a theory of the moon's internal structure.

The lunar materials seem to fall into three fairly distinct groups. First there are crystalline rocks of igneous origin. These rocks resemble terrestrial basalt, but the detailed chemical compositions are different. These are believed to be the base rocks which solidified in the early history of the moon. The second group of rocks are the breccias, which are composed of smaller rock particles cemented into a conglomerate. The particles in these rocks may have been blown out of the surface when meteorites crashed and exploded to form the craters. The broken materials might then have been formed into the breccias by the surface shock waves of later meteorite explosions. The third basic type of lunar material is the coarse, sandlike "soil" covering large areas of the moon. This is the loose, pulverized rock which has resulted from numerous meteorite explosions throughout the history of the moon. Some of this is in the form of a fine, dark powder which clings to the space suits of the astronauts.

The detailed analyses have revealed the relative abundances of the various chemical elements in the lunar materials. As in the case of the earth, silicon and iron are the most abundant elements. The percentages of these and other elements in the two bodies are not the same, however, and this has been interpreted as an indication that the origins of the two bodies were not the same. No evidence of water has been found, so water has had no role in the chemical history of the lunar rocks. It follows also that water cannot have had a role in shaping the lunar topography.

6.8 Effects of the Moon on the Earth

In our everyday language, words such as *lunatic* and *moonstruck* recall the ancient beliefs that the moon had great influence over men's lives and that certain practices and rituals were more meaningful if carried out when the moon was in one particular phase or another. There is no scientific basis by which special power can be attributed to moonlight, but the moon is an awesome and beautiful sight. It is easy to see how such notions could have begun.

In a very positive way the moon does have a profound effect on the earth through the **tides.** The most familiar are certainly the ocean tides, which range from high to low and back to high twice each day. There are also earth tides in which the flexible crust of the earth rises and falls very slightly. More recently discovered are atmospheric tides, whose effects are not well understood as yet.

The close resemblance of the tidal cycle and the synodic month suggested that the tides were in some way caused by the moon. Isaac Newton worked on the problem and computed that the tide should rise and fall about 35 cm if the earth were a smooth sphere covered with a uniform layer of water. The tide-raising force is the difference between the gravitational attraction of the moon on the earth and on the waters of the earth, which are free to flow. A point on the earth at which the moon is directly overhead is closer to the moon than any other point on the earth. At this point the gravitational attraction of the moon should be greatest. Directly opposite this point the attraction should be least. Along a circle around the earth 90° from these two points, the attraction of the moon exerts a nearly horizontal force. The result of these forces, if the water covered a smooth earth, would be an almost egg-shaped covering of water as indicated in Figure 6.17. The water would be deepest directly under the moon, almost as deep at the opposite side of the earth, and shallowest along the circle 90° from these two points. Along this circle, water would have flowed toward the point of deepest water. The actual picture is somewhat like this but is complicated by the presence of the various continents and the friction between the water and the ocean basins. Nevertheless as the earth rotates, the observer experiences two low tides and two high tides each day. Furthermore, one of the high tides is usually a little higher than the other each day.

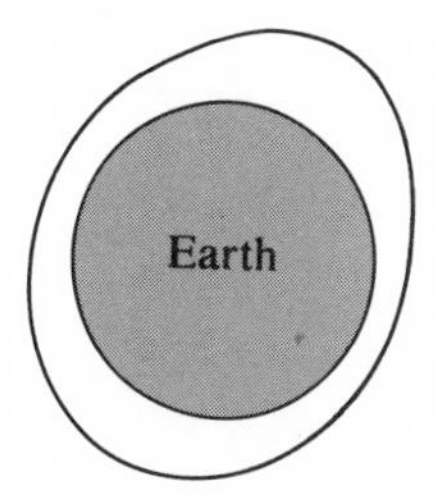

Figure 6.17. The moon's gravitational attraction for the earth and for the water surrounding the earth results in a tidal bulge.

From one place to another there is no particular pattern for the interval between the passage of the moon over the meridian and the arrival of high tide. The local tide times and the actual range from high to low tide depend most on the size, shape, and contours of the body of water in question. Points far up a tidal river may have their high tide much later than points near the mouth. A famous example of the effect of the shape of the body on the tides is the Bay of Fundy in Nova Scotia, where the range from high tide to low is in the neighborhood of 12 meters. Here a more or less normal ocean tide is squeezed into a tapering estuary.

Even though it is quite far from the earth, the immense mass of the sun exerts a considerable tidal force on the earth in the same manner that the moon does. In the course of the month, therefore, the sun and moon alternately act together and at right angles to each other. At new moon and at full moon the tide-raising force is greatest, and the range of the resulting **spring tides** is large. At the quarter phases the sun acts to lessen the effectiveness of the moon, and the range of the tides is somewhat less. These are referred to as **neap tides.** During severe storms the wind may actually push the water against the shore. Combined with natural high tides, such conditions often cause bad flooding along the coastlines.

There is a good deal of energy lost in the form of frictional heat as the tidal bulge meets the continents each day. This slows down the rotation of the earth by the very slight amounts mentioned earlier in this chapter. This loss of angular momentum from the earth-moon system is balanced by a small increase in the distance between the two bodies. Thus the total angular momentum of the system stays constant.

QUESTIONS

1. Define the horizontal equatorial parallax of the moon and explain the observations from which this quantity is calculated.
2. Explain the reasons why the time of moonrise depends upon the phase of the moon.
3. Why is the moon's synodic period longer than the sidereal period?
4. Why is it that before the advent of space vehicles, observers on the earth had seen only one side of the moon?
5. What causes the librations which enable the earthbound observer to see slightly more than half of the moon's surface?
6. Explain the circumstances of an eclipse of the moon. Why is such an eclipse not seen at least once a month?
7. How is the mass of the moon estimated and what is the approximate value?
8. What sort of observations confirm the notion that the moon has no atmosphere?
9. What are some of the possibilities discussed as explanations of the origin of the lunar craters?
10. Describe the present knowledge of the detailed nature of the lunar surface and the manner in which this knowledge has been gained from the Ranger, Surveyor, and Orbiter vehicles.

11. Why are there two high tides and two low tides each day?

12. How does tidal friction affect the rotation of the earth and the distance from the earth to the moon?

7

Lesser Bodies in the Solar System

Besides the sun and the moon the most conspicuous and influential bodies in the solar system are the planets. They are easily and regularly seen, and they affect each other's motions in a systematic manner. There are also, however, thousands of lesser objects which are sometimes hard to find and whose behavior is sometimes hard to predict. From these lesser bodies we have, nevertheless, learned a great deal not only about our own planet but about some of the other planets and the space between the planets as well. These bodies are the asteroids or minor planets, comets, meteorites, and meteors.

7.1 Discovery of Asteroids

In the latter half of the eighteenth century there was devised in Germany a number scheme which we know today as Bode's law. By writing down the series of numbers 0, 3, 6, 12, 24, 48, 96, adding four to each and dividing each sum by ten, one could obtain another series of numbers which were quite close to the distances of the planets from the sun in astronomical units:

0	3	6	12	24	48	96
4	4	4	4	4	4	4
4	7	10	16	28	52	100
0.4	0.7	1.0	1.6	2.8	5.2	10.0

◄**Figure 7.1.** Phobos, one of Mars' two satellites, pictured by Mariner IX from a distance of 5540 km. It is likely that the asteroids when photographed from short range will have an appearance somewhat similar to this. (NASA photograph.)

The actual distances of the planets Mercury, Venus, Earth, Mars, Jupiter, and Saturn are

0.38 0.72 1.0 1.52 5.20 9.54

The agreement between the two sets was good enough to make the scheme interesting even if there occurred a blank at 2.8. Then in 1781 Uranus was discovered and the radius of its orbit was calculated to be 19.19 astronomical units. The scheme of Bode's law was naturally carried one step further, giving the number 19.6. Uranus' newly found distance fitted so well that many astronomers became convinced that the number 2.8 in Bode's law meant that there must surely be an undiscovered planet between Mars and Jupiter. There was some talk of an organized search on an international scale, but before this could begin, the body suggested by Bode's law was found accidentally. In January, 1801, the Italian astronomer Piazzi found in a well-mapped star field a starlike object which was new to the area. He saw it move against the background of stars and thought that it might be a comet. Toward the end of the month he saw it make a small loop in the sky as the earth passed, and it displayed the typical retrograde motion of a planet. When Piazzi's object was relocated at its opposition the next year, there was no further doubt that this body was in fact a planet. Piazzi himself selected the name *Ceres* for his discovery. Ceres was the first of the **asteroids** or **minor planets** and its orbit was soon well known. At one favorable opposition a very large telescope was used to resolve the new planet's disc so that its diameter could be computed. The diameter of Ceres is 700 km, the largest discovered so far for a body of this type.

In the next five years three more similar bodies, Pallas, Juno, and Vesta, were found. Their diameters, derived from their angular diameters, are 460, 220, and 380 km, respectively. During the remainder of the nineteenth century the discovery of asteroids continued at a modest pace, and a total of several hundred became known. With the development of photography near the beginning of this century, the rate of asteroid discovery increased tremendously and today thousands have been catalogued.

The reason for the great upsurge in the number of minor planets discovered lies partly in the ability of the photographic emulsion to record faint objects and partly in the fact that the minor planet in its orbit is moving against the background stars at a rather noticeable rate of speed. On a long-exposure photograph, then, the stars appear as points and the minor planet appears as a streak (Figure 7.2). Today so many of these telltale streaks are found that only a few astronomers take the time to follow the asteroids and compute their orbital elements. The chances for simply rediscovering an old asteroid are pretty good; nevertheless, the persevering individual who wishes to immortalize his own or his sweetheart's name has the privilege of naming an asteroid if he discovers it, computes its orbital elements, and observes it on one or more subsequent oppositions. This is not so easy as it might seem because these small bodies are greatly disturbed by the gravitational effects of the earth, the moon, and the other planets if they happen to pass within a few million miles. The orbits are thereby changed to such a degree that the asteroids often become difficult to sight again at later times.

The orbits of most of the minor planets lie close to the plane of the ecliptic, and with some exceptions the orbits occupy the wide space between the orbits of Mars and Jupiter. All of the minor planets revolve in the direct direction just as the planets do. Typical periods range between 3.5 and 6 years.

Figure 7.2. The asteroid Icarus left this trail in a long-exposure photograph made on June 26, 1949. (Hale Observatories photograph.)

7.2 Probable Nature of the Asteroids

As we have said already, none of the asteroids is very large. For only the first four and a few others that have come particularly close to the earth can angular size and diameter be measured directly. For any of these we may compute an albedo from the observed brightness, the sizes, and the distance of the asteroid from the sun. If we assume the same albedo for the smaller ones, we may compute their diameters from their observed brightness. Observations of this sort were used to confirm the size of Pluto. From such estimates we know most asteroids are probably only 1 km or so across. Only a few dozen are more than 50 km in diameter.

For some asteroids the observed brightness fluctuates by as much as a factor of 2 in a period of a few hours. This variation is interpreted as an indication of an irregular body spinning and tumbling as it travels (Figure 7.1). On the occasion of one of the close approaches of Eros, a cigar-shaped form was actually observed. Careful photoelectric observations of Ceres also show fluctuations in brightness. It seems likely that none of the asteroids is very nearly spherical.

Many people have often been interested in the possibility that the minor planets are the broken pieces of a larger single body which once existed in the orbit at 2.8 A.U. There is, unfortunately, no way to prove or disprove this. We can try, however, to see what sort of a planet this might have been. Even with a very generous allowance for undiscovered asteroids and for material lost to the solar system in the explosion, the total volume of the lost planet would be much less than the volume of our moon. If a density is assumed comparable to the density of the moon, the mass of an asteroid may be computed. The total mass of all the asteroids is less than one-hundredth the mass of the moon. If the observed asteroid material ever comprised a single body, it is quite probable that it must have been quite small. Perhaps we really are seeing the wreckage of two or more Ceres-type objects.

7.3 Asteroids of Special Interest

The run-of-the-mill asteroid offers little of interest to the scientist. None is massive enough to hold an atmosphere. Once the orbital elements have been calculated and the size has been estimated, there is not much more that can be done except to try to detect it telescopically the next time the earth passes it. There are a few asteroids, however, which for any one of several reasons are of special interest. From some we have obtained information which was difficult or impossible to obtain in any other manner.

In Chapter 2 the problem of converting the orbital radii of the planets from astronomical units to kilometers was mentioned. It was pointed out there that triangulation techniques were employed by observers widely separated on the earth to measure the earth-Mars distance at a time when this distance was a minimum. At such times Mars is still some 40 million km from us and measuring its distance with any degree of accuracy is difficult. If there were another sun-orbiting body closer to earth at opposition, its distance could be measured with greater precision and thus the length of the astronomical unit in kilometers could also be known more accurately. The asteroid Eros is such a body. Its orbit is well known, and at the opposition of 1931, when faster-moving earth passed Eros, the minimum distance was only 22 million km. This favorable situation brought forth a world-wide cooperative effort among astronomers. At observatories in many countries, Eros was photographed against its background of distant stars as near as possible to the time of opposition. From hundreds of photographs the carefully measured positions of Eros against its background were obtained (Figure 7.3). These positions were then combined with other quantities such as time and latitudes and longitudes of the observatories into one grand and complex mathematical solution from which came the desired quantity, the parallax of the sun. Eros was ideal for this program not only because of its close approach but also because of its small size. The photographic image of Eros was similar to the images of the stars, and so the measurements on the photographs could be made with great accuracy.

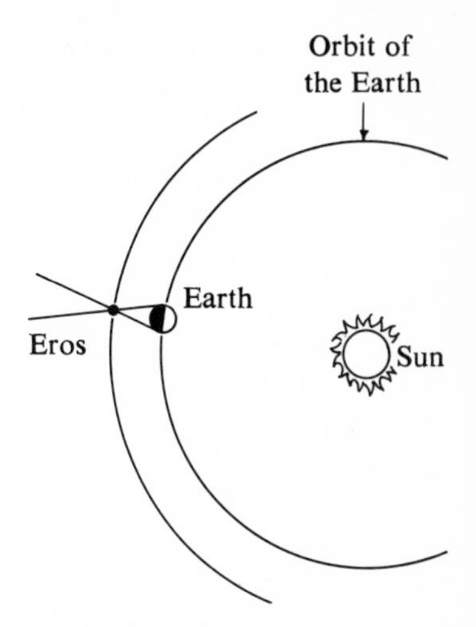

Figure 7.3. At opposition in 1931, Eros was only 22 million km from the earth. Its measured distance was used to calibrate the scale of the solar system.

Two other asteroids, Icarus and Geographos, also make periodic close approaches to the earth. Since the minimum distances for these two asteroids are 6.5 and 9.6 million km, respectively, from the earth, astronomers will have future opportunities for new and presumably better determinations of the length of the astronomical unit.

Of all the known asteroids, Icarus has the smallest mean distance from the sun and the most eccentric orbit. It approaches to within only 27 million km of the sun, while at the other end of its long slender orbit Icarus is well beyond the orbit of Mars (Figure 7.4). Crossing the orbits of the four inner planets as it does, Icarus

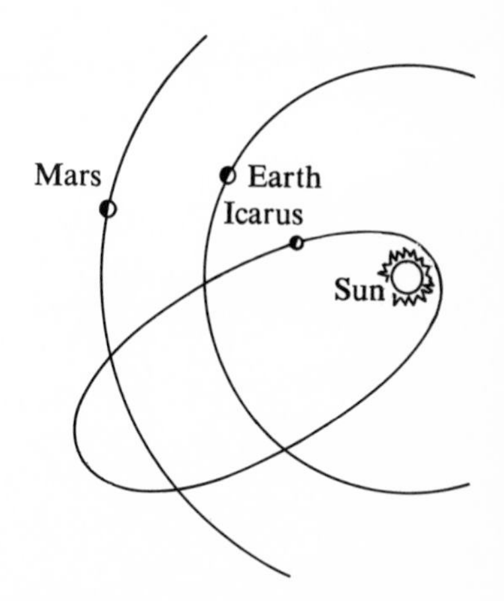

Figure 7.4. Orbit of Icarus brings that asteroid into the sun's vicinity once every 408 days.

is bound to come close to Mercury, Venus, earth, and Mars on rare occasions. A close approach to any one of these will mean that the gravitational attraction of the planet will cause perturbations or changes in Icarus' orbit. Observations of such perturbations caused by Mercury or Venus will be especially significant because they will help to improve our estimates of the masses of those two planets. The greater the amount of the perturbation, the greater must be the mass of the attracting body. (Icarus' mass is so small as to be negligible.)

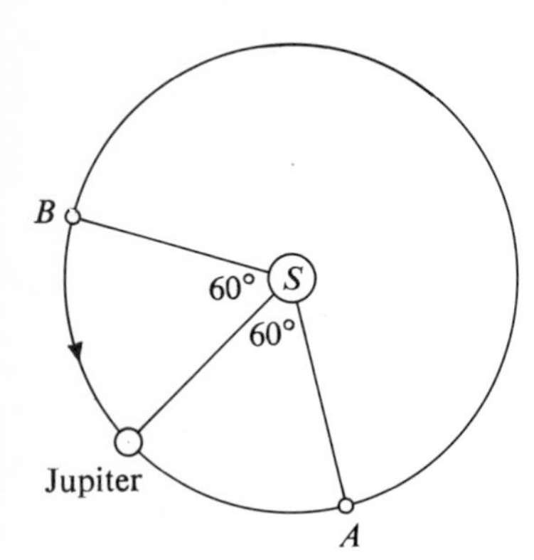

Figure 7.5. Points *A* and *B* are the Lagrangian points at which the Trojan asteroids are held.

Since the orbits of the majority of asteroids lie between the orbits of Mars and Jupiter, and since Jupiter has such a prodigious mass, it is not surprising to find Jupiter's influence apparent in several ways. Sixty degrees ahead of and behind Jupiter in its orbit are two small groups of asteroids. Many years ago the mathematician Lagrange showed that bodies of small mass could be maintained near the vertices of two equilateral triangles formed with Jupiter and the sun. (See Figure 7.5.) The bodies in the group to the east of Jupiter have been named for Greek heroes in the Trojan War, while those in the group to the west of Jupiter have been named for the Trojans. Together, all 14 bodies are referred to simply as the Trojan asteroids.

In the broad asteroid belt there should be a variety of revolution periods since the period is related to the asteroid's distance from the sun. Some of these periods should be simple fractions (1/4, 1/3, 1/2) of Jupiter's revolution period. When this is the case, after regular intervals of, say, 4, 3, or 2 revolutions, Jupiter exerts the same gravitational pull on the asteroid. The regularity causes these effects to be cumulative, and soon the asteroid's orbit and revolution period have been changed. Study of the data for a large number of asteroids does in fact confirm this theory. Asteroids simply are not found at distances from the sun which would result in periods related to Jupiter's. The resulting spaces in the asteroid belt are known as **Kirkwood's gaps** and are named for their discoverer. This is the same effect described on page 119 to account for the gaps in the rings of Saturn.

7.4 Appearance and Development of Comets

Every few years a bright comet appears in the sky and for a few weeks captures the attention of people all over the world. Such was the case with Comet Ikeya-Seki 1965(a), pictured in Figure 7.6.

Figure 7.6. Comet Ikeya-Seki 1965(a). Note the structure in the tail. (NASA photograph from Wallops Station.)

Even though only a few comets present the spectacle of this one, there are from seven to ten comets seen each year. Regardless of how bright they actually become, comets almost always follow a fairly definite pattern while they are near the sun.

A typical comet is usually discovered first as a simple diffuse spot. Watched from night to night, this spot changes its position and becomes brighter. Gradually a tail develops. As it nears the sun, the comet picks up speed, and the tail becomes longer and brighter. Rounding the sun the comet is for a short time invisible in the sun's brilliant glare. Then it reappears and moves away from the sun along its orbit. The tail shrinks and the head becomes less bright. At last the tail disappears and the comet soon fades from view altogether. Throughout the entire period the tail points away from the sun, as indicated in Figure 7.7. A few rare comets have been so bright that they could be seen in daylight, and it is hardly any wonder that primitive peoples have been amazed and terrified by such spectacles.

Detailed telescopic examination shows that the head of a comet is composed of a small bright **nucleus** surrounded by a rather diffuse cloud or **coma.** The material in the coma streams back away from the head to form the tail. The nucleus is probably never more than a few kilometers in diameter. The visible coma can be hundreds of thousands of kilometers in diameter and actually appear to grow smaller in the immediate vicinity of the sun. The tail is usually not very conspicuous until the comet approaches the orbit of Mars. After that the tail can become millions of kilometers in length. Comet tails extending for a full astronomical unit have been known.

On the basis of their periods and the shapes of their orbits, comets fit well into two groups. First there are the more typical comets whose orbits seem to be almost parabolic. If the orbits were really parabolic, the comets would make just one pass near the sun and then disappear into space. It is more probable that the orbits are really very much elongated ellipses. Observations show that orbits for bodies with periods greater than 1000 years are difficult to distinguish from a parabola. With periods ranging up to several thousand years there could be only one recorded passage for each of these comets. Comet orbits may cross the plane of the earth's orbit

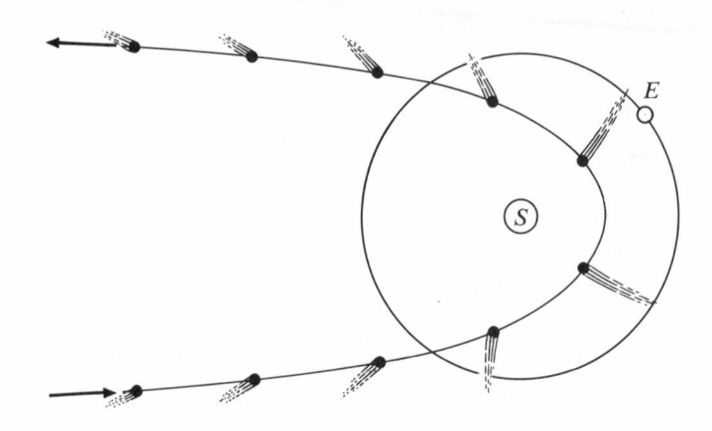

Figure 7.7. Changes in size and orientation of a comet's tail as the comet rounds the sun.

in any orientation and comets can proceed around the sun in either the direct direction (as do the planets) or in the retrograde direction.

The second group of comets are the **periodic** comets or those which return to the sun at relatively frequent intervals. Periods for those comets of shorter period range from 3.3 years for Encke's comet to 76.2 years for Halley's comet. The orbit planes of these comets all lie within 45° of the plane of the earth's orbit, and few of them travel in the retrograde direction. Most of the periodic comets may also be classed as members of **Jupiter's family of comets.** The gravitational attraction of Jupiter can have striking effects on the orbit of any comet passing within about 0.3 A.U. of Jupiter. In some situations Jupiter's pull makes the orbit hyperbolic and the comet leaves the solar system completely. At other times the effect is to make the orbit into a much smaller ellipse with its greatest distance from the sun somewhere near the orbit of Jupiter. When this happens, the chances of future encounters between Jupiter and the comet become pretty good. Further changes in the orbit are to be expected. It is highly probable that all of the periodic comets were "captured" by Jupiter, Uranus, and Neptune in this manner.

Although general statements may apply to all comets, there are never any two that are quite alike. Many extremely bright comets have been seen. Some have broken into several parts as they rounded the sun. Some have appeared to have double tails or a spike pointing toward the sun. Of all the comets, however, the most consistently spectacular one has been Halley's comet (Figure 7.8). Edmund Halley, for whom this comet was named, was a contemporary of Isaac Newton, and he showed that the comets seen in 1531, 1607, and 1682 were probably successive reappearances of a

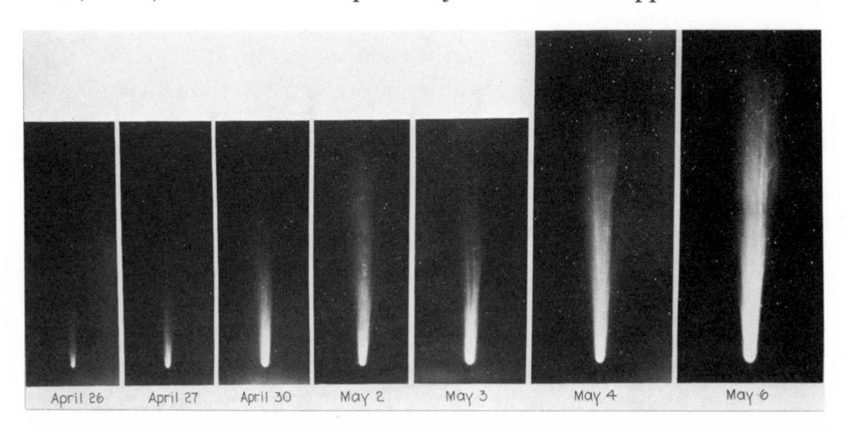

Figure 7.8. Changes in Halley's comet during its appearance in 1910. (Hale Observatories photographs.)

single comet with a period of 76 years. In accordance with Halley's prediction, the comet reappeared in 1758, just after his death. The comet has returned to the sun on schedule ever since and has continued to be a spectacular sight. People throughout recorded history have apparently been impressed enough by older appearances of Halley's comet to have made note of it as far back as 467 B.C. Over the centuries, the period has ranged from 74 years to as much as 79 years, due to the perturbations by Jupiter and the other planets.

7.5 Spectra of Comets

The uniqueness of individual comets extends from their appearance and behavior to the spectrum as well. There are, however, general spectrum characteristics which seem common to most comets. A comet's spectrum usually shows both the continuous spectrum of reflected sunlight and the emission features or bright lines of its gases. The continuous spectrum is seen as the comet first becomes visible, and indicates that the comet at this stage consists of a cloud of solid particles which reflect sunlight. These particles surround the solid central nucleus, and probably were once part of the nucleus itself. The presence of gases is revealed as the comet nears the sun and solar ultraviolet radiation is able to excite the gases. From the resulting pattern of bright lines and bands, the gases present are all seen to be composed of molecules of the more abundant chemical elements. C_2, CN, OH, NH, CH, CO^+, and N_2^+ have all been identified in the spectra of the heads of comets. In the tails of comets the gases are chiefly the ionized molecules N_2^+, CO^+, OH^+, CO_2^+. The intense solar radiation is sufficient to break up the normal neutral molecules, and the extremely low density of the gases in both the coma and the tail makes it unlikely that the ions will recombine. For this reason we see the cometary gas in a state which would be almost impossible to duplicate in the laboratory.

If a comet approaches close enough to the sun, the radiation will be sufficiently concentrated to excite the atoms of some of the metals, and bright-line spectra of sodium, iron, and a few other elements have been seen.

7.6 The Sun's Effect on Comet Tails

The manner in which comet tails point away from the sun was noted by many observers before the time of Kepler. Only in modern

times has a fairly complete explanation based on solid physical evidence become available. There seem to be two influences which combine to blow the tail radially outward from the sun. The first of these is **radiation pressure,** a weak force applied on small particles of more than atomic or molecular dimensions by the outflowing radiant energy of the sun. The second is the **solar wind**,* which was experimentally established by Mariner II and other satellite investigations that could be performed outside the influence of the earth's magnetic field. This solar wind consists of ions and electrons emitted from the sun. These particles travel outward at velocities of thousands of kilometers per second. This outflow is not uniform either in time or in direction and will be mentioned again in the next chapter. When these outflowing particles encounter the ions in a comet's coma and tail, combinations can occur and the cometary material is swept along with the solar wind. Sudden fluctuations in the solar wind are known to occur and these may be the cause of sudden changes in brightness and structure in the comets (Figure 7.9).

Figure 7.9. Rapid changes in the head of Comet Humason are seen in these photographs made two days apart in 1962. (Official U.S. Navy photographs.)

*It was a study of comet tails by Biermann in 1952 that first led to the hypothesis of the existence of the solar wind.

7.7 Meteors and Meteor Showers

On a clear moonless night, the observer who has taken a few minutes to become dark-adapted will begin to see **meteors.** The appearance of meteors may vary from short, barely detectable streaks lasting about one-tenth of a second and extending only a few degrees on the celestial sphere to the rarer bright **fireballs** which exceed Jupiter in apparent brightness and may streak through the sky for more than one second, covering a path of 20° or more. Exploding fireballs are known as **bolides.**

By careful coordination of observations from two stations, it is possible to photograph specific individual meteors simultaneously. By noting the directions from each observer to the beginning and ending points of the meteor's track in the sky, the heights of the meteors may be determined (Figure 7.10). Very significant work in this area has been pioneered by Fred L. Whipple of Harvard, who used unusual wide-angle cameras in his work in northern Arizona. Whipple's cameras were equipped with shutters that interrupted the meteor's streak at some known frequency (Figure 7.11). From the length of the streak between interruptions and the computed distance of the meteor from the observer, the velocities of meteors were determined. Analysis shows that the velocities of meteors range from about 11 to 73 km/sec. The faster ones are usually seen in the early morning. Heights at which meteors first appear are usually in the neighborhood of 100 km. Faint meteors are usually burned up by the time they are about 85 km high, while brighter

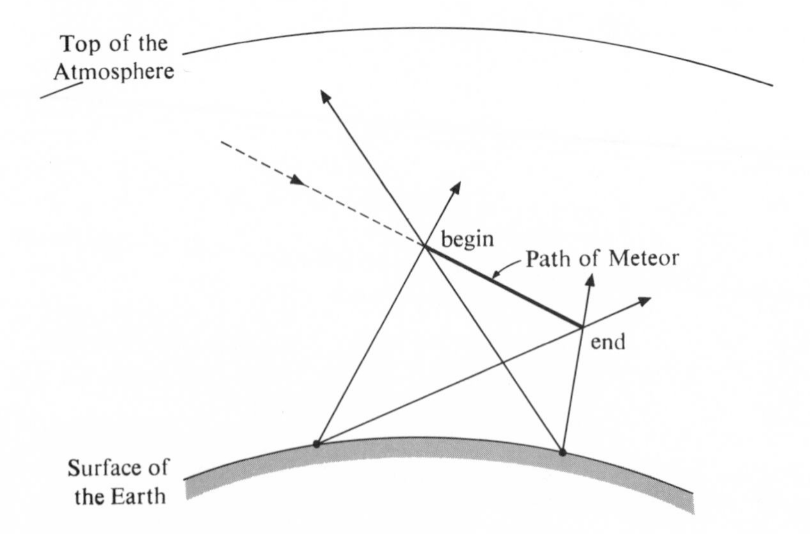

Figure 7.10. Simultaneous observations from two stations can give knowledge of the heights of the beginning and end of the meteor.

Figure 7.11. The trail of a spectacular meteor near the bowl of the Big Dipper. A rotating shutter interrupted the trail 60 times per second. The photograph was made with a Super-Schmidt meteor camera. (Harvard Meteor Project photograph.)

ones may persist as low as 60 km. The first estimates of the height of the earth's atmosphere were actually based on visual observations of meteors.

Once the distance from the observer to the meteor and the duration of its flash have been determined, it becomes possible to compute from the observed brightness of the flash the actual amount of energy involved. The energy released in any meteor's flash depends upon the **mass of the meteor** and the velocity with which it enters the atmosphere. The brighter meteors are either the faster-moving ones or the larger ones. Carried out completely, calculations show that meteors are quite small. The spectacular meteor flashing across a long arc in the sky might have a mass of less than one gram. The more typical meteors just visible to the eye probably have masses less than one-hundredth of a gram and are bits of material the size of a grain of sand or a piece of rice.

Modern radars have been able to detect meteors in the daytime as well as at night, since the trail of ionized gas in a meteor's wake reflects the transmitted radar pulse. The rate of daytime meteors is comparable to the rate of night ones. A patient observer on clear, moonless nights ought to see on the average about 10 meteors per hour, and the number of those too faint to be visible to the unaided eye is much higher.

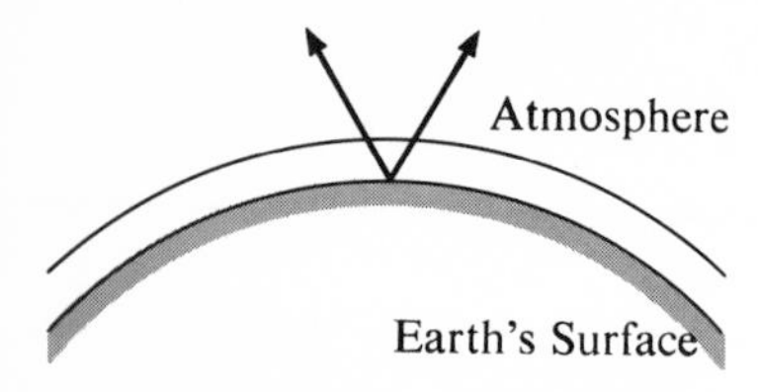

Figure 7.12. Because the atmosphere is such a thin layer around the earth, an individual observer can see meteors in a very small volume of the the total atmosphere even though he can see nearly half of the celestial sphere.

In order to estimate in a meaningful manner the total number of meteors entering all parts of the earth's atmosphere, the observer must know how much of the earth's atmosphere he is actually seeing. Even though one can see about a third of the celestial sphere at a given time, only a very small fraction of the atmosphere can be seen from any particular spot on the earth (Figure 7.12). The little cone of atmosphere watched by a single individual is only about one hundred-thousandth of the whole area to be patrolled. At the rate of 10 per hour for one observer, one million visible meteors per hour reach the earth. If each of these had a mass of one gram, the earth would be collecting about 24 tons of meteoric material per day. When the vast numbers of faint meteors are considered, the total daily accumulation is more than one thousand tons! This seems like a lot of material, but spread over the entire earth it is insignificant. Even in a million years the total mass of the meteoric fragments would be an insignificant fraction of the earth's mass.

The type of observations mentioned above and pictured in Figure 7.10 gives us the actual direction in space of the meteor's path as it entered the atmosphere. Knowing this direction, the velocity and the distance of the meteor from the sun (1 A.U.), one can calculate the trajectory that the meteor would have had if it had not collided with the earth. All the meteors studied by radar methods have been found to have had closed elliptical orbits around the sun. None of these objects could have been swept up from interstellar space as the sun moved and carried the planets with it. Each tiny individual meteor must be a true member of the solar system.

During certain fairly well-defined periods each year, the number of meteors per hour may be somewhat higher than usual. Anywhere from 15 to 50 meteors per hour might be counted. These increases in activity are referred to as **meteor showers,** and at such times the meteors may seem to enter the atmosphere from the direction of some particular constellation. Information on principal meteor showers is given in Table 7.1. Thus in early August the Perseid meteors seem to emanate from the constellation Perseus, and in December the Geminids are seen in the direction of Gemini. Figure 7.13 is a time-exposure photograph made during the Leonid meteor shower in November, 1966. If extended backward, the meteor trails seem to converge toward a common point. This is referred to as the **radiant point** of the meteor shower. This is the effect that should be noted if the meteoric particles are travelling

Table 7.1. Principal meteor showers.

Name	Dates	Constellation	Approx. Radiant R.A.	Approx. Radiant Decl.	Approx. Hourly Rate
Quadrantids	Jan. 1–4	Bootes	15^h20^m	49°	30
Lyrids	Apr. 20–22	Hercules	18^h00^m	33°	8
η Aquarids	May 2–7	Aquarius	22^h20^m	0°	10
δ Aquarids	Jul. 20– Aug. 14	Aquarius	22^h40^m	−10°	15
Perseids	Jul. 29– Aug. 18	Cassiopeia	3^h00^m	58°	40
Draconids	Oct. 10	Draco	17^h40^m	54°	*
Orionids	Oct. 16–26	Orion	6^h20^m	15°	15
Taurids	Oct. 20– Nov. 25	Taurus	3^h30^m	16°	8
Leonids	Nov. 14–19	Leo	10^h10^m	22°	*
Geminids	Dec. 8–15	Gemini	7^h30^m	32°	50
Ursids	Dec. 19–23	Ursa Minor	14^h00^m	78°	12

*In some years the Draconids and Leonids are much more spectacular than in other years.

Figure 7.13. Leonid meteors seen in November, 1966. The two circular images near the radiant point are the trails of meteors heading directly toward the camera. The brighter stars here are the principal ones of the constellation Leo. (Photograph by Dennis Milon.)

in parallel paths. (Recall the example in Chapter 2 of the car driving in a snowstorm.) During the period of a meteor shower, the earth must be crossing a broad **meteor stream.** The direction of the radiant point gives the orientation of the meteor stream. A number of such streams are recognized from the annual meteor showers, and there may well be others through which the earth does not pass. The complete paths of these streams are elliptical orbits around the sun.

7.8 Relationship between Comets and Meteors

In a surprising number of cases the orbits of meteor streams are the same as the orbits of particular comets. In some cases the meteor stream was noticed after a periodic comet had failed to reappear when predicted. It seems quite clear today that the meteoroids or particles which cause the visible meteor were once part of a comet and became separated from the comet. Each particle has a small random component of velocity, so it continues around the sun in its own individual orbit. The small differences in the orbits cause the particles to become more and more separated from the parent comet. Eventually there will be particles along most of the orbit in a

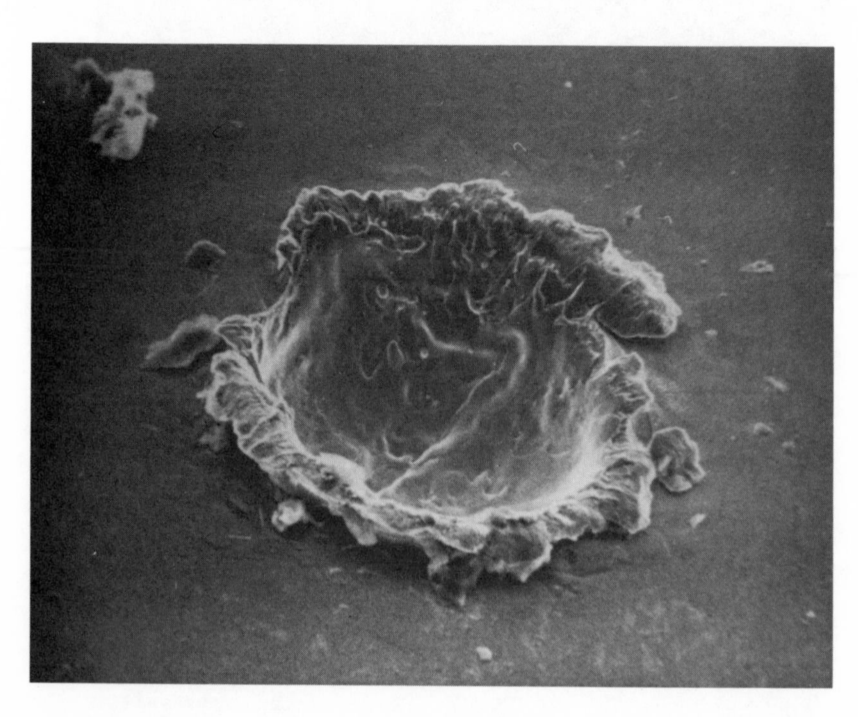

Figure 7.14. This crater was formed when a micrometeorite struck a stainless-steel plate attached to an earth-orbiting Agena rocket. This photograph was made with an electron microscope since the crater is less than 0.2 mm in diameter. After four months in orbit the plate was retrieved during the Gemini X mission. (Dudley Observatory photograph.)

wide path. As the earth crosses the path we see a meteor shower. The fainter sporadic meteors which are not related to any shower may have been in well-defined streams which over the years have lost their identities (Figure 7.14).

The relationship of meteor streams to comets has led to the concept of the comet as a mixture of frozen material and small solid particles. As this "icy conglomerate" approaches the sun, the frozen material evaporates from the solid surface to form the gaseous coma and tail of the comet. As the icy nucleus gradually becomes smaller, the solid particles are first exposed at the surface and then detached from it. The particles may spread out within and beyond the coma, and they help to explain the appearance of reflected sunlight in cometary spectra. This model of a comet requires that the comet lose some of its material on each trip around the sun. Thus the comet must necessarily have a limited existence. When all of the gas has been evaporated and dissipated into space, only the solid particles are left to orbit the sun along the old path.

7.9 Meteorites

Not many years ago the distinction between meteors and meteorites was simply that the meteor was the flash in the sky and the meteorite was the material which, in a few rare cases, was large enough to survive the flight through the atmosphere and land on the earth. It now appears that meteorites are completely unrelated to meteor streams and cometary debris. There is no known case of a meteorite (or even a fireball) which can be traced to a meteor stream.

Reports of "stones from the sky" have been recorded for hundreds of years, but only in the past 200 years or so were a few falls sufficiently documented to offer convincing proof of the extraterrestrial nature of these bodies. Every year in a few spots on the earth meteorites are recovered after an observed fall. Allowing for a surface area covered with oceans and unpopulated areas, it has been estimated that there may be two meteorites per day large enough to be seen and retrieved. Known cases of bodily injury and property damage from meteorites have fortunately been very few. In a few cases objects have been identified as meteorites even though the falls were not observed. This is possible because meteorites have certain rather definite characteristics. A meteorite usually has a dark, shiny, sometimes glasslike surface. This is the result of surface melting and ablation due to friction with the air while the meteorite speeds through the atmosphere. As the surface becomes molten, droplets of material fall off. Before striking the earth the meteorites usually have been slowed to the point where melting no longer occurs and some of the molten material can solidify leaving this thin crust. Inside the crust most meteorites are composed of **stony** material containing small bits of iron and nickel. About 5% of all meteorites are almost completely composed of **iron,** and a very few percent are best described as **stony-iron.** Of the meteorites found but not observed to fall, most are iron because stony meteorites are more easily confused with terrestrial rocks. When the interior section of an iron meteorite is polished and etched with acid, it usually shows a characteristic pattern of large crystals (Figure 7.15). To the geologist this means that the material has cooled slowly. It has been suggested that the iron meteorites may once have been in the interior of a planetlike body.

Chemists have detected many chemical elements in meteorite samples and, as one might expect, no elements unknown on the earth have ever been found. In detailed analyses chemists have also measured the relative amounts of uranium, lead, and helium.

Figure 7.15. Polished section of a meteorite showing the characteristic pattern of large crystals. (American Museum of Natural History photograph.)

Since uranium decays spontaneously into lead and helium at a known rate, the present ratios of these elements may be used to determine the age of a meteorite sample. The best estimates suggest that meteorites are some five billion years old. This figure is reasonably close to the age of the earth. On the basis of very precise chemical analysis, meteorite samples seem to fall into a limited number of well-defined groups. This has been taken as an indication that meteorites originally came from a small number of parent bodies.

7.10 Significant Meteorite Falls

It was pointed out in Chapter 6 that on the earth there is ample evidence of large meteorite falls. The craters left by these impacts and explosions are to be found in most areas of the earth even though recognition is difficult after many centuries of erosion. Large meteorites lose almost none of their orbital speed in their rapid pass through the atmosphere. Their tremendous kinetic energies must be dissipated very quickly and so a violent explosion results. Smaller ones may be slowed in the atmosphere enough so that they simply burrow into the ground. (See Figure 7.16.) In other cases meteorites have been known to explode in the air.

Figure 7.16. The Ahnighito meteorite on display at the American Museum–Hayden Planetarium. (American Museum of Natural History photograph.)

In recorded history large meteorite falls near populated areas have indeed been rare. The two largest, in fact, both occurred in Siberia during this century. On June 30, 1908, near the Tunguska River in Siberia, a body which may have weighed several hundred tons apparently exploded above the earth's surface. No large crater was left, but many fragments did land in the forests. Over an area of hundreds of square kilometers around the explosion, trees were knocked down with their branches pointing away from the explosion zone. Earth tremors were felt in western Europe more than 1600 km away.

On February 12, 1947, a second major meteorite fall occurred in Siberia, this time farther east, near Vladivostok. This time the "body" seemed to be composed of many lesser parts, for more than 100 individual craters were found. The larger ones showed definite signs that explosions had occurred.

The most thoroughly studied large crater of definite meteoritic origin is undoubtedly the famous Barringer crater near the town of Winslow in northern Arizona (Figure 7.17). The top of the rim is about 50 meters above the level terrain in the area and the bottom of the crater is some 180 meters below the rim. From rim to rim the diameter of the crater is 1280 meters. Extensive studies carried out about 1900 revealed much meteoritic iron scattered over a wide area and suggested that the underlying rock may be badly shattered. For years it was imagined that the main mass of the body

Figure 7.17. The Barringer meteorite crater near Winslow in northern Arizona. (Photograph by John Farrell.)

might be buried under the crater, but it seems more probable that the original body was completely destroyed in the explosion. The meteorite may have weighed many thousands of tons. The age of the crater is difficult to establish with any degree of certainty, but the crater is almost certainly more than 30,000 years old.

In the northern region of the province of Quebec in Canada there is a circular lake about 3 km in diameter. The great central depth of 240 meters and the raised rim some 90 meters high suggest the profile of a meteorite crater. The true nature of the lake has not yet been confirmed by the discovery of positive meteoritic samples in the area. This crater and a number of other probable meteorite centers are very likely as old or older than the Arizona crater. The fact that no large young crater has been discovered on the earth means that the present rate of arrival of "asteroid-sized" meteorites is fortunately very low.

7.11 Orbits of Meteorites

Since large meteorites are rather rare, there are few cases for which many reliable observations are recorded. Nevertheless, from well-documented reports of fireballs it has been possible to compute the orbits of some meteorites before their collisions with the earth. These studies indicate two things: (1) the meteorites travel around the sun in the same direction as the planets and (2) the orbits are more nearly circular than are the orbits of the comets. This kind of evidence supports the idea that meteorites are perhaps related to the asteroids but bear no relation to the comet-related meteors.

7.12 Interplanetary Dust

In addition to the types of bodies already discussed, there is also a significant amount of dust in the solar system. In this usage, "dust" describes particles too small to be observable as meteors but larger than molecules. Since there is some range in the size of the particles which are seen as meteors, it is not surprising to find large numbers of these dust-sized particles. Just like their larger counterparts, each dust particle must be in its own individual orbit around the sun.

The presence of interplanetary dust and its distribution in the solar system are inferred from a phenomenon known as the **zodiacal light.** At twilight on a clear spring evening, the zodiacal light may be seen as a faint glow in the western sky. The glow is centered near the point at which the sun sets and forms a narrow triangle extending upward some 20° above the horizon. The ecliptic or zodiac passes through the center of this luminous triangle, hence the name which has been used to describe the glow. Since the orientation of the ecliptic with respect to the horizon depends upon the season and the observer's position on the earth, the zodiacal light is sometimes more nearly vertical than at other times. This general appearance suggests that the sun is surrounded by a flattened cloud of particles which becomes thinner in the plane of the ecliptic as the distance from the sun increases.

The spectrum of the zodiacal light is the same as the spectrum of sunlight, indicating that we are actually seeing sunlight scattered by dust particles and electrons. Since this light is not appreciably redder than sunlight, we know that the scattering was not by gas molecules which scatter the blue light more strongly relative to red light. The zodiacal light is slightly polarized, and this is attributed to a scattering by free electrons which must also be present in the interplanetary medium.

Under extremely favorable conditions, the zodiacal light may be seen all the way around the ecliptic. This suggests that the vast disc of interplanetary dust extends well beyond the earth. At such favorable times the **gegenschein,** or counterglow, may also be seen. The gegenschein may subtend an angle of 20° or more, but it is so faint that it is extremely difficult to see. There is no complete explanation of the gegenschein, but it has been suggested that we are actually seeing a higher concentration of dust particles than in other directions. This may come about through some interactions between the solar wind and the earth's magnetic field. It is also possible that the gegenschein is simply a phenomenon resulting from the angle at which the dust particles are viewed from the earth. When we look in the same direction as the sunlight, we see no shadows as we do in other directions. A similar effect causes the moon's surface to appear brightest at full moon.

QUESTIONS

1. Describe the role of Bode's law in the discovery of the asteroids.
2. Why did the rate of discovery of asteroids increase dramatically after the introduction of photography into astronomy?

3. What is the evidence for the belief that many asteroids are quite irregular in shape?
4. How has the earth-sun distance been determined from observations of the asteroid Eros?
5. Describe the changes in the appearance of a comet as it approaches the sun.
6. Why is the tail of a comet always directed away from the sun?
7. What happens to the orbit of a comet if the comet happens to pass close to Jupiter or one of the other large planets?
8. How can observers determine the heights of meteors above the earth's surface and what does this tell us about the earth's atmosphere?
9. What sort of orbits do meteors have? Does this prove that they are truly members of the solar system?
10. How did the observance of meteor showers lead to the icy-conglomerate theory of the nature of comets?
11. Differentiate between meteors and meteorites.
12. Why are there so few large meteorite craters on the earth when there seem to be so many on the moon?

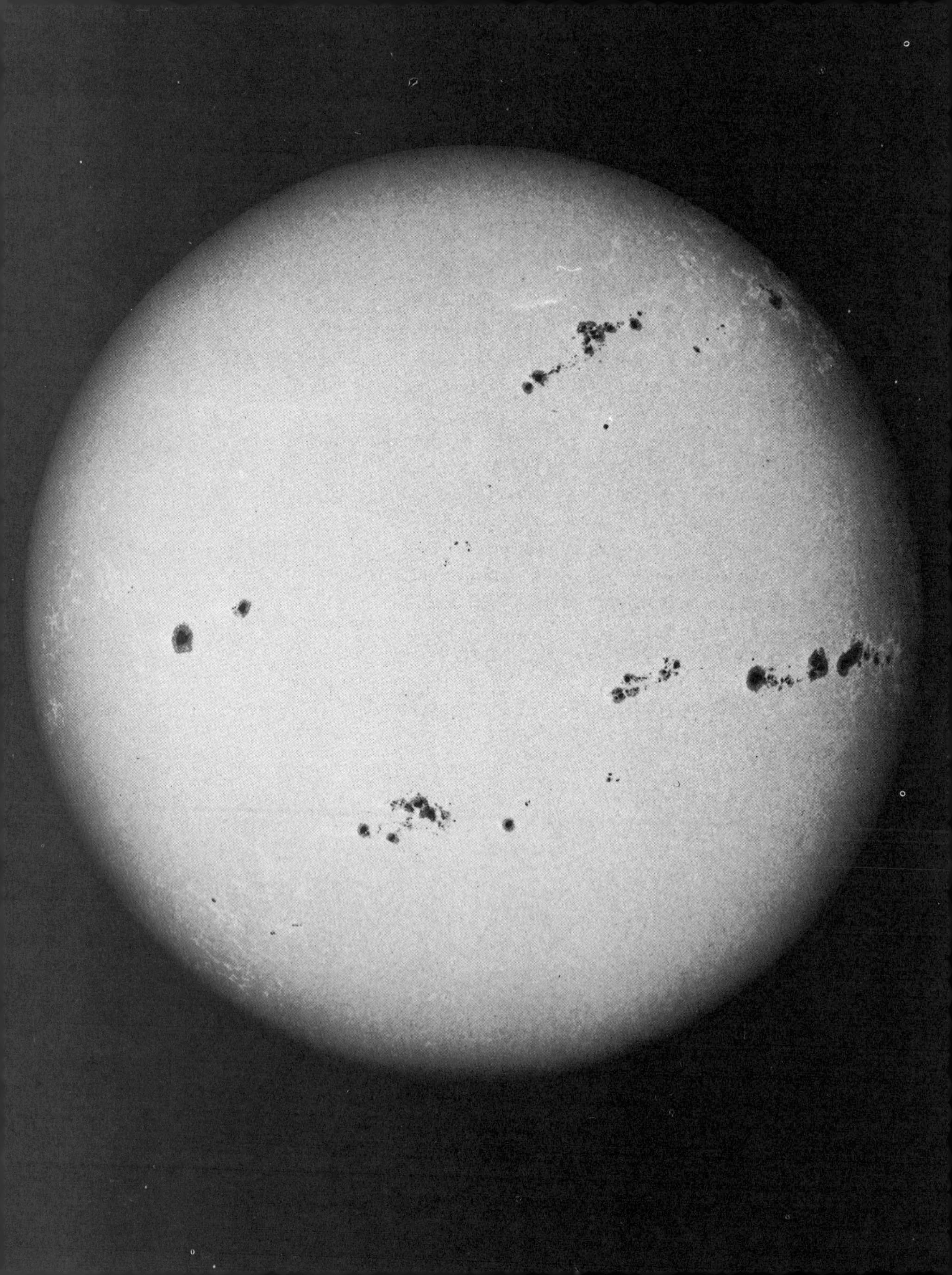

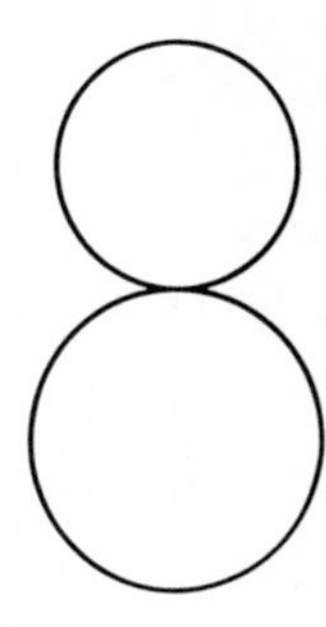

The Sun

As the gravitational center of our planetary system, the sun controls the motion of all of its family of orbiting bodies. As the central source of radiative energy, the sun is responsible for the light and heat of the planets. In this chapter we shall examine the sun as a physical body and discover what its sources of energy are. In later chapters we shall compare the sun to other stars and see how the wealth of solar information may be extended to stars of other masses and other temperatures.

8.1 Physical Properties

By this time the student should know how the sun's diameter can be determined. The observational quantity needed is the sun's angular diameter, for which the accepted value is just under 32′ when the earth is at its mean distance from the sun. (Compare this with the moon's angular diameter of 31′5″.) Since we know also the distance from the earth to the sun, we can compute the diameter of the sun by the same method described in Chapter 5.

$$\frac{32'}{3437'.7 \text{ per radian}} = \text{sun's ang. diam. in radians} = \frac{D}{150{,}000{,}000}$$

$$D = 1{,}396{,}282 \text{ km}$$

◄**Figure 8.1.** Photograph of the sun near sunspot maximum showing a large number of sunspots. (Hale Observatories photograph.)

When the exact values of angular diameter and earth-sun distances are used, the sun's diameter is actually closer to 1,392,000 km. The sun's angular diameter is the same along all diameters, indicating that for all practical purposes the sun is spherical.

In Chapter 5 the procedure for determining the temperature of the sun was inferred. Several approaches based on the radiation laws may be used. Recalling the Stefan-Boltzmann law, we know that the total energy radiated from a unit area of a hot surface is proportional to the fourth power of the temperature and that this energy is radiated in all wavelengths, not just the visible ones. This total energy is related to the solar constant, which was also mentioned in Chapter 5. The solar constant is the measured energy falling on each square centimeter at the earth's distance from the sun. The total energy output of the sun may then be found by multiplying the solar constant by the number of square centimeters on the surface of an imaginary sphere with a radius equal to the earth-sun distance. We may then divide this vast amount of energy in ergs by the number of square centimeters on the surface of the sun to find the number of ergs per square centimeter actually radiated at the solar surface. It is this quantity which is related to the absolute temperature in the Stefan-Boltzmann law. When the computation is carried through completely, a temperature of about 5800°K is obtained.

Good agreement for the solar temperature is obtained when the other two radiation laws are applied. Using Wien's law, we look for the wavelength at which the sun's energy is most intense. On this basis the temperature is about 5600°K. To use Planck's law we must measure the radiant energy at many wavelengths so that a spectral-energy distribution curve may be plotted. The temperature is then found by comparing this observed energy distribution with the theoretical energy distributions for black bodies of various temperatures. The actual temperature should be close to that of the theoretical curve which most nearly fits the observed data. Since the curve for the sun does not fit any theoretical curve exactly, the sun may be described as a **gray body** rather than a black body. Some of the factors causing the sun to radiate as an imperfect radiator will be brought out in our discussion of the solar spectrum. Although the methods do not agree exactly, the surface temperature of the sun may be taken as under 6000°K.

By all terrestrial standards, 6000°K is an extremely high temperature. At such a temperature all of the chemical elements exist

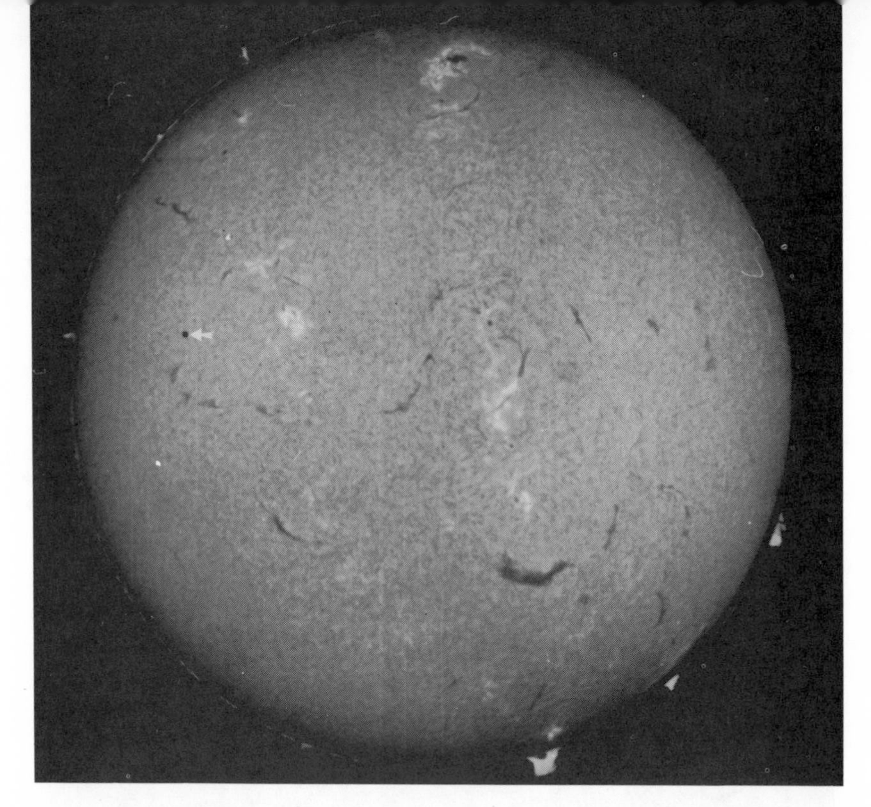

Figure 8.2. A composite photograph of the sun on November 7, 1960, taken while the planet Mercury was in transit. One photograph reveals the features of the photosphere; the other shows the prominences at the sun's limb. (U.S. Army Engineers photographs by R. Gerharz.)

only as vapors or gases. Furthermore, at such a temperature all but the simplest molecules will be separated into their component atoms. The sun we infer, then, is gaseous. Since the sun is held together by the mutual gravitational attraction of all of its parts, the gases must be under ever-increasing pressure at progressively deeper levels below the surface. In accordance with the gas laws, these greater pressures accompany greater temperatures and nowhere is it possible for the sun's material to be in anything but a gaseous state.

We have spoken of the sun's "surface," but if the sun is gaseous, it obviously cannot have any sort of conventional surface. Sometimes gas is transparent, as it is for the most part in our atmosphere. At other times, as in a hot flame, for example, the gas may be more or less opaque. Looking into the gas of the sun, one can see to the depth at which the hot solar gases become opaque, and no deeper. The level beyond which the light is blocked has been defined as the **photosphere,** and it is the sun's apparent surface. (See Figures 8.1 and 8.2.) In reality the photosphere is a layer of hot gas about 200 km thick. Atoms in all parts of this layer emit photons of many energies. Together these photons give rise to the continuous spectrum which we see in normal sunlight. Temperature in the photosphere increases with depth, so the cooler atoms near the top of the region are able to absorb photons and produce the absorption lines of the solar spectrum.

Since the sun is spherical, our line of sight strikes the photosphere at progressively smaller angles as we look from the center of the disc toward the edge or limb. The opacity of the gas is such that it limits the depth from which photons can escape to the earth (or the depth to which our view can penetrate into the gas). Figure 8.3 indicates that when we look at the center of the sun's disc we see to a certain depth in the gas. Near the limb the identical path length ends closer to the top of the photosphere layer. Thus, near the limb the gas into which we look is cooler than the gas into which we look near the center of the disc. The result of this is that the brightness of the disc decreases as we look from the center toward the limb. This phenomenon is evident in Figure 8.1 and is referred to as solar **limb darkening.** By studying this effect, astronomers have been able to understand the variations of temperature with depth in the photospheric gases.

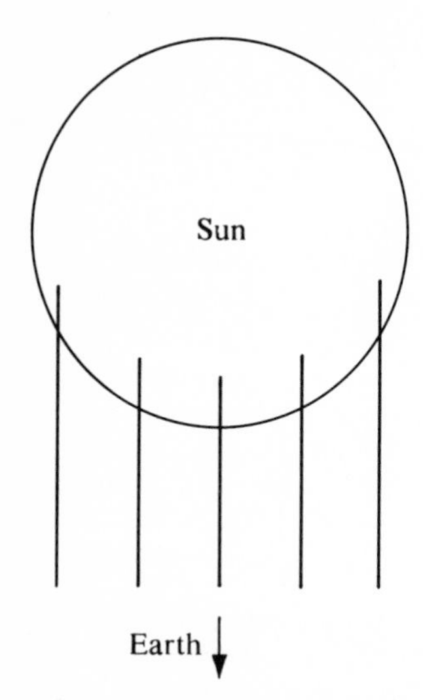

Figure 8.3. The limb-darkening effect is caused by the fact that we are looking into gases cooler near the sun's limb than near the center of the sun's disc.

Figure 8.4. The solar corona photographed at the eclipse of June 8, 1918. (Hale Observatories photograph.)

From the appearance of a red ring around the sun at solar eclipses, the layer of gas above the photosphere is called the **chromosphere.** Above the chromosphere and also visible during solar eclipses is the extensive and beautiful **corona** (Figure 8.4). These regions of the sun's atmosphere will be discussed more fully in a few pages.

On page 97 it was stated that the mass of the sun is found from the effect that the sun's gravitational attraction exerts on the earth. The sun's mass turns out to be 330,000 times the mass of the earth or 1.989×10^{33} g (2.2×10^{27} tons).

8.2 Sources of Solar Energy

The fossil record here on earth undeniably reveals that the sun has been radiating at more or less its present rate for many millions of years. Astronomers have quite naturally been concerned with the question of how the sun could shine so brightly for so long a time. The gases of the sun might react with each other chemically much the same as gas and air react and produce the heat in a Bunsen burner flame. The energy released in such exothermic reactions is well known, and it is fairly easy to compute the amount of gas which would be needed to produce chemically the proper number of ergs per square centimeter per second on the earth for so long. Chemical reactions in a body the size of the sun simply could not produce energy at the required rate. Astronomers realized fairly early that the sun could not be "burning" as we generally use the term. Today we know that the sun is too hot to be burning chemically, anyway.

For a time astronomers thought that they might have the answer when in 1853 it was suggested by Helmholtz that the sun's energy could come from the sun's own gravitational collapse. As the sun contracted, heat would be released. The potential energy of the gas in a large volume would be converted to kinetic energy as the gas "fell" slowly toward the sun's center. Here was a source of incredibly large amounts of energy—enough to keep the sun shining as it now does for millions of years. The sun would only have to shrink in radius by about 25 meters a year to release energy in the observed amounts. Working backward from this, astronomers were able to establish the maximum age of the earth—that time since the sun's surface reached as far as the earth—as about 25 million years. This, however, was much too short a period to be acceptable to the scientists who believed that the sun had shone much as it does today for a considerably longer period of time. The excitement and energy expended in the development and defeat of the contraction theory were not wasted, however, and eventually astronomers realized that contraction did play an important role in the early existence of the sun, as it does with other stars, even though other energy sources must be more significant during most of their remaining "life."

Long before atomic energy and nuclear explosions had become familiar to most of the world's population, astronomers and physicists had begun to understand the basic concepts in the release of energy through nuclear reactions. In 1939, the physicist Hans Bethe described the first series of reactions which could, on the basis of established theory, be considered a source of energy in the sun. Bethe's theory required the presence of carbon atoms mixed with the basic hydrogen which was already known to be the most abundant element in the sun and the stars. The complete series of reactions is known as the **carbon-nitrogen cycle** and is given below. It should be noted that the temperature required for these reactions is of the order of 20,000,000°K, and that all of the atoms involved have lost all of their electrons.

$$
\begin{aligned}
{}_6C^{12} + {}_1H^1 &\rightarrow {}_7N^{13} + \text{gamma ray} \\
{}_7N^{13} &\rightarrow {}_6C^{13} + {}_1e^0 \\
{}_6C^{13} + {}_1H^1 &\rightarrow {}_7N^{14} + \text{gamma ray} \\
{}_7N^{14} + {}_1H^1 &\rightarrow {}_8O^{15} + \text{gamma ray} \\
{}_8O^{15} &\rightarrow {}_7N^{15} + {}_1e^0 \\
{}_7N^{15} + {}_1H^1 &\rightarrow {}_6C^{12} + {}_2He^4
\end{aligned}
$$

In the first reaction a carbon nucleus combines with a proton (or hydrogen nucleus) to form a nitrogen nucleus. In the process some energy is radiated as a gamma ray (a high-energy photon). The numbers below and above the chemical symbols refer respectively to the charge and the mass of the nuclei. Thus the carbon nucleus, for example, has a charge of six units, having lost six electrons, and a mass of 12 units. This means that the nucleus must contain six protons and six neutrons. The hydrogen nucleus contains one proton and no neutrons. The nitrogen-13 is unstable and spontaneously decays into carbon-13 and a positron. When the positron combines with one of the abundant electrons, another gamma ray is produced.

The other reactions proceed in the same manner until in the end a carbon nucleus and a helium nucleus are left. Four protons have entered the reactions and have been combined into one helium nucleus.

The combined mass of the four protons is 4.0304 mass units, and the resulting helium nucleus has a mass of 4.0027 units. The small amount of mass lost in the process, 0.0277 units, has been converted into energy according to Einstein's famous equation $E = mc^2$. Since c is the velocity of light, which is very large, even a small mass m can be converted into tremendous amounts of energy E.

Earlier in this chapter we stated that the total amount of energy radiated per second from the sun could be known if the solar constant and the earth-sun distance were known. Modern observations show that the sun radiates 4×10^{33} ergs per second from its total surface area. Each time that a helium atom is created, 4.2×10^{-5} erg is released, and so it is a straightforward matter to compute the number of reactions per second which must occur in the sun's interior. The amount of mass converted to energy follows also. To account for the observed solar energy, the sun must lose 4.3×10^{12} grams of its mass each second. This, at first thought, seems to be a truly prodigious amount of material. A second look at the sun's total mass, 2.2×10^{33} grams, makes it clear that the 4.3×10^{12} grams per second is quite insignificant even over very long periods of time.

Soon after Bethe's work, nuclear physics progressed to the point where another important nuclear reaction was recognized as a possible source of solar energy. This one is known as the **proton-proton reaction,** and it is believed to be the chief source of energy in the sun. It is now recognized that the carbon-nitrogen cycle would be the probable energy source only in stars of mass much larger than that of the sun. Under the conditions in the sun's interior, the following reactions produce energy. First, two protons combine to form heavy hydrogen or deuterium.

$$_1\mathrm{H}^1 + {_1\mathrm{H}^1} \rightarrow {_1\mathrm{H}^2} + {_1e^0} + \text{neutrino}$$

The deuterium nucleus then combines with another proton to form a light isotope of helium.

$$_1\mathrm{H}^2 + {_1\mathrm{H}^1} \rightarrow {_2\mathrm{He}^3} + \text{gamma ray}$$

$$_2\mathrm{He}^3 + {_2\mathrm{He}^3} \rightarrow {_2\mathrm{He}^4} + {_1\mathrm{H}^1} + {_1\mathrm{H}^1} + \text{gamma ray}$$

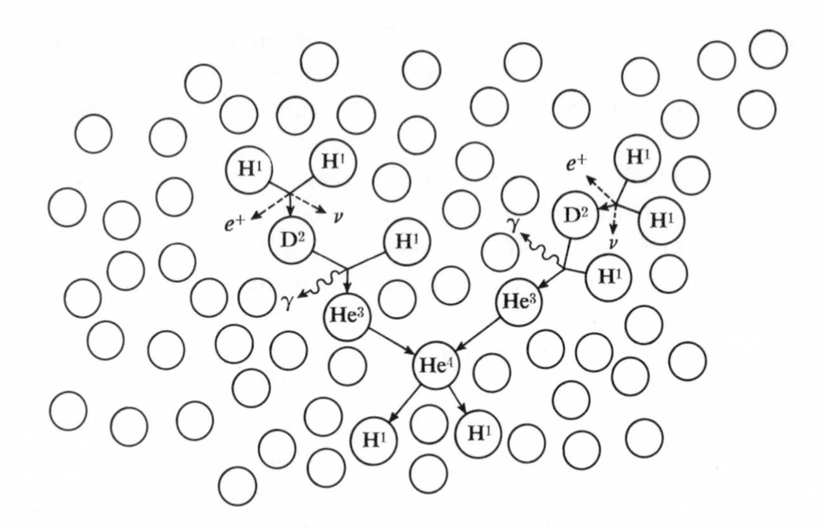

Figure 8.5. The proton-proton reaction. At a temperature of 10 million degrees Kelvin and a density of 100 grams per cubic centimeter, hydrogen nuclei fuse to produce helium as the end product. Here the symbol D^2 is used to indicate deuterium. (After a diagram by W. Fowler.)

This process is shown schematically in Figure 8.5. The net result is the same as that for the carbon cycle in that four hydrogen atoms have been combined into one helium atom, with a slight loss of mass which appears as radiant energy. The proton-proton reaction produces maximum energy release at a temperature lower than for the carbon-nitrogen cycle. However, we shall see later that during some phases of star development there are times when the carbon-nitrogen cycle plays the dominant role. It has become customary to refer to the conversion of hydrogen to helium by either of these means as **hydrogen burning,** but the student should be careful to remember that this does not refer to any sort of chemical burning or oxidation in the familiar sense.

8.3 Internal Structure of the Sun

Having once accepted these nuclear processes as the source of solar energy, one may consider in some detail the conditions in the sun's interior and the problem of transfer of the sun's energy from its center to its surface. Because of the nature of a self-gravitating gaseous mass, the pressure and temperature of the gas must increase at progressively greater depths in the interior. Starting with the known dimensions, mass, and surface temperature of the sun, and making some assumptions about the characteristics of the gas, the temperatures and pressures at various depths may be calculated. Knowing also the temperatures and pressures required for the conversion of hydrogen to helium, one realizes that everywhere below a certain depth the conditions are such that hydrogen burning can occur. Furthermore, at depths beyond this limit, hydrogen burning goes on at progressively more rapid rates. This implies, then, the existence of a rather well-defined core in which most of the sun's energy production takes place. Within the sun, approximately 90% of the energy is produced in a region with a radius equal to 0.2 of the sun's radius (Figure 8.6).

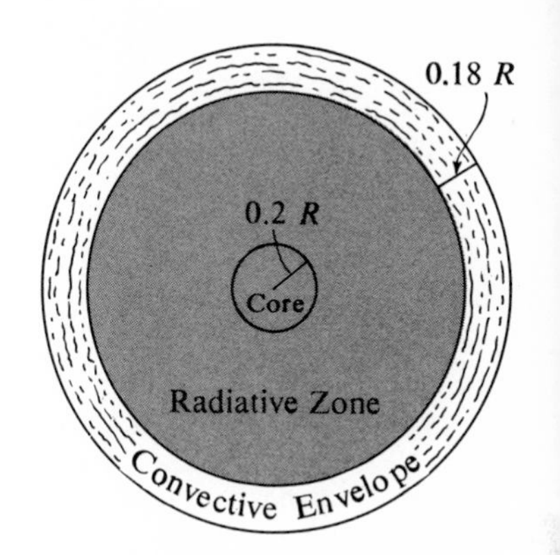

Figure 8.6. Internal structure of the sun.

The solar energy originates as gamma rays, which are high-energy photons. In order that the sun may radiate as it does, the energy of these photons must find its way through the layers of gas overlying the core. The original photons never make it, but the energy of the original photons does reach the surface after countless absorptions and re-emissions by the atoms above the core. Because the processes of absorption and radiation represent the dominant energy transport means, this region is referred to as the **radiative zone.** In this zone another important process is taking place as well. After absorbing a high-energy photon, an excited atom can lose energy, either by emitting a single energetic photon or by cascading from one state of excitation to another and emitting a number of less energetic photons. Because this cascading takes place extensively, the original high-energy photons are gradually replaced by larger numbers of less energetic photons. Even though the solar energy is produced as gamma rays, the photosphere radiates most strongly the low-energy photons which we see as visible light. Within the radiative zone there is relatively little mixing of the gas.

A second mechanism for the transport of energy in a fluid or gas is **convection.** In this case there is an actual circulation of gas within the sun. The hot gas rises from lower levels to the surface and cools off as it radiates. The cooler gas then sinks to lower levels, where it again becomes heated. It has been possible to calculate the relative importance of both convection and radiation as energy transport mechanisms within the sun. Only in the outer layers to a depth of about 0.18 of the radius is convection important. There is now firm observational evidence for the presence of a convective layer above the radiative zone. The photosphere observed from high-altitude balloons and from ground-based observatories under excellent "seeing" conditions shows a rapidly changing pattern of **granules.** The average granule has a diameter of about 1000 km and maintains its identity for only about six minutes. The granules appear as bright irregular areas, as may be seen

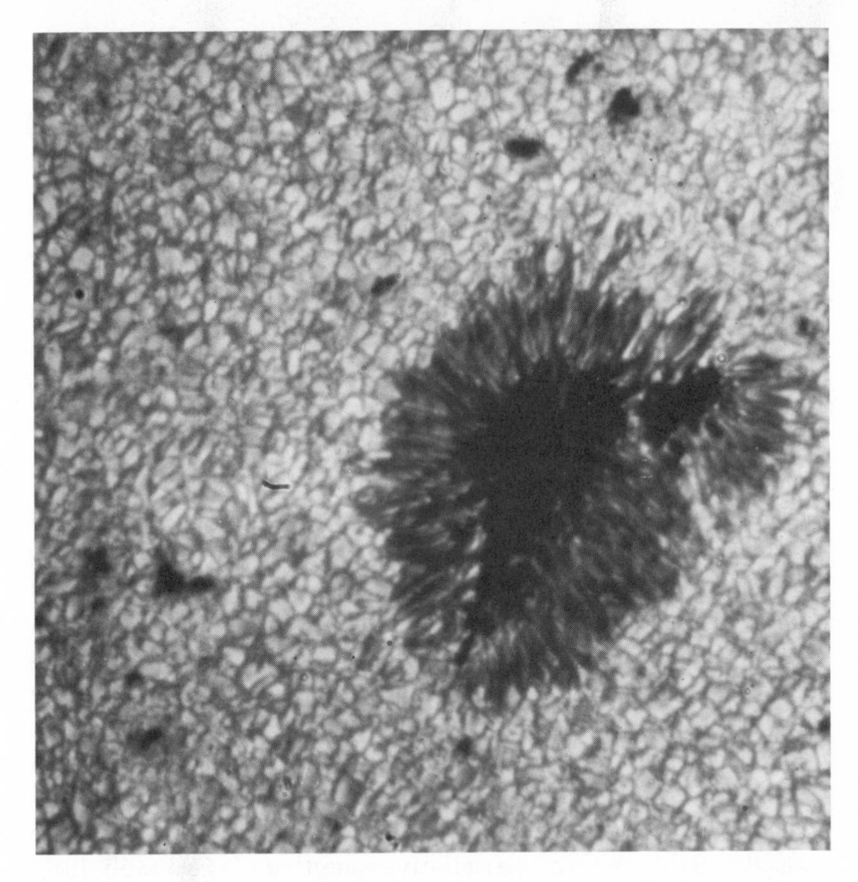

Figure 8.7. The granular appearance of the photosphere photographed from a ballon-borne telescope high above most of the earth's atmosphere. (Project Stratoscope of Princeton University, sponsored by the Office of Naval Research, The National Science Foundation, and the National Aeronautics and Space Administration.)

in Figure 8.7. It is thought that these are places in which hot gas is rising to the surface. As each area cools, new areas of hot gas appear nearby in a continuously changing pattern.

The theoretical analysis of the sun's internal structure is actually rather complex. Here we have sketched only the broadest sort of reasoning which gives rise to these ideas. In actual practice the theoretical astronomer begins with a series of observed characteristics and, with the aid of a large electronic computer, attempts to calculate what sort of a sun would result from an assumed chemical composition. If his computed sun matches the actual sun, he concludes that his theoretical model is reasonable.

8.4 Sunspots

The safest way to try to look at the sun is to adjust a small telescope so that it projects an image of the sun onto a card or screen placed a short distance behind the eyepiece. Filters in front of the eyepiece may be used for direct viewing. This can be very dangerous unless the observer has made certain that the filter removes all of the harmful radiation from the light. Since the sunlight is concentrated on the filter, the observer's eye may be exposed to the full intensity of the sun's radiation if the filter cracks. When the sun is observed with proper precautions, **sunspots** may be regularly observed on the photosphere. Sunspots appear as dark areas simply because the gases in these areas are cooler than the gases around them. Temperatures of spot areas are about 4000°K, compared with the 6000°K of the normal photosphere. Galileo saw sunspots with his telescopes, and on rare occasions some individual spots have been large enough to be visible to the unaided eye.

A large sunspot group may be seen in Figure 8.8. The sizes of some of the individual spots compared with the size of the sun are clearly seen. The spots may range from areas 20,000 kilometers across down to the limit of visibility. For comparison, the white

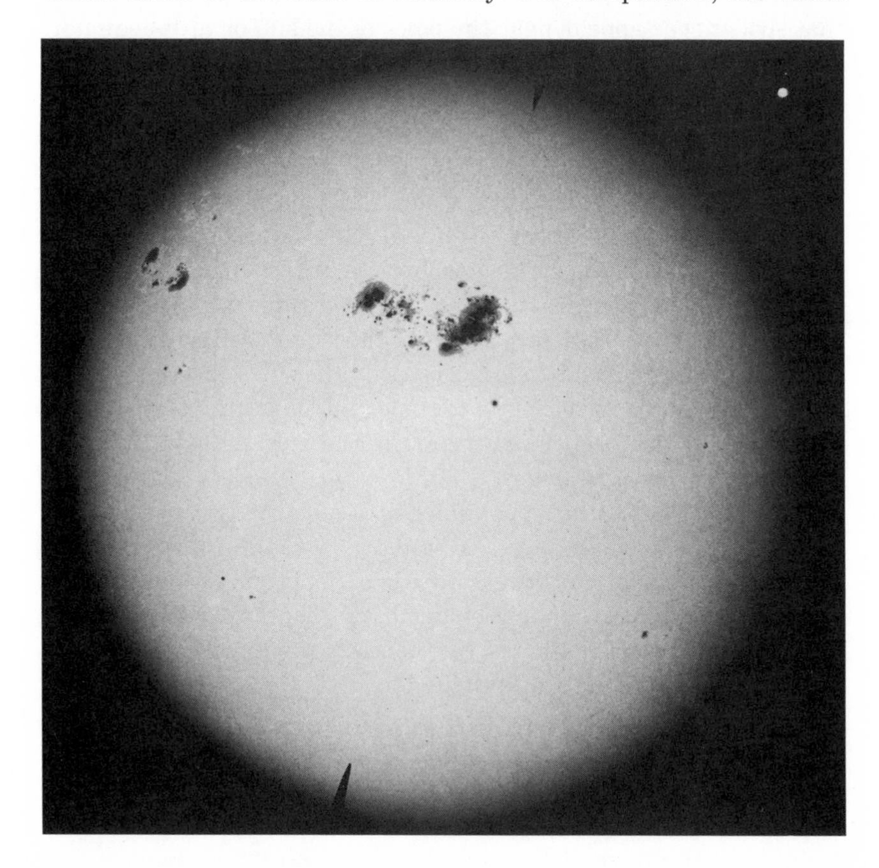

Figure 8.8. An unusually large sunspot group photographed on April 7, 1947. The white dot in the upper right corner of the photograph represents the earth on the same scale. (Hale Observatories photograph.)

spot near the limb of the sun represents the earth drawn to the same scale.

Although the appearance and lifetimes of no two spots are ever the same, there are certain typical patterns which are characteristic of most sunspots. First a small, single spot may appear and over a period of a few days begin to grow larger. Other small spots may then appear nearby, until the group may have 20 or more members of various sizes. Usually within the group there will be two particularly prominent spots. These prominent spots are known as the "leading spot" and the "following spot" with the direction of rotation of the sun as a reference. Gradually the spot activity in the area begins to decrease. The smaller spots disappear. The larger ones diminish in size and eventually they too disappear. Small spot groups may last days or several weeks, while large ones may persist in a slowly changing manner for as long as a month.

There is no way in which to predict exactly where or when sunspot activity will appear on the sun. It has been noted, however, that spots never appear near the poles of the sun or at its equator. All spot activity is confined to two **sunspot zones** extending from about 5° latitude to about 35° latitude on either side of the solar equator.

Observed from day to day, the spots always appear to move from west to east across the face of the sun. This apparent motion is due solely to the **sun's rotation.** Each individual sunspot is essentially fixed in its location on the photosphere. A surprising point may be noted when sunspots are watched for some time in order to measure the sun's rotation period. Near the equator the spots move faster than at higher latitudes. To specify the sun's rotation period, then, one must make reference to some specific latitude. At 16° latitude, for instance, the sun's rotation period is 25.4 days; at higher latitudes the period is longer.

Figure 8.9 shows the appearance of a sunspot group as photographed on successive days. As stated above, the rotation of the sun causes the spots to appear to move. Sometimes the apparent path of such a spot group is clearly curved. The curvature is due to the fact that the sun's axis of rotation is not perpendicular to the plane of the earth's orbit but is tilted by 7°15′.

8.5 Sunspot Cycle

The sunspot cycle was discovered in 1843 by S. H. Schwabe, a druggist in the town of Dessau, Germany. Sunspots had been ob-

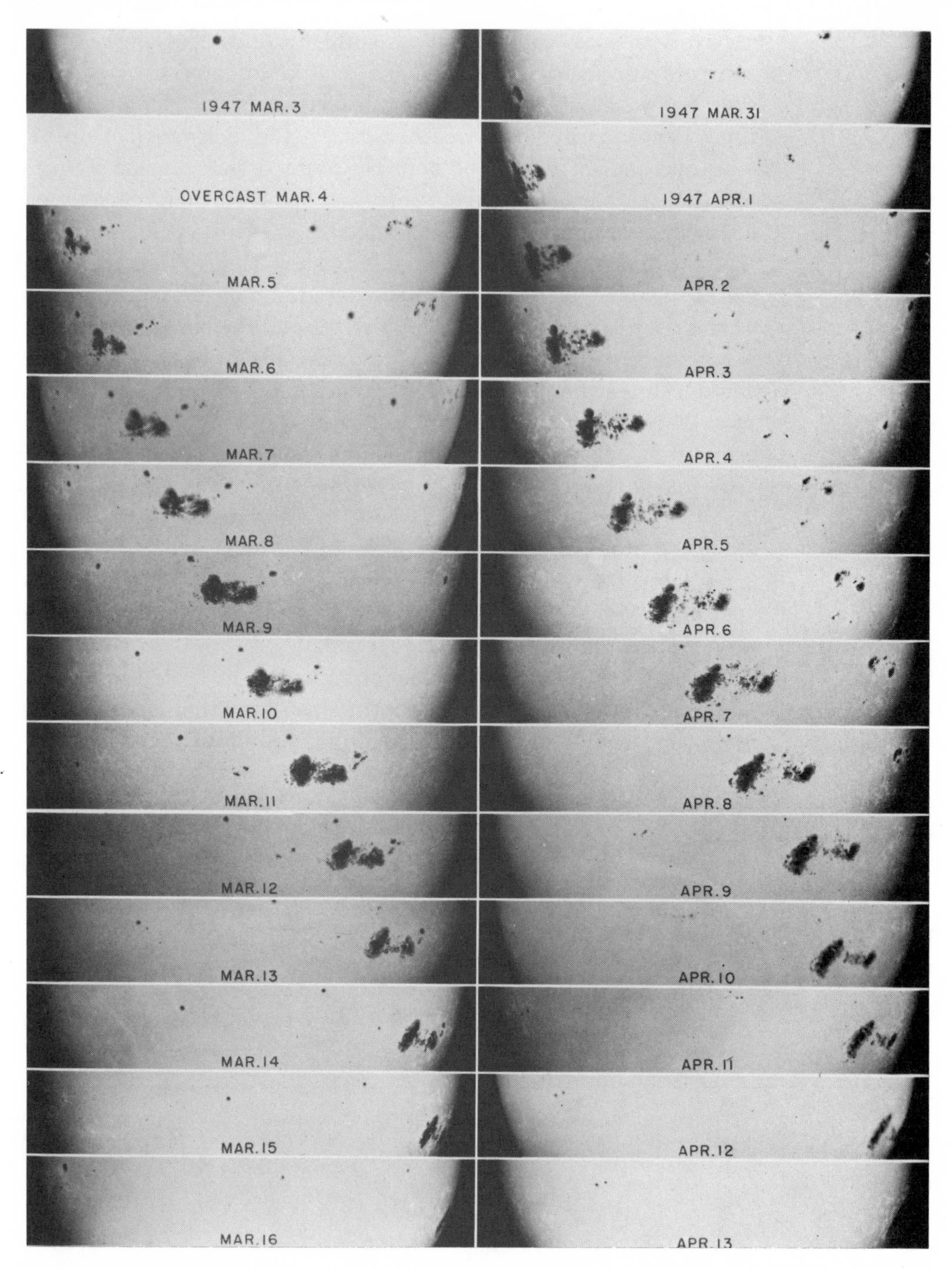

Figure 8.9. The appearance of the sun over a period of 42 days during 1947. The large spot group persisted long enough to be carried twice across the face of the sun. (Hale Observatories photographs.) Seen from the sun's north pole the sun rotates in the same sense as the earth, i.e. west to east.

served with some care ever since Galileo's telescopes showed that sunspots were common phenomena. Unlike his predecessors, however, Schwabe extended his observations over a very long period of time. Each day at noon, weather permitting, Schwabe set up his telescope and looked at the sun. He counted the number of spots and the number of groups of spots, and recorded them. Strictly on his own, Schwabe continued his diligent observations for 17 years. At the end of that time he was able to show clearly that the number of sunspots visible per year increased and decreased in a cycle of just over 11 years. Ample confirmation of the sunspot cycle has been achieved in the years since Schwabe's observations. The average interval from one year of maximum activity to the next year of maximum activity is 11.2 years. Individual cycles have been as short as nine years and as long as 14 years.

Counting the number of spots per year one would usually find that the numbers increase for about five successive years. Thereafter, the numbers counted decrease annually for six years. There is no particular value for the number of spots counted either in a year of maximum number or a year of minimum number, but the ranges shown in Figure 8.10 are typical.

Simply from the annual counts of sunspots the cycle from maximum to maximum seems to be continuous. Two other observations, however, indicate that each cycle from minimum through maxi-

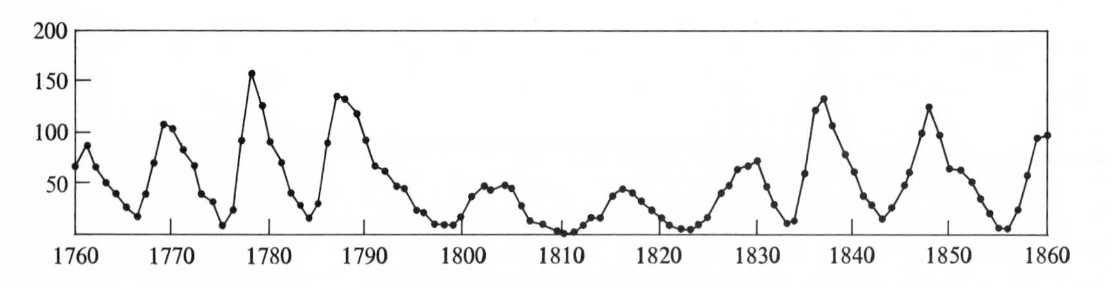

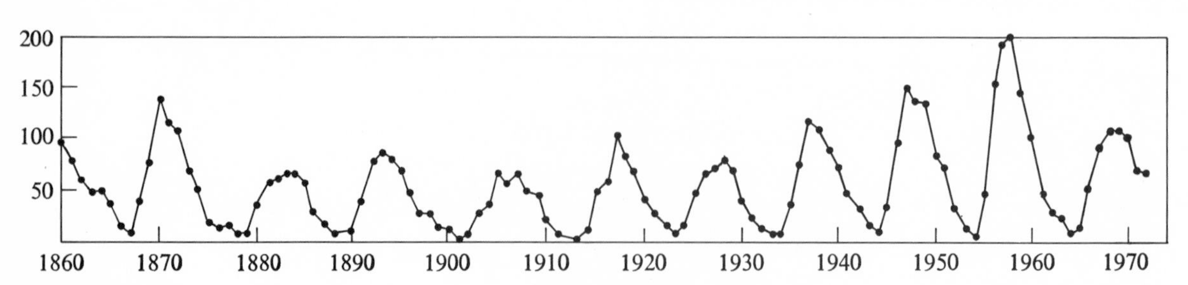

Figure 8.10. The sunspot cycle is seen in the variation in the number of spots seen per year on the photosphere.

mum to the next minimum is somewhat independent. The first of these lesser effects is related to the latitude of the spots, and the second is related to their magnetic polarity. At a time of sunspot minimum the few spots seen will appear near the high-latitude edges of the two sunspot zones. As the cycle progresses toward maximum, the numbers counted increase, and the spots appear at progressively lower latitudes in both hemispheres. At the time of sunspot maximum, the spots appear in a wide range of latitudes centered on the northern and southern zones. As the counted numbers decrease after maximum, their locations are progressively closer to the equator. In the final year of the cycle, only a few spots are seen, and these are relatively close to the equator in both hemispheres. As a cycle ends, however, the first few spots of the next cycle begin to appear in high latitudes again. Throughout each cycle, then, the spots are progressively closer to the equator as they appear and persist for their brief lifetimes.

The second lesser effect is the change in magnetic polarity of sunspots from one cycle to the next. Physicists know from theory and laboratory practice that a strong magnetic field associated with a gas will cause the spectrum lines to split. This phenomenon is known as the **Zeeman effect** or Zeeman splitting. The spectra of sunspots show this effect. Furthermore, the spectra show that in any sunspot pair the leading and following spots have opposite polarities. If the leading spot is a north pole, then the following spot has the opposite or south polarity. All of the spot groups in a given hemisphere will have the same polarity for leading and following spots, and in the opposite hemisphere the polarity will be reversed. If the leading spots in the northern hemisphere are north (+), then the leading spots in the southern hemisphere are south (−). (See Figure 8.11.) These polarities will be constant throughout a given 11-year cycle. At the beginning of the next cycle, however, the polarities in both hemispheres will be reversed. In a sense, then, the cycle of magnetic polarity lasts for 22 rather than 11 years.

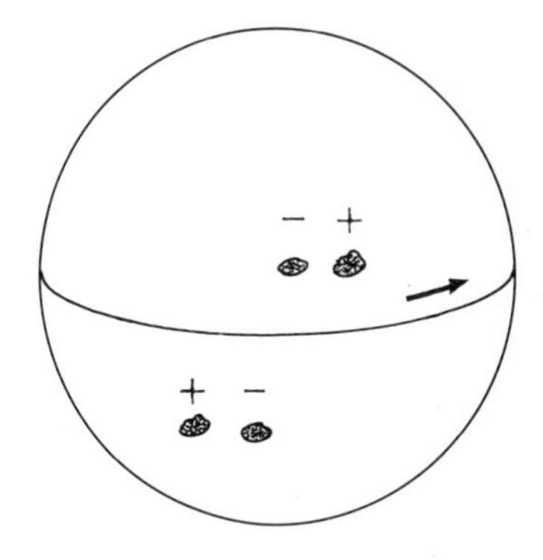

Figure 8.11. The magnetic polarity of the leading and following spots is opposite in the two hemispheres and reverses after each 11-year cycle.

In summary the three parts of the sunspot cycle are:

1. a variation in the mean number of sunspots per year;
2. a variation in latitude of spots throughout the cycle;
3. a reversal of the magnetic polarities of leading and following spots at the beginning of each new cycle.

8.6 Atmosphere of the Sun

The presence of an extensive gaseous region above the photosphere was mentioned earlier in this chapter. It should hardly be surprising that most of our information regarding this atmosphere is obtained from studies of the solar spectrum. The spectrum of the sun shows thousands of absorption lines superimposed on the continuous spectrum of the photosphere. These lines are, of course, characteristic of the gases in the photosphere.

Of the more than 20,000 absorption lines in the solar spectrum, the vast majority have been identified. By "identified" we mean that their wavelengths have been measured and have been found to coincide with the wavelengths of known chemical elements or molecules previously studied in the laboratory. Absorption lines of more than 66 of the 92 natural elements have been found in the solar spectrum, indicating that these elements at least are present in some degree in the sun. The rest of the elements are probably in the sun also, but in such small quantities that their absorption lines are too weak to be found. Only once have astronomers and physicists been able to associate some of the sun's absorption lines with a chemical element unknown on the earth. This element was given the name *helium* to remind us of the place of its discovery, and it was later found to occur in small quantities here on the earth. Since that one occasion, however, no elements not known here have been found. Furthermore, our laboratory knowledge of the elements, when extrapolated to the solar temperature and pressure conditions, predicts that the elements in the sun's atmosphere should show the absorption features which are observed.

By comparing the absorption lines of one element with those of another in width and intensity, astronomers have learned to gauge the relative abundances of the elements in the sun. In spite of the fact that its lines are not particularly conspicuous, hydrogen is by far the most abundant element in the sun. About 63% of the sun is hydrogen, and about 36% is helium.

The gas in the sun's atmosphere is very hot. If it were not for the presence of the very bright photosphere behind it, the atmospheric gases would show an emission-line spectrum instead of absorption lines. There are moments at the time of solar eclipses when some of this atmospheric gas may be viewed by itself with rather interesting effects. Just before the moon completely covers the disc of the sun, there will be a small crescent of the sun's atmosphere exposed.

This is indicated in an exaggerated way in Figure 8.12. For only a few moments we are able to see the light emitted from the gases in this part of the sun's atmosphere. The spectrum of this light is referred to as the **flash spectrum,** since it is visible for a short time. Figure 8.13 shows three photographs of such a flash spectrum. The emission lines show up as short arcs because no slit was used in the spectrograph. In an interesting way the flash spectrum demonstrates the effects mentioned in the earlier discussion of spectra in Chapter 4. From the fact that at different wavelengths the arcs are not all of the same length astronomers can tell a great deal about the conditions of excitation and ionization at various levels in the gas. The H and K lines of ionized calcium are the longest arcs on the spectrum. We infer that at great heights above the photosphere the temperature is high enough to maintain a large number of calcium atoms in an ionized state and to cause these atoms to emit.

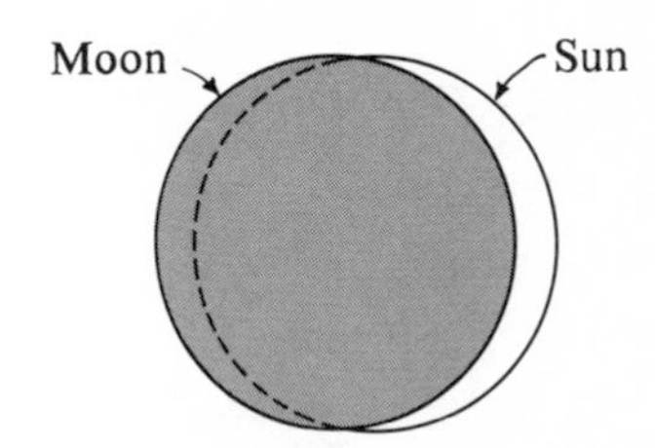

Figure 8.12. Just before the sun becomes completely covered by the moon, the hot gases of the chromosphere may be seen in emission.

Just after the flash spectrum is seen during an eclipse, the moon covers the sun more completely and the period of total eclipse begins. During totality a bright reddish ring may often be seen around the moon's disc. Because of its brilliant color, the upper layer of atmosphere which the ring represents has been called the **chromosphere.**

Visible in the photograph of the flash spectrum are some irregular extensions of the chromosphere. These are called **prominences**

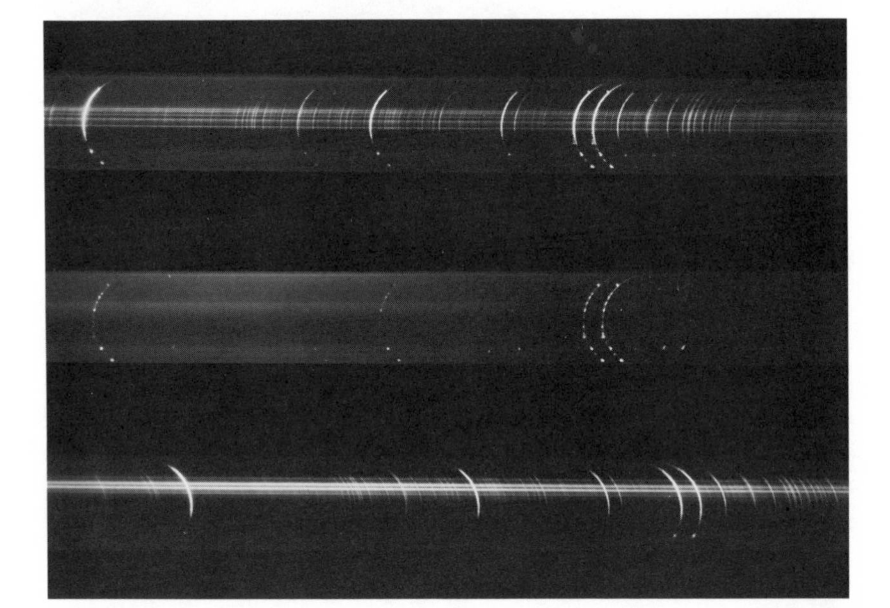

Figure 8.13. Three flash spectra of the sun photographed at the eclipse of January 24, 1925. (Hale Observatories photographs.)

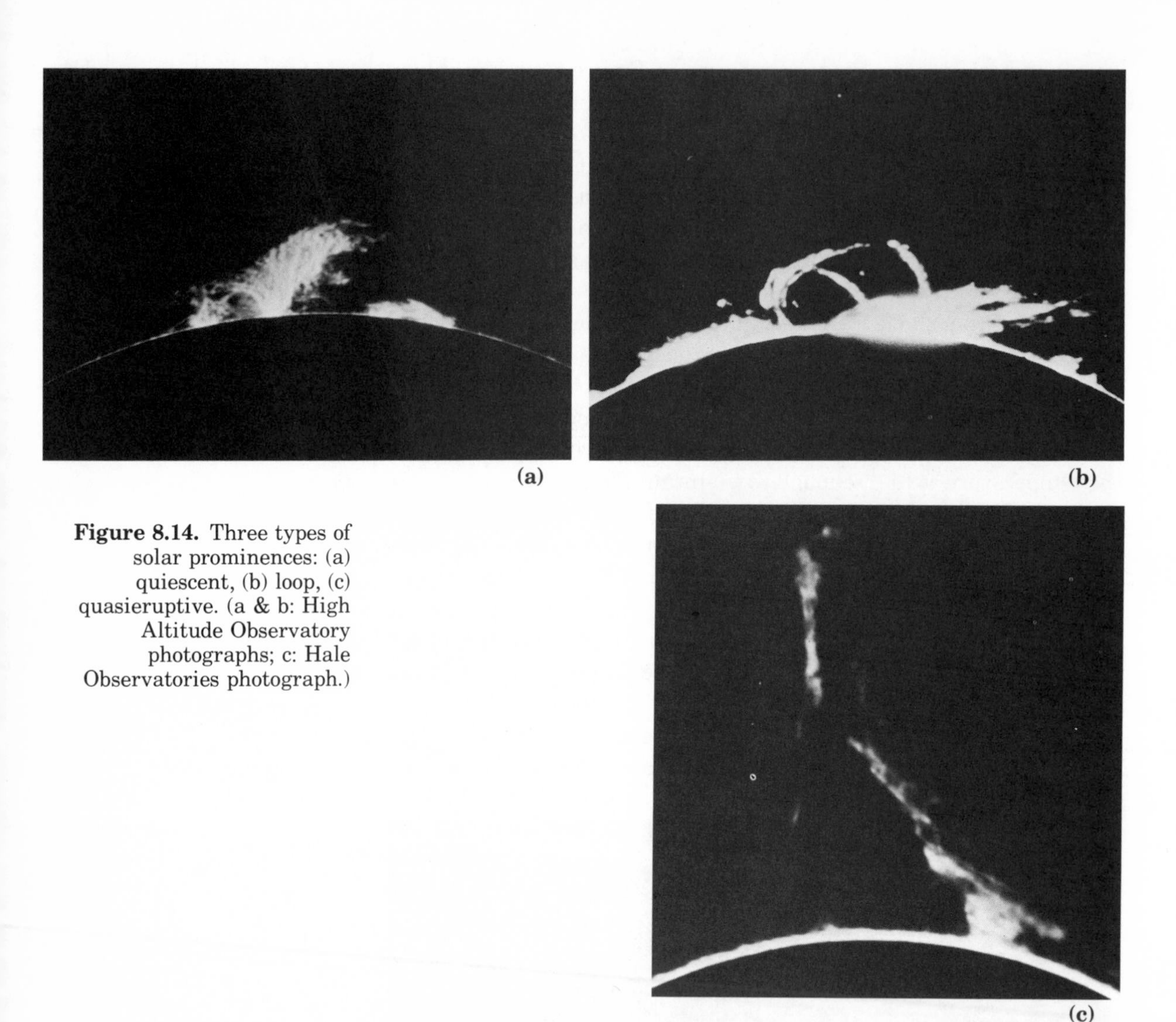

Figure 8.14. Three types of solar prominences: (a) quiescent, (b) loop, (c) quasieruptive. (a & b: High Altitude Observatory photographs; c: Hale Observatories photograph.)

and may take a variety of forms as shown in Figure 8.14. In most prominences hot gases appear to rain down into the sun from condensations of material above the chromosphere. Sometimes the prominences are associated with sunspots, and the gas seems to be constrained by the magnetic field of the spots into loops. On other occasions, the prominence material rises in huge arches thousands of kilometers long. Prominences were first seen along the rim of the sun at the times of solar eclipses, and for many years it was possible to study them only at eclipses. Today there are several specialized types of telescopes which permit observation of prominences at almost any time. These instruments are always located on high mountains away from most of the dust and moisture which might otherwise scatter unwanted light into the telescope. Solar telescopes at appropriate locations in Europe, Asia, and North America attempt to maintain an almost constant watch on the sun in order to record the motions of the prominence material.

Some rather interesting views of the chromosphere are seen when the sun is viewed through **narrow-band filters.** Such filters are designed to transmit light of only a short range of wavelengths centered at some specified location in the spectrum. The filter might transmit only the light of the Hα line at 6563 Å, or it might transmit only at the wavelength of the K line of ionized calcium. All other wavelengths are blocked. The absorption lines in the solar spectrum are normally dark only in contrast to the regions on either side of the lines. When the narrow-band filter is used there is plenty of light in which the whole sun may be seen. Figures 8.15 and 8.16 show the sun photographed in the light of hydrogen at 6563 Å and in the light of ionized calcium at 3934 Å. Since these two gases are excited at great heights in the chromosphere, we see in these two photographs the appearance of the top of the chromosphere. The mottled appearance means that the chromosphere is not homogeneous. In the bright areas the gas is hotter than in the other areas. These bright areas are known as **faculae**, or plages, and are almost always in roughly the same areas as sunspot activity.

Also to be seen in Figure 8.15 are some irregular dark lines known as **filaments.** Because these areas are dark, the gas in them is cooler than that in the neighboring areas of the chromosphere. The filaments are actually prominences seen from the top instead of in silhouette at the sun's edge or limb. This has been confirmed from numerous examples in which a prominence appeared as the sun rotated and carried a large filament to the limb.

Perhaps the most significant phenomena observable in the narrow-band photographs are the **solar flares.** Flares are sudden intense chromospheric brightenings which may last an hour or more. They usually occur in the vicinity of active sunspots and are limited to small areas near the spot. The flare is very bright in ultraviolet light as well as in visual light, and it has been well established that flares also emit streams of charged particles, principally electrons and protons. Some of these particles become trapped in the earth's magnetic field to form the Van Allen radiation belts. High above the atmosphere, the magnetic field causes the particles to travel rapidly from one pole to the other along spiral paths. In the vicinity of the magnetic poles the field lines converge, and the par-

Figure 8.15. The sun photgraphed in the light of the Hα line at 6563 Å. (Manila Observatory photograph.)

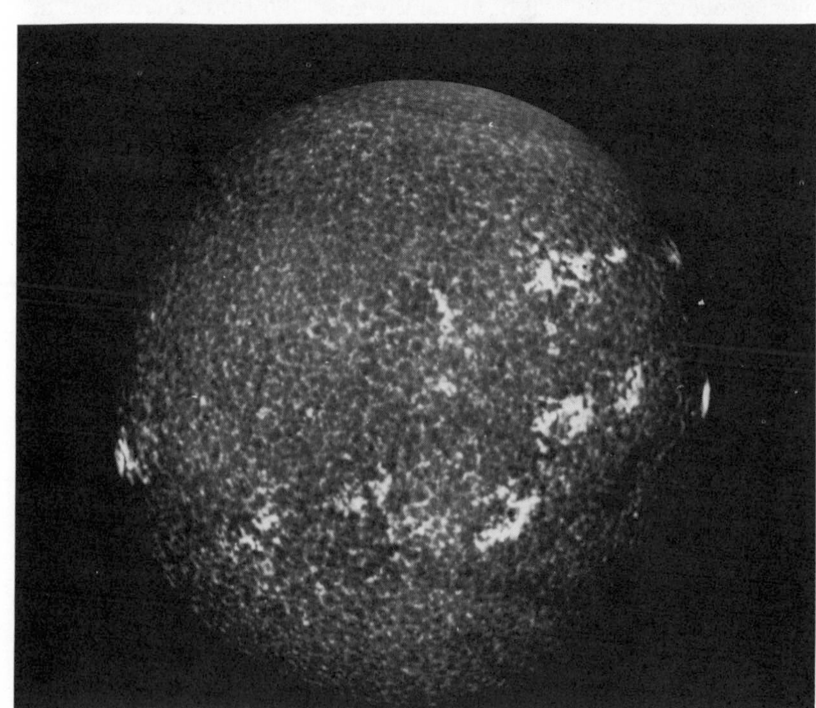

Figure 8.16. The sun photographed in the light of CaII at 3934 Å on the same day as Figure 8.15. (Manila Observatory photograph.)

ticles approach the earth's surface. Some of the particles come so low before they change direction that they strike atoms or molecules of the atmospheric gases. Through these collisions the gases become excited and emit light. We see this emission as an **aurora.**

8.7 Solar Corona

Extending outward from the photosphere to distances easily as large as the solar radius is the **corona.** The full extent of this pearly white halo can only be seen during the period of totality when the sun is in eclipse. Observations have shown that the corona is more or less symmetrical at the time of a sunspot maximum and considerably extended in the equatorial plane at sunspot minimum. As may be noted in Figure 8.4, streamers are often seen extending from the directions of the poles. See also Figure 8.17.

The spectrum of the corona is continuous in the inner regions. Here it is believed that light from the photosphere is being scattered by free electrons moving at high speeds. We know that the velocities of the particles in a gas depend upon the temperature of the gas. Knowing the velocities of the electrons in the corona, then, we may speak of a temperature which would be required to give the electrons these velocities. This is best described as an **electron temperature,** and in the lower levels of the corona the electron temperature is 300,000°K. The density of the gas in the corona is so low that the temperature could not be recorded on a thermometer. There would not be enough collisions per second to warm the fluid in the tube.

Farther out in the corona are emission lines which come from highly ionized atoms of iron, nickel, and calcium. The iron atoms

Figure 8.17. The solar corona at the eclipse of November 12, 1966, photographed from the mountains of Bolivia. A special filter, decreasing in density radially from the center, was used to record the faint outer parts of the corona without overexposing the inner parts. The bright spot at the left (NE) of the sun is an overexposed image of Venus. (High Altitude Observatory, Boulder, Colorado.)

have lost 13 electrons and by the sort of reasoning applied above, this suggests a temperature of about one million degrees Kelvin.

8.8 Cyclical Sun-Earth Effects

In spite of the extensive studies of sunspots and the sunspot cycle, astronomers really do not yet know why sunspots appear on the sun or why the occurrence of sunspots follows the 11-year cycle as it does. It can be said, however, that the sunspot cycle reflects variations in the level of solar activity. One striking manifestation of the variation of solar activity is seen in the annual growth rings in certain kinds of trees. When the cross section of a large tree trunk is examined, it is easy to see the 11-year cycle in the width of the rings. When the sun is active at the time of a sunspot maximum the growth rings are thickest. It is certainly doubtful that there is any direct relationship between sunspots and biological effects. It is much more likely that the sunspot cycle and the many variable effects which are in phase with it are actually manifestations of more general fluctuations in the overall activity of the sun. Thus when the sun is most active, the number of sunspots increases. Also, because of the increased activity, there are more solar flares and auroras, and the trees show greater growth. Again, the reasons for such variations in solar activity are completely unknown.

8.9 Origin of the Solar System

There is ample geological evidence to support an age of the earth somewhere in the neighborhood of four billion years. Furthermore, fossil records show plainly that the sun must have been radiating much as it does today for many millions of years. Beyond these questions of age lie the questions of origin of both the sun and the planets. In seeking answers to these questions, there is not much with which the astronomer can work. He must consider the present condition of the sun, earth, and solar system as well as the probable condition of the material from which these bodies may have formed. He must then use known physical laws and processes to proceed from the original material to the system as it exists at the present time. There can be no fossil record of the events that occurred at the earth's beginning, and we cannot hope to see many evolutionary processes at other locations in the universe. Nonetheless, self-consistent theories of the origin of the solar system, though largely speculative, are interesting and worthwhile. It has long been held that any theory that seeks to explain the origin of the sun, earth, and planets ought to be general enough to permit the formation of a planetary system around another star. This approach is necessary in order to avoid the view that the earth and planets are in any way unusual. If we believe that the earth is a unique body and that planets are extremely rare in the universe, then intelligent life is rare or perhaps unique to the earth. This point of view is almost as "earth-centered" as the views of Ptolemy. Astronomers have tried to make their theories such that planets might be expected in company with large numbers of stars. There is always the possibility, however, that the sun as a star is unique in having a system of planets around it. The detection of planets in orbit around other stars is an extremely difficult problem, as the reader will realize after studying subsequent chapters in this book.

Most of the modern theories of the sun's origin fit somewhere under the general heading of the **nebular hypothesis,** a name derived from the idea that stars are formed from a nebula or large cloud of gas and dust in space. Such clouds are known to contain enough material to form a great many individual stars. In the earliest stages, a condensation begins at some localized point in the cloud. The density of the gas and dust at this point becomes great enough so that the surrounding material is gravitationally attracted toward this center. If the original cloud or nebula is small, it may begin to contract under the gravitational attraction of all of its parts for each other. If the cloud is large, the contracting mass may be only a part of the total, and there could be several simultaneous contractions in various parts of the cloud.

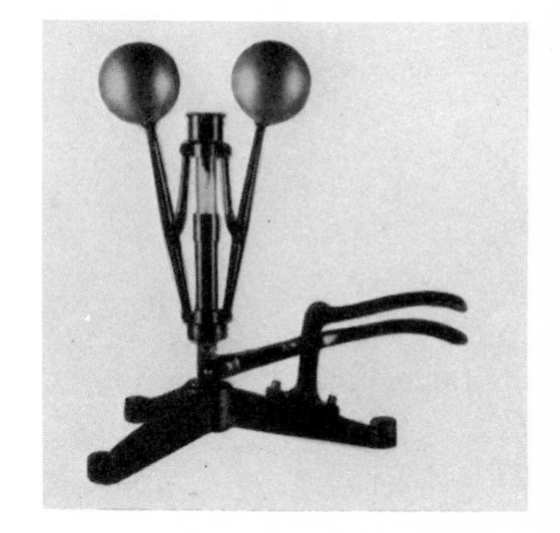

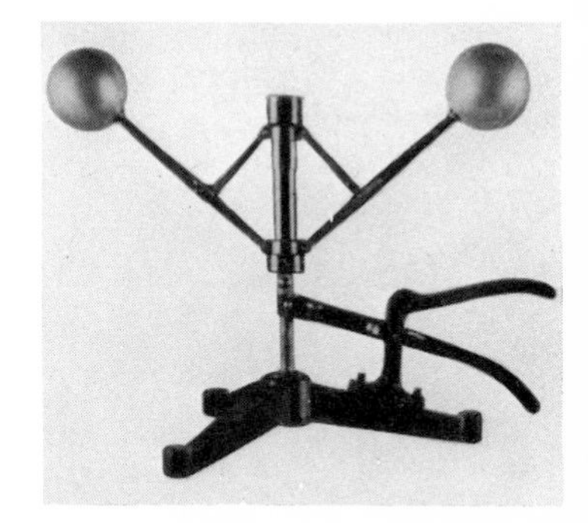

Figure 8.18. Apparatus for demonstrating the conservation of angular momentum.

A nebula is likely to be rotating as it contracts, with the rotation rate continuously increasing. The rotation rate must change because the laws of physics require that the total angular momentum in a rotating system must remain constant. The angular momentum of each particle in the cloud depends, then, upon the mass of the particle, its velocity, and its distance from the axis of rotation. Since the mass of a particle does not change, a reduction in distance from the axis must be accompanied by an increase in velocity. This velocity increase may be demonstrated by the simple apparatus pictured in Figure 8.18. Two spherical weights are arranged in such a way that they are brought closer to each other when the handle below them is compressed. The weights may be given an initial spin by hand so that they revolve around the central shaft. When the handle is compressed to bring the weights closer together, the angular velocity of the weights increases dramatically. When the handle is released, the weights move apart, and the angular velocity decreases. This change in speed is necessary in order that the original angular momentum may be conserved. Another example is that of the ice skater who begins to spin herself rapidly in one place. If her arms are outstretched when she begins to spin, she can pull her arms in close to herself and spin even more rapidly.

As the large low-density cloud rotates, therefore, its contraction will cause the rotation to become more rapid. The more the gas and dust contract, the more rapidly the whole mass rotates. The contracting mass which began with a spherical shape soon becomes flattened by the rapid rotation. Eventually a relatively thin disc, somewhat larger than the solar system, is left surrounding a large central mass which will become a star. Up to this point there are no serious problems in the theory. Beyond this, however, one must attempt to explain the formation of the planets in the rotating disc of gas and dust. It has been suggested that eddies form in the disc, and that these eddies can foster the accumulation of material into the planets. A possible pattern of such eddies is indicated in Figure 8.19. The broader aspects of this theory can account for the fact that the orbital planes of the planets are nearly coincident and that most of the planets rotate in the same direction. Whatever the details of the actual formation of the solar system, it is very likely that the process may have followed this general pattern.

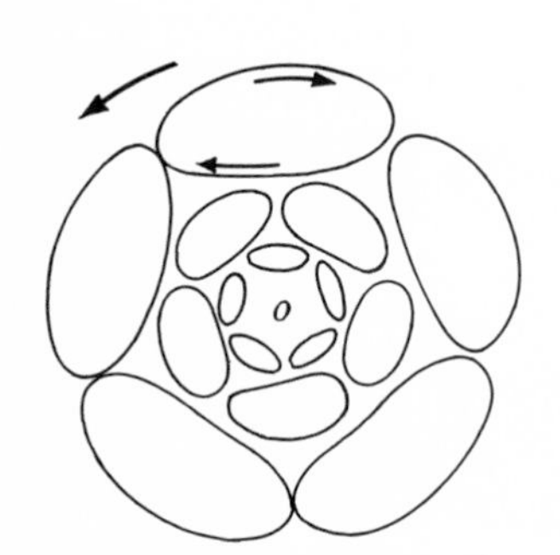

Figure 8.19. A possible pattern of eddies in the rotating disc surrounding the proto-sun.

The proto-planets which formed in the nebular disc may all have been rather similar to each other in the earliest stages of evolution. The particular distance of each one from the sun would affect the subsequent evolution of each planet, so that they eventually became the objects we see today. Let us imagine a proto-planet in orbit at Mercury's distance from the sun. Within this large body we would expect the heavier materials to settle toward the center, so that a rather dense core would form. Around this core there would then be an extensive atmosphere of light gases. So near to the sun, the proto-planet would be subject to strong tidal effects caused by the sun's large mass and would become an ellipsoid oriented as in Figure 8.20. If the proto-planet should have an original rapid rotation, there should be considerable tidal friction between the small core and the extended atmosphere. The central core would then slow its rotation considerably. At the same time the high temperature, the solar wind, and the radiation pressure would cause the

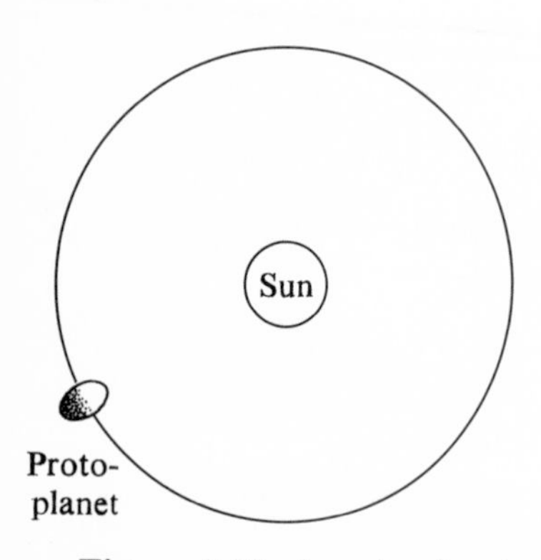

Figure 8.20. A proto-planet distorted by the strong tidal forces of the sun.

light gases in the atmosphere to be blown off into space. By this rather simple mechanism we would be left with a rather dense, slowly rotating planet which had lost essentially all of its atmosphere. In the extreme, the rate of rotation could be equal to the revolution period since this would also have been the rotation period of the tidal bulge of the proto-planet.

The situation for a proto-planet at Jupiter's distance from the sun is much less extreme. Very little of the original rotational momentum is lost due to tidal friction. Likewise, very little of the atmosphere is lost because of the lower temperature and the lessened effects of the solar wind at these distances. As a result, the outer planets are able to maintain their light gases and their original short rotation periods. We see the outer planets, therefore, much as they were long ago when they first formed from the vast disc surrounding the proto-sun. So far, there seem to be no serious difficulties with the general form of this theory; however, many of the details are not yet well worked out. Future study may show the necessity for changes in parts of this theory within the general framework.

QUESTIONS

1. What is the approximate surface temperature of the sun and how is it determined?
2. Why is the sun believed to be completely gaseous in spite of the high pressures in its interior?
3. What is the definition of the photosphere?
4. What is the principal source of energy in the sun?
5. By what sort of reasoning has the internal structure of the sun been inferred?
6. What quantities are observed to vary during the 11-year sunspot cycle?
7. Describe the appearance and changes in appearance of a typical sunspot group.
8. Where do the dark lines in the solar spectrum originate? How many chemical elements have been identified in the solar spectrum?
9. Account for the brief appearance of the flash spectrum seen at solar eclipses. What does the flash spectrum tell us about the conditions in the gas above the photosphere?
10. Define prominences, filaments, and solar flares.

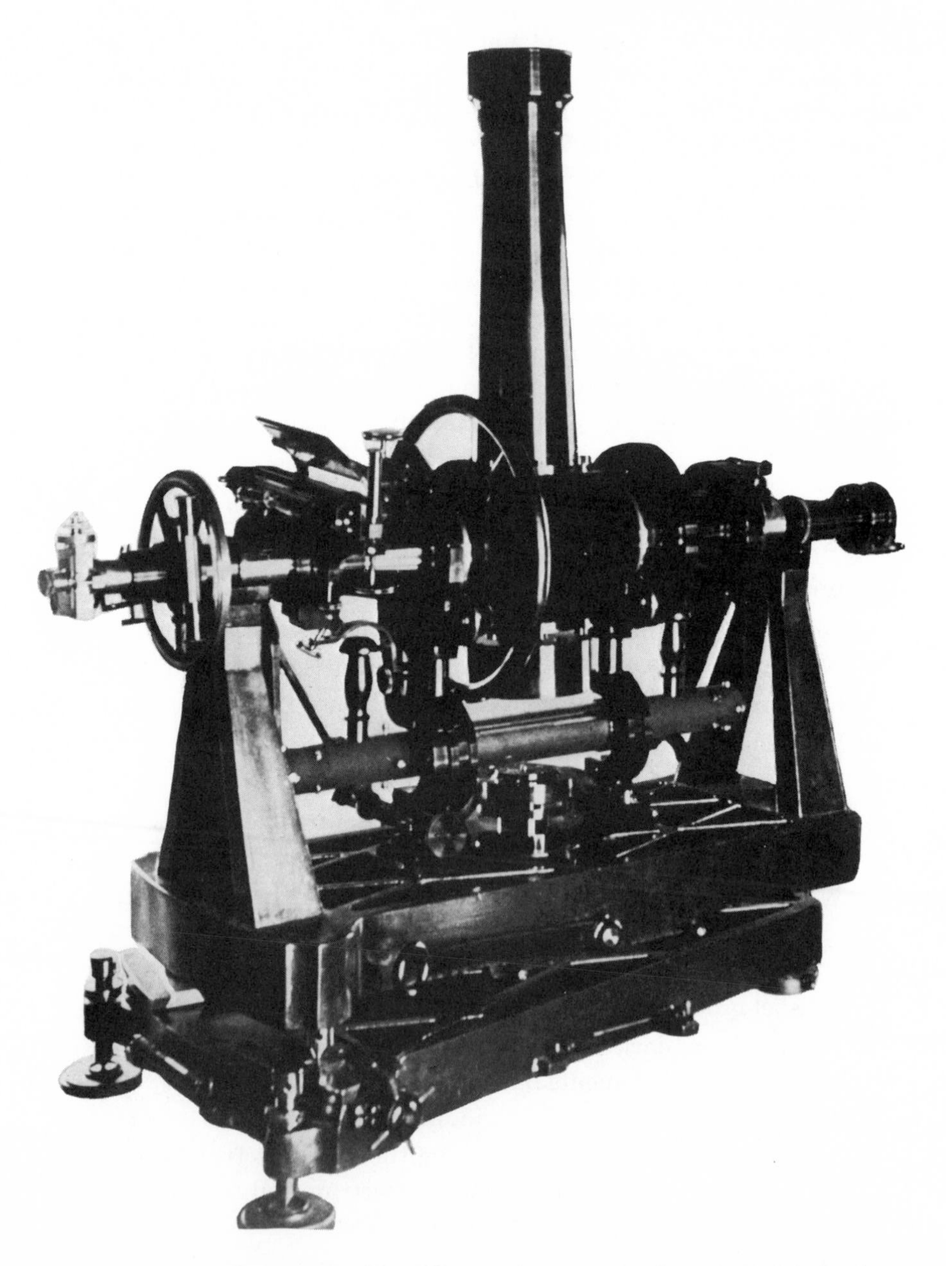

Figure 9.1. A transit telescope of 3-inch aperture. The telescope is constrained to move only in the north-south plane along the local meridian. The time for the longitude of the observatory may be determined by noting the instant that a star of known coordinates passes through the field of view. Larger instruments of this type are used in the reverse sense to determine the coordinates of stars when the local time is known. (Wellesley College photograph.)

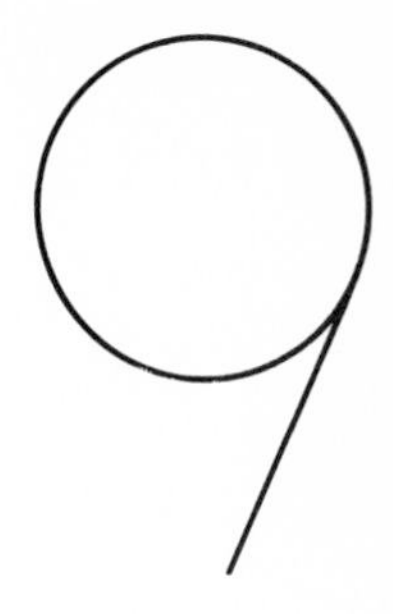

Positional Astronomy

In the previous chapter, the sun was discussed in detail as the source of all of the light and heat in the solar system as well as the star most accessible to us for detailed study. In this and the chapters to follow the stars in general will be considered. We shall begin with the appearance of the stars on the sky, noting certain facts about their positions on the celestial sphere and their brightnesses with respect to each other.

9.1 Celestial Coordinate System

There are some general rules by which the location of a point on any sphere may be specified. The same rules may be applied to determining the stars' positions on the celestial sphere.

A rotating sphere is indicated in Figure 9.2. The axis of rotation of the sphere is shown, and this axis passes through the center of the sphere. Imagine a plane passing through the sphere in such a way that it passes through the center of the sphere and is at right angles to the axis. The intersection of this plane with the surface of the sphere may be defined as the **equator.** Other planes which pass through the center of the sphere but are at right angles to the equator will include the axis of the sphere (for example, circle *PSF*

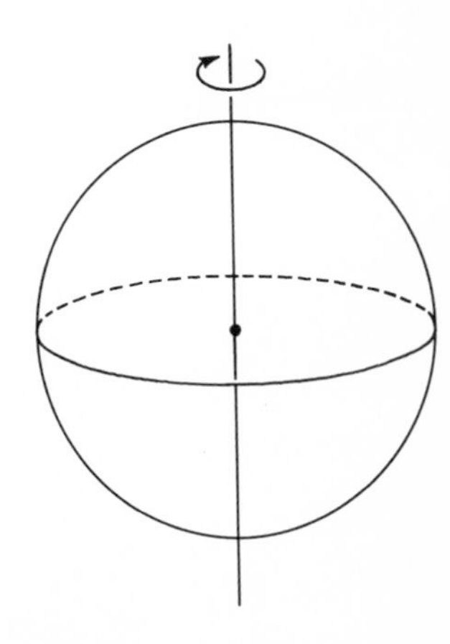

Figure 9.2. A hollow sphere rotating about an axis.

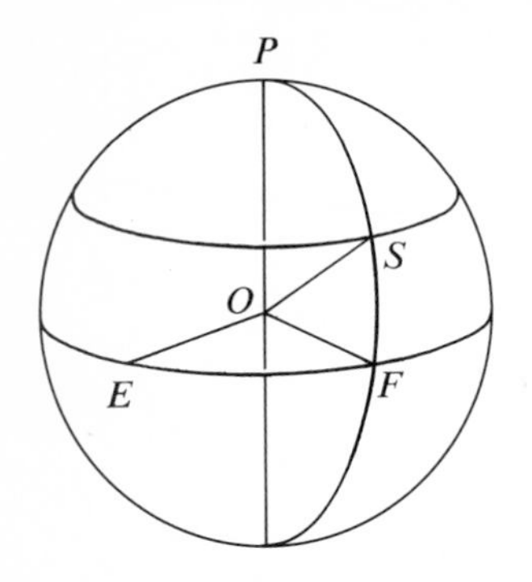

Figure 9.3. Planes may be passed through a sphere in such as way as to include the axis. Such planes will all be at right angles to the equator.

in Figure 9.3). The portion of such a circle on one side of the sphere is called a **meridian.** A meridian may be drawn through any point on the surface of a sphere. In Figure 9.3 a line has been drawn from the center of the sphere to the intersection of a meridian with the equator, and a second line has been drawn from the center to point *S* on the meridian. The angle *SOF* between these two lines locates point *S* on its meridian quite specifically, but the same angle would apply also to a similar point on any other meridian. In order to identify point *S* uniquely, we must identify the meridian through *S* in some convenient manner. If some arbitrary point on the equator is selected as a reference point, then any meridian may be positively identified simply by the angle between the reference point and the meridian measured in the plane of the equator. In Figure 9.3 this angle is indicated as angle *EOF*. In a general way, then, two angles may be used to specify the location of a point on the surface of a sphere.

Although we seldom think of it in quite this manner, we are using two angles to find a point when we determine the locations of places on the earth in terms of their latitudes and longitudes. Latitude locates a place at some distance north or south of the equator and longitude locates the meridian of the place east or west of an arbitrarily chosen zero meridian that passes through Greenwich, England. Actual determinations of latitude and longitude are made by indirect methods on the surface of the earth. Nevertheless, these coordinates really represent angles measured at the center of the earth.

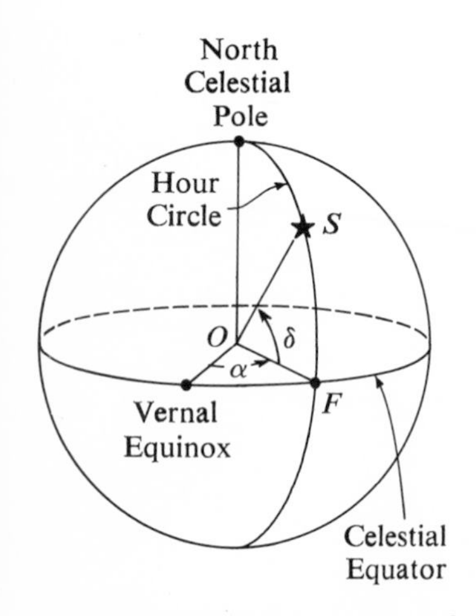

Figure 9.4. The equatorial coordinate system for specifying the positions of stars on the celestial sphere.

On the celestial sphere the situation is analogous, but specialized names are assigned to the various circles and angles, as shown in Figure 9.4. Point *S* now represents the location of a star on the sky. The celestial equator is simply the intersection of the earth's equatorial plane with the celestial sphere. Circles at right angles to the celestial equator which pass through the north and south poles of the sky are called **hour circles,** and an hour circle may clearly be passed through any star located on the sky. The angle *SOF* is indicated in the figure as δ, and is referred to as the **declination** of the star. The point on the equator which serves as a reference for the identification of hour circles is the **vernal equinox.** The angle measured from the vernal equinox to a star's hour circle is called **right ascension** (abbreviated R.A.), and is indicated in the figure as α. Declination is always measured in degrees north or south of the equator. Right ascension is most conveniently mea-

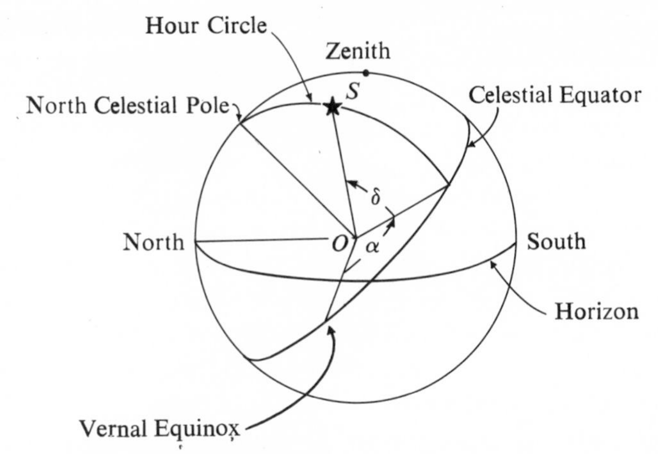

Figure 9.5. The orientation of the celestial sphere with respect to the horizon depends upon the observer's latitude and sidereal time.

sured in hours, minutes, and seconds (24 hours is equivalent to 360°) and is measured only toward the east from the vernal equinox. Thus, a star slightly to the west of the hour circle through the vernal equinox might have a right ascension of 23^h56^m. A star on the opposite side of the sky from the vernal equinox would have a right ascension of 12^h. The declination of any point on the celestial equator is 0° and the declination of the north celestial pole is +90°. Since this system is based on the celestial equator, it is generally referred to as the equatorial coordinate system. (See Figure 9.5.)

9.2 Determination of Right Ascension and Declination

It is useful to define one other circle on the sky. This is the **observer's meridian**—an imaginary circle running across the sky from the north point on the horizon through the zenith to the south point on the horizon. The **zenith** is simply a point on the sky that is directly above the observer. The celestial meridian is clearly fixed with respect to the observer, and as the celestial sphere appears to turn, the stars must each cross the observer's fixed meridian once in each 24-hour period. The passage of a star across the meridian is called a **transit.** Observations of the transits of stars provide the basic data for the determination of the right ascensions and declinations of stars.

Observatories engaged in measuring the coordinates of stars make use of specialized telescopes known as transit telescopes or meridian telescopes. Transit telescopes are usually refractors mounted to move around a fixed axis carefully oriented in an east-west direction. The telescope is thus able to point only toward the

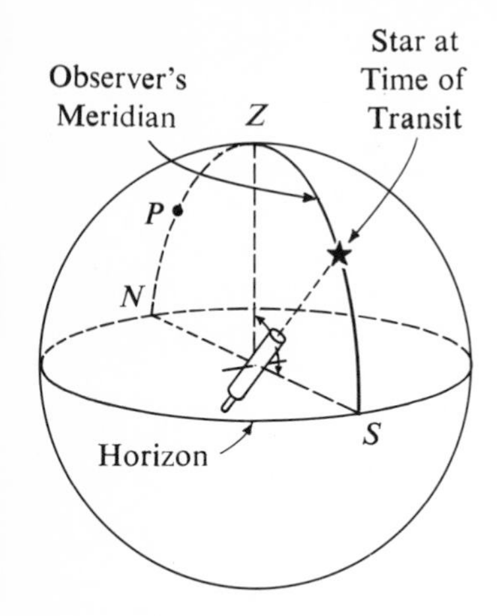

Figure 9.6. The transit telescope is constrained to move only along the observer's celestial meridian.

meridian, but can swing around its axis toward any point on the meridian. Continuous observations of individual stars are impossible, but once each day any star may be observed as it makes its daily transit. With the telescope pointed at the proper region on the meridian, the star appears to move slowly through the field of view. A graduated circle is attached to the axis of the telescope and is adjusted to read exactly 90° when the telescope is directed toward the celestial north pole. When the telescope is swung through an angle of 90°, it must point to the celestial equator. The graduated circle must now read 0°. With this type of instrument both the right ascensions and declinations of stars are readily measured. (See Figures 9.1 and 9.6.)

When the transit telescope has been adjusted so that a particular star passes through the center of the field, the declination of the star may be read directly from the graduated circle. The right ascension of the star may be computed from the time of the star's meridian passage if this has been carefully noted during the observation. Observatories are always equipped with a clock which has been set to keep local sidereal time. The clock will read 0^h at the instant the vernal equinox makes its transit of the meridian. Since right ascension is measured from the vernal equinox, the local sidereal time at the moment of a star's transit is equal to the right ascension of the star. This helps to explain why astronomers prefer to measure right ascension in hours, minutes, and seconds rather than in degrees.

The nature of transit observations is such that the coordinates of the sun may be determined as the sun crosses the meridian each day at noon. As mentioned in Chapter 2, the day-to-day changes in the sun's position reveal the sun's steady motion along the ecliptic.

One of the few observatories in the world at which astronomers make fundamental observations of stellar coordinates on a routine basis is the U.S. Naval Observatory in Washington, D.C. (Figure 9.7). Some of the transit instruments there are used in the manner described above, but others are used in a reverse sense. Instead of finding right ascension from the observed time of a meridian passage, the Observatory's telescopes are used to find the time itself from meridian passages of stars of previously known right ascension. Time determined in this manner is converted to Eastern Standard Time and is distributed throughout much of the world on the familiar radio time signals. It should be evident that the longitude of a place may be determined from observations of this sort.

Figure 9.7. The United States Naval Observatory in Washington, D.C. (Official U.S. Navy photograph.)

The observer is able to determine his own local time and compare it with Eastern Standard Time. The difference is equal to the difference between the observer's longitude and the longitude 75°W, which is the center of the Eastern Standard Time zone.

Observations made with a meridian telescope are somewhat laborious and are usually restricted to the brighter stars. In many situations the coordinates of large numbers of faint stars are needed, and these may be determined from photographs. If accurate right ascensions and declinations of several stars on one photograph are known, they can serve as a reference frame for the determination of the coordinates of the fainter stars on the same photograph.

9.3 Changes in the Coordinates of Stars

Thousands of years ago the planets became known as special objects from their motions against the fixed background of stars. To the ancient observers the relative positions of the stars remained unchanged over the long periods of their observations. Only with the development of techniques for very accurate measurements of coordinates have astronomers been able to note the very small changes in position which actually do occur. There are three important factors which cause changes in right ascension and declination: precession, parallax, and proper motion.

Figure 9.8. A gimbel-mounted gyroscope by which the precession effect may be demonstrated. (Wellesley College photograph.)

Precession

In physics courses an effect known as **precession** is customarily demonstrated with a spinning gyroscope (Figure 9.8). When a force is applied to the axis of the gyroscope, the axis is not deflected in the direction of the force but in a direction at right angles to the force. When a spinning toy top begins to slow down, gravity acts to lay the top on its side. This is the same as a force applied to the axis of the top, and this force causes the axis to move at right angles to the direction of the force. The result is that the axis of the top traces out a cone. Since the earth is rotating on its axis, one would expect it to behave in the same manner if a force should be applied to its axis.

Recalling that the earth is not spherical but flattened through the poles, let us look at the effects of the moon's gravitational attraction for the earth. Part of the earth's mass is in the equatorial bulge, and one side of this bulge is slightly closer to the moon than is the center of the earth. Consequently, the attraction of the moon for the material in the bulge is a little stronger than for the material at the earth's center. Since the orbit of the moon is not in the plane of the earth's equator, the moon pulls on the bulge as if to try to pull the bulge into the plane of its orbit. The same effect would occur if a force were applied to the earth's axis. As a result of the moon's attraction, the earth's axis moves in the expected conical path (Figure 9.9). Because of the rather large distance from the

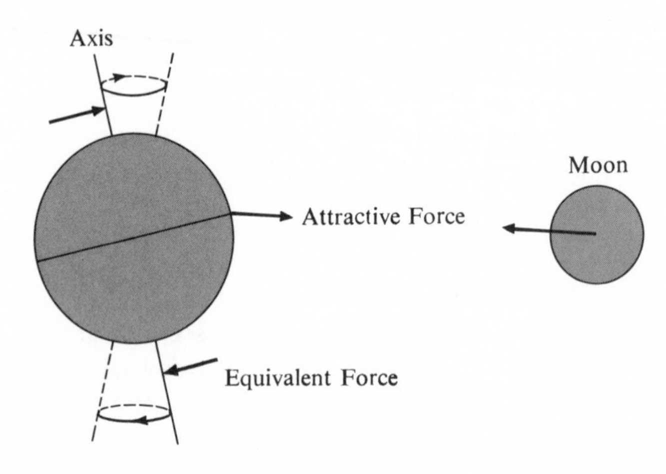

Figure 9.9. The moon's gravitational attraction acting on the earth's equatorial bulge causes the precessional motion of the earth's axis.

earth to the moon, and because the earth is only very slightly flattened, the force causing the precession of the earth's axis is small. The resulting precessional effect is also small, but it is quite significant.

If precession causes the earth's axis to move, the north pole of the sky must move also, since it is defined by the direction in which the earth's axis points. A new north celestial pole means a new celestial equator and a new frame of reference for the coordinates of all of the stars (Figure 9.10). Due to precession, then, all right ascensions and declinations are changing in a progressive manner. As viewed on the sky, the north celestial pole moves in a counterclockwise direction around the pole of the ecliptic. The equinoxes must therefore move in a westward direction along the ecliptic. As mentioned above, this affect is very small, and it will take 26,000 years for the earth's axis to make a complete circuit and return to the present pole.

The builders of the great Egyptian pyramids were the first to notice the progressive change in stellar coordinates which we now attribute to precession. The Egyptians had noticed that the annual flood of the Nile River came just a week or so after the day on which the sun and the bright star Sirius rose above the eastern horizon at the same time. Since the flood was crucial in their agricultural cycle, the Egyptians kept a close watch on Sirius in order to know when to expect the flood. For hundreds of years the priests watched Sirius rise, and gradually it became apparent that the point at which Sirius rose was moving very slowly. The reasons behind this movement were unknown to the Egyptians, but the cause was precession of the earth's axis.

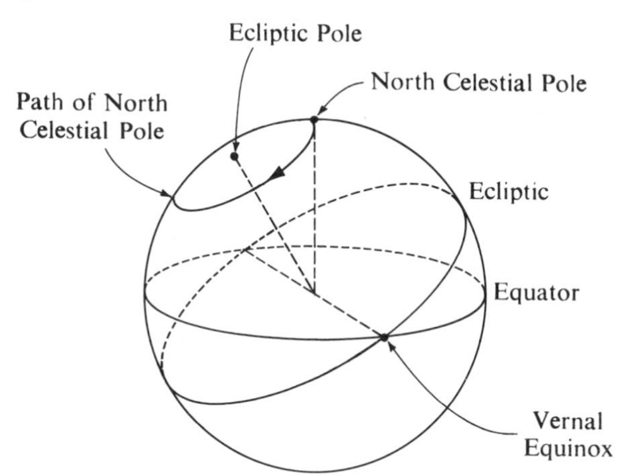

Figure 9.10. As the earth's axis moves, the plane of the earth's equator must also move. Since the reference frame moves against the celestial sphere, the coordinates of all stars must be changing continuously.

During the lifetime of one man, precession will cause no obvious changes in the appearance of the sky to the naked eye. The largest precession effect will be a change in a star's right ascension of less than 2^m over a period of 10 years. Such changes are of little consequence as far as the constellations are concerned. In order to study the other factors which are detected as changes in star positions, however, astronomers must understand and make corrections for precession. In addition, astronomers must often set their telescopes on very faint objects. This is done most quickly and efficiently when a star's exact coordinates are known. It is interesting to realize that in 13,000 years the earth's axis will have moved through one-half of the precession cycle and that the seasons on the earth will be reversed. (Refer to Figure 2.14, page 41.)

Parallax

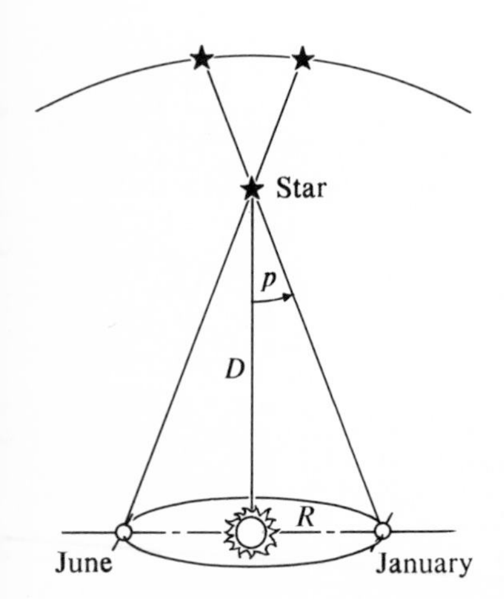

Figure 9.11. The parallax of a star is the angle p subtended by the radius of the earth's orbit as seen from the star.

Parallax as a proof of the earth's orbital motion is discussed in Chapter 2, where the stars are described as appearing to move in small circular or elliptical paths (on the ecliptic the ellipses become straight lines) with a one-year period. In terms of a star's coordinates, parallax results in small periodic changes in right ascension and declination. In addition to parallax serving as a convincing proof of the earth's revolution, the parallax angle p, indicated in Figure 9.11, is a measure of a star's distance from the sun. The larger the measured parallax, the nearer the star. It is for this reason that the study of stellar parallaxes is a basic and important aspect of astronomy.

All of the stars are so far away that the angle p is never very large. For the nearest star other than the sun, Proxima Centauri, visible in the southern hemisphere, the parallax is only 0.765 arcsec. Parallaxes as small as 0.01 arcsec are measurable, but when the parallax is smaller than this, the errors of measuring become large compared to the size of the angle being measured.

The first parallax determinations were made in the first half of the nineteenth century, and the principal observers devised ingenious methods for measuring the small displacements. When first-class meridian telescopes became available, they were used to find the parallaxes of the nearest stars. The greatest advances came just after 1900, however, with the development of photographic parallax methods much the same as those in use today. A series of photographs of a desired star and its background of faint stars is made over a period of at least two and one-half years. The photographs are usually restricted to seasons about six months apart when the displacements are largest. The position of the parallax star is then carefully measured with respect to the background stars on each photograph. After a lengthy series of calculations, the parallax is determined. The parallaxes of some of the nearest stars are listed in Table 9.1. To appreciate the problem of measuring very small parallax angles, one may note that the scale of the long-focus McCormick refractor (Figure 4.1) is such that an angle of 1 arcsec is represented by a measured distance of only 0.05 mm on the photograph.

In discussions of the distances to stars, astronomers often refer simply to the parallax angle. To most people, however, this is rather awkward, and it is advantageous to convert parallax to more familiar units customarily used to express distances. Referring again to Figure 9.11, one may note that $\tan p = R/D$. Since R, the radius of the earth's orbit, is known, the distance D is readily found after the parallax has been measured. If the parallax is converted to radians from seconds of arc, then $p = R/D$. If R is one astronomical unit, D will also be expressed in astronomical units. Since one second of arc is equivalent to 1/206,265 radian, the distance to a star with a parallax of one second of arc would be 206,265 A.U. It is unnecessarily awkward to deal with such large numbers when the precision of the data really does not warrant it.

Table 9.1. Some of the nearest stars.

	Parallax (arcsec)	Distance (pc)	Distance (light-years)
Proxima Centauri	0.763	1.31	4.27
α Centauri	0.752	1.33	4.34
Barnard's Star	0.545	1.83	5.97
Wolf 359	0.425	2.35	7.66
Lalande 21185	0.398	2.51	8.18
Sirius	0.377	2.67	8.70
Ross 154	0.350	2.86	9.32
ϵ Eridani	0.303	3.30	10.76
61 Cygni	0.292	3.42	11.15
Procyon	0.287	3.48	11.34
τ Ceti	0.275	3.64	11.87
Kapteyn's Star	0.251	3.98	12.97

Astronomers have, therefore, invented a new unit, the **parsec** (abbreviated pc), to express stellar distances. The parsec is defined simply as the distance at which a star would have a *par*allax of one *sec*ond of arc. From this definition the distance in parsecs is equal to $1/p$ (where p is expressed in seconds). The parsec is the distance unit most used among astronomers, but it requires too much explanation to be meaningful to laymen. As a result, another unit, the **light-year,** is used in most popular astronomical writing. A light-year is the distance that light travels in one year. For purposes of conversion, 1 parsec = 3.26 light-years. Since the parallax of the nearest star is less than one second, there is no star within one parsec of the sun. Table 9.1 lists not only the parallaxes of the nearest stars, but their distances in parsecs and light-years as well.

Proper Motions

When the coordinates of stars have been determined on two or more occasions widely separated in time, significant differences are often detected. For example, Edmund Halley in 1718 redetermined right ascensions and declinations for a number of the brighter stars in the catalogue Ptolemy had made more than 1000 years earlier. The differences between the two determinations were too large to be attributed to errors in Ptolemy's observations after taking out the precession effect. Halley concluded that the stars in question had actually changed their positions by significant amounts. Subsequent comparisons between positions given in older star catalogues with those in more recent ones have revealed the progressive changes in coordinates for thousands of stars. Such changes are known as **proper motions,** and are simply the angular velocities of the stars as seen from the sun. Proper motion is indicated by the arrow marked μ in Figure 9.12. The most convenient units for proper motion are seconds of arc per year or seconds of arc per hundred years. A few stars of large proper motion are listed in Table 9.2.

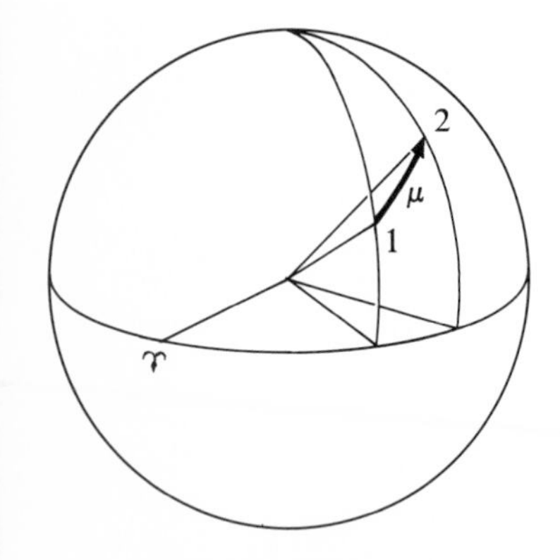

Figure 9.12. The proper motion of a star is the star's angular velocity as it moves slowly and steadily on the celestial sphere.

Table 9.2 shows that displacements due to proper motions are relatively large compared with those caused by parallax. Also, since proper motions are cumulative, it is not surprising that proper motions were discovered before parallaxes. This helps to explain why a series of photographs for a parallax must extend for at least several years. The continued observations are necessary in

Table 9.2. Stars of large proper motion.

	Proper Motion (arcsec/year)	Position Angle (degrees)	Parallax (arcsec)
Barnard's Star	10.27	356	0.545
Kapteyn's Star	8.73	131	0.251
BD −36° 15693	6.90	79	0.273
BD −37° 15492	6.08	113	0.219
Ross 619	5.40	167	0.151
61 Cygni	5.20	52	0.292
Lalande 21185	4.78	187	0.398
Wolf 359	4.71	235	0.425
ϵ Ind	4.69	123	0.286
BD +44° 2051	4.53	295	0.172
40 Eridani	4.08	213	0.201
Proxima Centauri	3.85	282	0.763

order to separate the proper motion and parallax effects from each other. The numerical solution for parallax actually gives a star's proper motion as a by-product. It should be clear that if two stars have the same actual velocity through space, the nearer of the two will have the larger proper motion. It is for this reason that stars with large proper motion were selected for study in the first attempts to detect parallax. Generally speaking, large proper motion is an indication of a relatively near star.

The complete specification of a star's proper motion requires two quantities: the angular velocity of the star, indicated as μ in Figure 9.13, and the position angle θ, which is measured between the direction toward north and the direction in which the star is moving. An alternative procedure, used in some catalogues, is simply to list the proper motion components in right ascension and declination as μ_α and μ_δ.

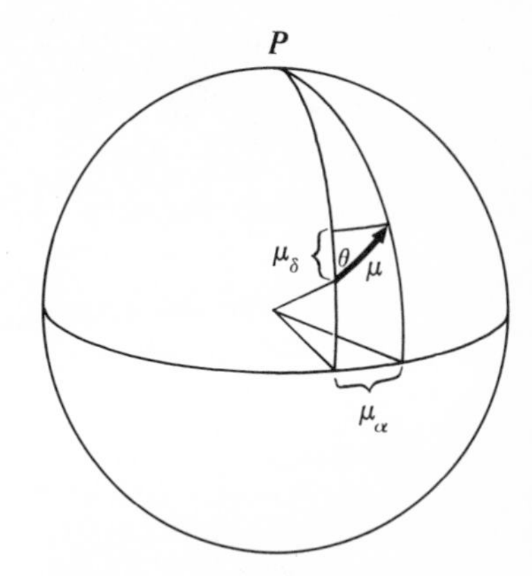

Figure 9.13. The proper motion of a star may be divided into components in right ascension and declination.

Only the brighter stars can be studied with meridian telescopes; therefore, proper motions of faint stars must be obtained photographically (Figure 9.14). To do this one simply determines the motions of the nearer stars with respect to the more distant ones whose motions are hoped to be very small and random. When the interval between observations is long, even the distant stars will have detectable motions and may not be used as a reference frame. Fortunately, most photographs made with large astronomical

Figure 9.14. Proper motion of Barnard's Star from July 31, 1938 to June 24, 1939. Photographs from the two dates are shown together, with the earlier one to the left. Barnard's Star is the bright one near the center. (Sproul Observatory photographs.)

cameras show an extremely distant background of faint galaxies which furnish the necessary fixed system to which the motions of the stars may be related.

In later chapters we shall see that much of our detailed knowledge of the physical nature of the stars and their arrangement in our vast stellar system is derived from basic observations of parallaxes and proper motions. Some of the most interesting challenges in astronomy lie in taking the data derived from the region within about 100 pc of the sun and applying them to the extremes of our galaxy and to other independent systems of stars.

QUESTIONS

1. Show with a sketch how, in the general case, two angles may be used to define the location of a point on a sphere. What names are

given to these angles if the point is on the spherical earth and what names are used if the point is a star on the celestial sphere?

2. How are transit telescope observations used to determine the coordinates of stars in the sky?
3. What is the cause of the precession effect and why does this result in changes in star coordinates?
4. Draw a labeled sketch to show what the parallax of a star actually represents.
5. Why are there limits to the measurement of stellar parallaxes and what are these limits?
6. How is parallax related to the actual distance to a star? Explain the units commonly used to express stellar distances.
7. Define *proper motion* and explain how the proper motions of stars are determined.
8. Why is it that two quantities are actually required in order to specify the proper motion of a star?
9. Which of the three effects, precession, parallax, and proper motion, would be the most noticeable over a period of 50 years?

6
1
2
3
4
5

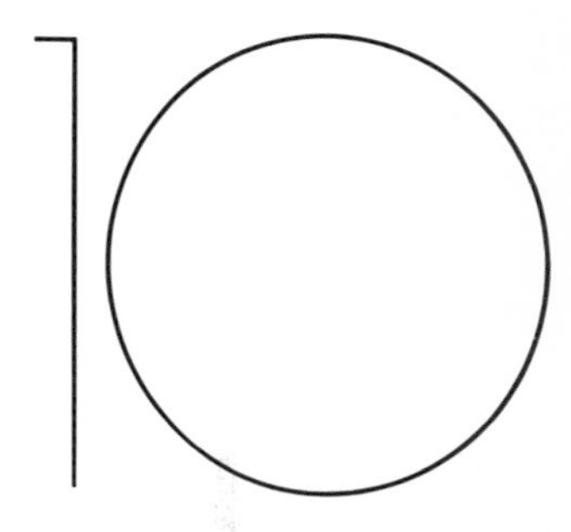

Photometry and Spectroscopy

Along with the data on positions and motions of stars, some fundamental observations relate to the measurement of their brightnesses. Studies in the broad field of astronomical photometry range from crude visual estimates to remarkably accurate measures made with sensitive photoelectric devices. Less obvious to the layman than differences in brightness are the differences in the colors of stars. Star colors are precisely described by the astronomer's photometric data. Combining the parallax methods of the preceding chapter with stellar photometry, astronomers are able to correct the observed brightnesses for the effects of distance and make meaningful comparisons of the true luminosity of stars.

Perhaps the greatest insight into the true nature and variety among stars comes from the careful analysis of photographs of stellar spectra. Temperature, composition, and radial velocity are the basic data derived from spectra, but information regarding rotation, diameter, and magnetic fields is also attainable. When all types of observational data are combined, two great pictures begin to emerge—the life cycles of the stars and the pattern of their distribution in the immense volume of space around the sun.

◄**Figure 10.1.** Objective-prism spectra of a representative group of stars. After studying Section 10.6, the student may try to classify the numbered spectra. The correct classifications are given at the end of this chapter. (Warner and Swasey Observatory photograph.)

10.1 Stellar Magnitudes

The astronomer's conventional system of describing the relative brightnesses of stars is based on a scale which was first used in the star catalogue of Hipparchus, a Greek astronomer of the second century B.C. Hipparchus called the brightest stars "first magnitude" and those barely visible to the unaided eye "sixth magnitude." This convention has been in use ever since. Through the years it has become possible to make measurements of stellar brightness with greater and greater precision. Today the magnitude system has been formalized and extended to the sun in one direction and to very faint stars in the other direction. Measurements ranging from crude visual estimates to rough, whole-number classes have been replaced by magnitude determinations accurate in the third decimal place.

The formal expression of the magnitude scale began with the recognition that a first-magnitude star was just 100 times brighter than a sixth-magnitude star. Thus, a ratio of 100/1 in brightness is equivalent to a difference in magnitude of 5. From this, one may derive the following expression which relates any ratio of brightness to a difference in magnitude.

$$b_1/b_2 = 2.512^{(m_2 - m_1)}$$

A slightly more usable formula for actual computations is obtained by taking the logarithm of both sides and writing

$$\log (b_1/b_2) = (m_2 - m_1)0.4$$

or

$$2.5 \log (b_1/b_2) = m_2 - m_1$$

In these expressions, b_1 and b_2 are the brightnesses observed for two stars whose magnitudes are respectively m_1 and m_2. The quantity 0.4 is simply the logarithm of 2.512, and 2.512 is an approximate value of the fifth root of 100.

Two important points must be noted in connection with these basic formulae which define the magnitude scale. First, no zero point for the magnitude system is implied in these formulae; second, only differences in magnitude are measurable. Since ratios of brightnesses are being measured, the brightness units are not important. Furthermore, since differences in magnitude are being measured, there is no way to find the magnitude of a single star by

Table 10.1. Magnitude differences and equivalent brightness ratios.

Difference in magnitude	1	2	3	4	5	10	15	20
Ratio of brightness	2.5	6.25	16	40	100	10,000	1,000,000	100,000,000

itself. There was necessarily a time when astronomers had to assign arbitrarily a specific magnitude for one single star. The magnitudes of all other stars could then be determined in relation to this one. In Table 10.1 are listed the brightness ratios which are equivalent to several magnitude differences.

Since magnitude determinations must always be made relative to stars of previously known magnitude, astronomers in the early years of this century established a group of stars near the north celestial pole as international standards. Astronomers working anywhere in the northern hemisphere could make reference to these standards on any night of the year. In more recent years similar magnitude standards have been set up in all parts of the sky for more convenient and precise reference.

After the magnitude system had been mathematically defined and a zero point agreed upon among astronomers, it was apparent that some of the objects listed as first magnitude by Hipparchus were really too bright to be first magnitude. From the basic formula one may see that these very bright objects must have fractional or negative magnitudes. Thus Vega, in the constellation Lyra, has a magnitude of 0.14, and Sirius, the brightest star in the sky, has a magnitude of −1.47. The magnitude of Venus, when that planet appears brightest, is −4.3 and the magnitude of the full moon is −12.5. Still on the same scale, the magnitude of the sun is −26.7.

10.2 Magnitude Systems

In Chapter 4 it was stated that the color response of photographic materials is not the same as that of the human eye. The most common films are more sensitive to blue light than is the eye, and for this reason a blue star may appear brighter relative to some group of stars on a photograph than it would to the eye. Similarly, a red star might seem relatively faint on the photograph. It has been possible for photographic engineers to manufacture emulsions which, when used with certain filters, reproduce the sensitivity of the eye. In Figure 10.2 two photographs of a small section of the sky are reproduced. The left one was made with blue-sensitive materials

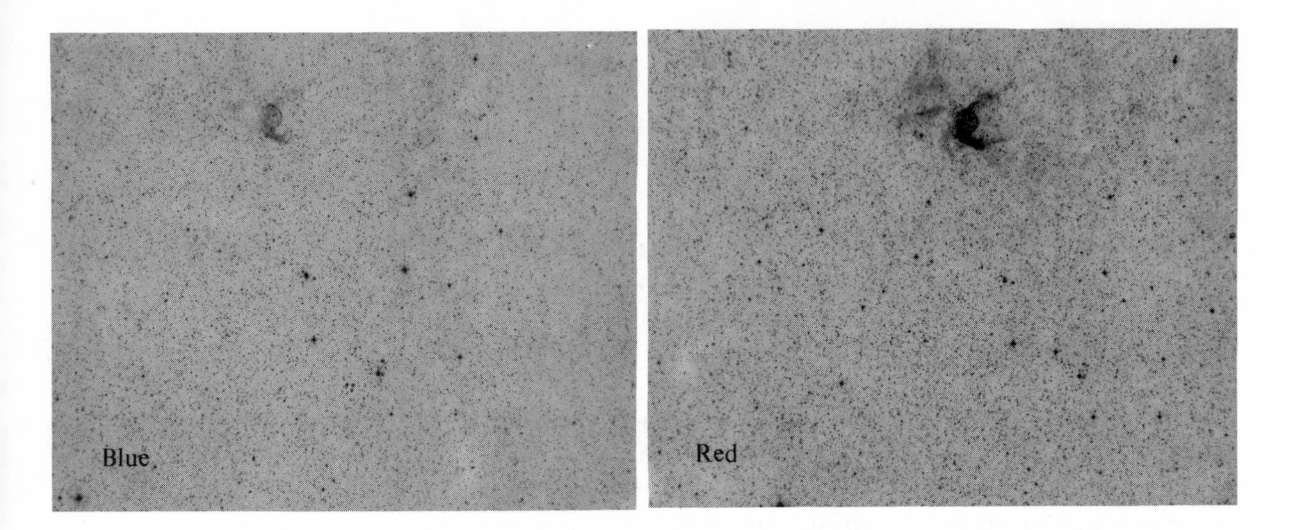

Figure 10.2. A star field photographed in blue light (left) and in red light (right). The colors of stars are indicated by the relative brightness in the two photographs. (Hale Observatories photographs, © copyright by National Geographic Society.)

and the right one with red-sensitive materials. There are distinct differences in the relative brightnesses recorded on the plates for a number of the stars seen here. This is, of course, an indication of the actual differences in the colors of the stars. The blue stars appear brighter in the left photograph and the red stars appear brighter in the one on the right.

It is, therefore, not enough to speak merely of the magnitude of a star. One must also speak of the **magnitude system** in which the observations were made. The two principal systems in use in the past were referred to simply as the visual and the photographic systems. The photographic equivalent of the visual magnitudes were referred to as photovisual magnitudes. The magnitude system used for some particular set of observations is usually indicated with an appropriate subscript. For example, $m_{pg}=3.9$ indicates that a star's photographic magnitude is 3.9 and $m_v=4.6$ indicates that the star's visual magnitude is 4.6.

In more recent times, with the great advances made in photoelectric methods of measuring star brightnesses, a number of other systems have come into use. For convenience, the newer systems were originally designed to duplicate the earlier photographic and visual systems. For greater versatility, a third system was added to measure brightnesses in the ultraviolet. Together, the three photoelectric systems comprise what has come to be called the UBV system, with the letters referring to ultraviolet, blue, and visual. Extensive lists of magnitudes of stars in the UBV system are now

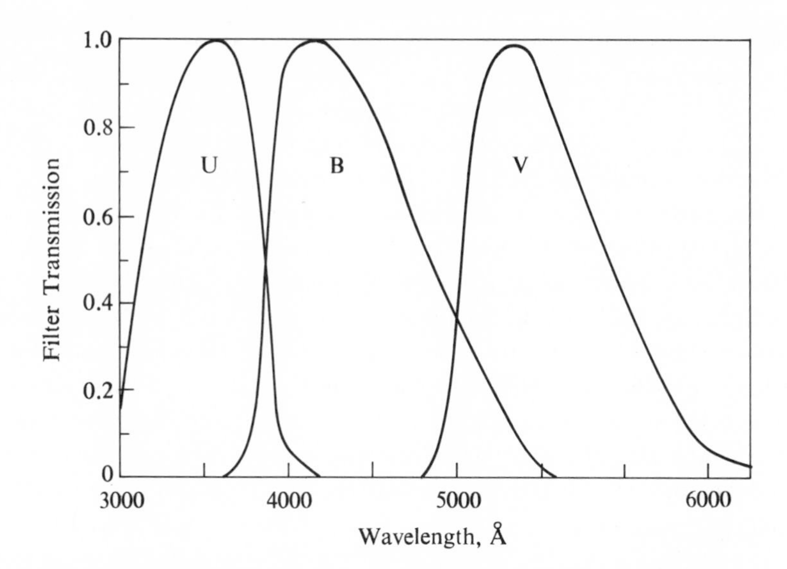

Figure 10.3. Curves representing the transmission of the UBV filters at various wavelengths.

available for the great convenience of astronomers, and the UBV system is the one most widely used today. An important extension of the UBV has been the addition of red and infrared systems designated by the letters R and I, respectively. In Figure 10.3 the wavelength bands covered in the UBV system are indicated.

In each system magnitude differences are determined from measured brightness ratios. The zero points of all the systems are completely independent of each other and must be related in an arbitrary manner. Thus at an early date it was decided that the photographic and visual magnitudes for certain approximately white stars should be the same. As newer systems have been added, the same convention has been maintained, so that for an ideal white star, the magnitude would be the same in each of the various systems.

10.3 Color Indexes

Differences in the colors of stars have been mentioned in Chapter 4 and elsewhere, and the student should make an effort on some clear night to look for these color differences specifically. Color variations among stars may be easily seen in the stars in and near the constellation Orion (see page 9). Betelgeuse (α Orionis) is quite definitely red and Rigel (β Orionis) is quite definitely blue. It is hard to describe the star colors in a quantitative manner since color differences are rather subtle. Beyond a mere statement that a star is blue, white, yellow, or orange, little may be said. When observations are available in two or more of the systems described above, however, a comparison of two magnitudes such as B and V may be made, and such a comparison is related to the star's color in a very simple way. Astronomers have defined a quantity called a **color index** simply as the difference between the magnitudes determined in two of the standard systems mentioned above. Thus, a color index might be $m_{pg} - m_v$, U − B, or B − V. Again from earlier definitions, the color index for a white star should be zero. A red star will be fainter in the blue system than in the visual, and so its blue magnitude will be a *larger* number than its visual magnitude. The color index for a red star is then positive, and the color index for a blue star must be negative. The most commonly used color indexes are U − B and B − V, and the B − V color index is very nearly the same as the older $m_{pg} - m_v$. Actual values of the B − V color index may range as high as +2 for the very red stars or as low as −0.4 for blue stars. Since magnitudes of stars are measurable with great accuracy, it is apparent that the color index is also measurable with great accuracy.

10.4 Absolute Magnitudes

Our discussion so far has been concerned with the magnitudes of stars as we observe them from the earth. Such magnitudes are referred to as **apparent magnitudes** and must depend upon the true luminosities of stars and upon their distances from us. In order to make meaningful comparisons among the physical properties of individual stars, the effect of distance must be taken into account. In the habit of thinking in terms of the magnitudes of stars, astronomers calculate for any star of known distance what the magnitude of that star would be if it could be seen from a point 10 pc from it. Such magnitudes are referred to as **absolute magnitudes** and are designated by the symbol M. Following the same convention used for apparent magnitudes, we may speak of absolute visual magnitude M_V, absolute blue magnitude M_B, and so forth.

Recalling that the observed brightness of a light source is inversely proportional to the square of the distance from the observer to the light source, one may derive in a fairly simple manner an expression by which the absolute magnitude of a star may be determined if the distance is already known. Begin with the expression

$$b = \frac{L}{r^2}$$

where b is the observed brightness, L is the true luminosity, and r is the distance to the star. If the star could be viewed from two points at distances r_1 and r_2, then the ratio of the brightnesses of the star as seen from the two points would be

$$\frac{b_1}{b_2} = \frac{L}{r_1^2} \bigg/ \frac{L}{r_2^2} = \frac{r_2^2}{r_1^2}$$

Now substitute this into the basic magnitude relation given above and write

$$2.5 \log (r_2^2/r_1^2) = m_2 - m_1$$

or

$$2 \times 2.5 \log r_2 - 2 \times 2.5 \log r_1 = m_2 - m_1$$

If r_2 is now taken as the standard distance of 10 pc, we may replace m_2 with M and write

$$M = m + 5 - 5 \log r$$

With this expression the absolute magnitude may be computed from the observed apparent magnitude and the measured distance to the star. This is an extremely important expression in astronomy and will appear repeatedly in later chapters.

Table 10.2 is a list of some of the bright stars and some of the near stars with their apparent magnitudes, color indexes, distances, absolute magnitudes and other data. In this brief list one may see most of the range of absolute magnitudes. Even in a volume of 10^6 cubic parsecs, it is rare to find a star brighter than $M = -5$, and at the faint end the limit seems to be in the neighborhood of $M = +15$. This range of 20 magnitudes is equivalent to a brightness ratio of 100,000,000/1. We can feel somewhat more certain that we have discovered samples of the brightest stars than we can that we have discovered samples of the faintest kinds of stars. This is because we are able to see the faint stars only in a relatively small volume of space close to the sun. We cannot be sure that all types of stars are represented in this small volume and that the faintest stars have been discovered.

Table 10.2. Properties of some bright stars and some nearby stars.

	App. Vis. Mag. (m_v or V)	Color Index (B − V)	Parallax (arcsec)	Distance (pc)	Proper Motion (arcsec/year)
Sirius	−1.47	−0.01	0.376	2.66	1.32
Canopus	−0.72	+0.16	0.018	55.50	0.02
α Centauri	−0.27	+0.71	0.761	1.31	3.678
Arcturus	−0.05	+1.24	0.091	10.99	2.285
Vega	0.03	0.00	0.123	8.13	0.346
Capella	0.09	+0.81	0.071	14.08	0.436
Rigel	0.11	−0.05	0.004	250.00	0.002
Procyon	0.36	+0.41	0.287	3.48	1.247
Achernar	0.49	−0.17	0.028	35.71	0.096
β Centauri	0.63	−0.24	0.008	125.00	0.036
Sun	−26.78	+0.62	...	...	...
Barnard's Star	9.53	+1.75	0.545	1.83	10.27
Lalande 21185	7.47	1.51	0.398	2.51	4.78
ϵ Eridani	3.74	0.87	0.303	3.30	0.97
61 Cygni	5.20	1.21	0.292	3.42	5.22
ρ Ceti	3.5	0.72	0.275	3.63	1.92
40 Eridani B	9.50	+0.03	0.205	4.87	4.08
van Maanen's Star	12.36	+0.61	0.236	4.24	2.98

*The luminosity of the sun is taken as 1.

If the earth-sun distance is expressed as a fraction of a parsec and then combined with the sun's apparent magnitude, the sun's absolute magnitude is found to be +4.8. This, it will be noted, places the sun near the middle of the absolute magnitude range. When the absolute magnitudes for a large number of stars have been determined, it may be seen also that there are larger numbers of faint stars than bright stars. More stars are fainter than the sun than brighter. It hardly seems necessary to mention that the range in apparent magnitude is limited only by the size of the telescopes in use. The faintest apparent magnitudes detectable with the 200-inch telescope are approximately +23.

10.5 Stellar Spectra

In general terms, the spectrum of a star is a continuous one on which a pattern of absorption lines has been superimposed. Just as was true for the sun, the continuous spectrum and the absorption lines originate in the star's photosphere. When in the last century

Abs. Vis. Mag. (M_V)	Luminosity*	Spectral Type	Radial Vel. (km/sec)
+1.41	22.5	A1 V	−8
−4.4	4742.4	F0 II	+20.4
+4.2	1.7	G2 + K1	−22
−0.2	99.1	K1 III	−5.2
+0.5	52.0	A0 V	−13.7
−0.6	143.2	G8 + G0	+30
−7.0	52,000	B8 Ia	+22
+2.7	6.9	F5 IV−V	−3.2
−2.2	625.2	B5 IV	+19
−5.0	8241.4	B1 II	−12
4.79	1.0	G2 V	. . .
+13.21	0.00044	M5	−108
10.42	0.0056	M2	−87
6.14	0.29	K2	+15
7.53	0.080	K5	−64
5.70	0.43	G8	−16
11.02	0.0032	wA2	−42
14.24	0.00017	wdF	?

the spectra of a large number of stars had been examined, it was seen that the absorption lines of hydrogen were present, to a degree at least, in the spectra of almost all of them. Lines of helium, calcium, sodium, and other elements were present in one star or another, but the hydrogen lines seemed to persist through most stellar spectra. The first classification scheme was based, therefore, on the relative strength of these lines. The spectra in which the hydrogen lines were most conspicuous were called type A. Those in which the hydrogen lines were weaker were called type B, and so on. Classes as far as M, R, and S were even listed. It was seen also that certain of these types were regularly described by certain colors. The B-type stars were always blue; the A stars were white; and the M stars were red. Eventually most of the classes were eliminated as being too similar to some other class, and the seven remaining ones were arranged in a new order: O B A F G K M. This provided a sequence in terms of color from blue to white to yellow to red and a sequence in terms of decreasing temperature. Examples of these types of spectra may be seen in Figure 10.4.

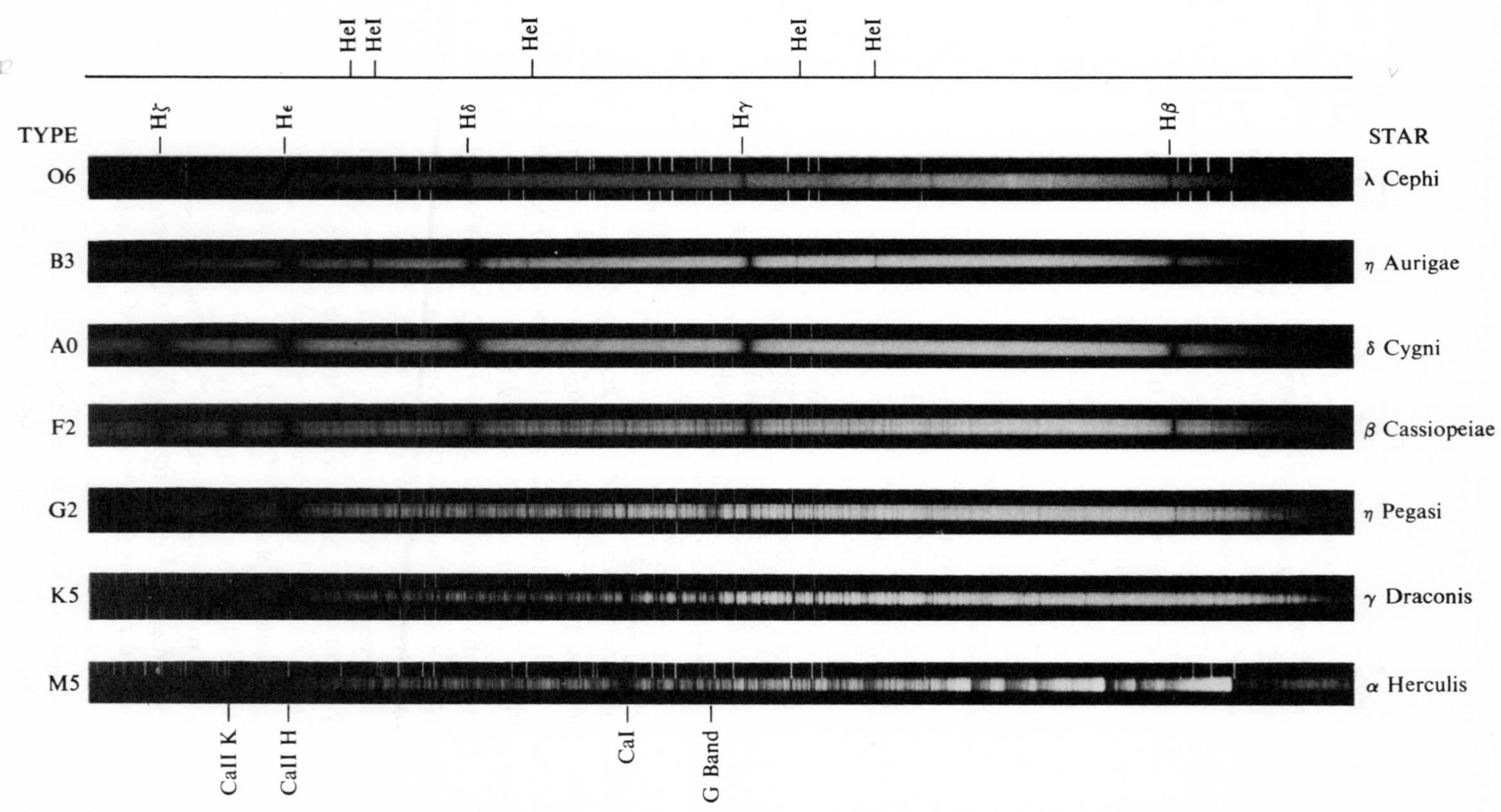

Figure 10.4. Representative samples of stellar spectra showing the principal classification features. The comparison spectrum is the iron spectrum. (Hale Observatories photographs.)

The major theoretical work which brought about this rearrangement of spectral classes was done by the physicist M. N. Saha in 1921. Saha showed that the appearance of the spectrum of a star should depend more on the star's temperature than on its actual chemical composition. Thus, if the compositions of all of the stars were the same, differences in temperature from one star to the next would cause differences in the observed spectra. Two important terms previously introduced in our discussion are *excitation* and *ionization*. Let us note the effect of a progressive increase in temperature on the excitation and ionization of the atoms in a gas, and then the related effect on the appearance of the spectrum.

Consider first a stellar atmosphere which is composed only of hydrogen and is at a temperature of about 4000°K. The hydrogen atoms can produce absorption lines in the Balmer series only when they are excited from the second energy level to a higher one. At 4000°K not many hydrogen atoms are in level 2 at a given time, so not many atoms are able to absorb the photons which will excite them from level 2 to some higher level. The absorption lines of hydrogen are therefore weak. If the temperature is now increased, there will be more atoms which have been excited to level 2 and are thus able to absorb the Balmer lines. As the temperature increases, the absorption lines should become stronger and stronger. At some point, however, the temperature becomes high enough to begin to ionize some of the atoms. For hydrogen this temperature is about 10,000°K – the approximate temperature of the type A stars mentioned above. The ionized hydrogen atom can neither absorb nor emit in the visible wavelength region, and so beyond this critical temperature the number of atoms capable of absorbing decreases as the number of ionized atoms increases. Since there are now fewer absorbers, the Balmer lines become weaker. Thus the increase in temperature first results in stronger hydrogen lines, and after some maximum intensity is reached, the lines become weaker. This effect is indicated in Figure 10.5.

The same reasoning applied to hydrogen may be applied to other gases as well. It becomes clear then that the absorption lines of certain elements will be strong at certain temperatures and weak at other temperatures. Since the atoms of the various elements are all quite different, the maximum strength of the absorption lines will be at different temperatures for each element. The work of Saha, however, permits one to calculate what the variations in appearance of spectra with temperature should be for any element.

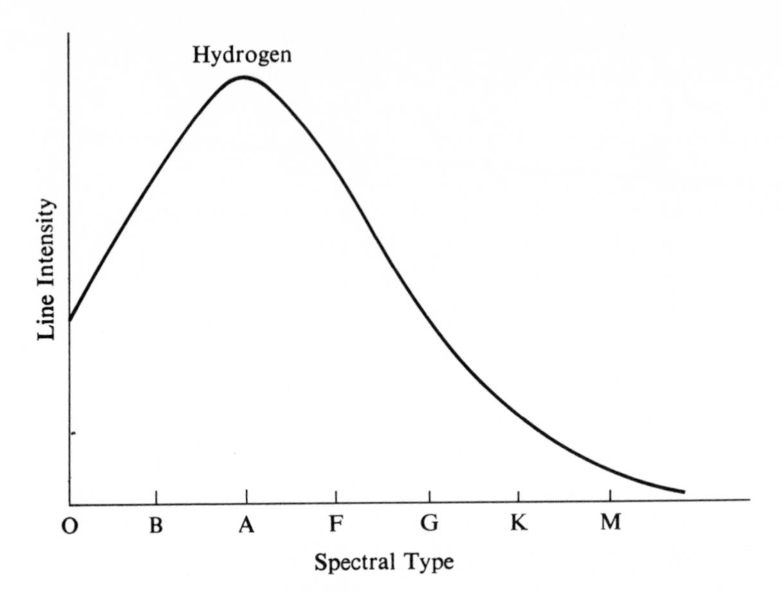

Figure 10.5. As the temperature increases from spectral type M to spectral type O, the absorption lines of hydrogen become stronger, then weaker.

In Figure 10.6 curves representing variations in strength of lines with temperature have been drawn for a number of elements common in stellar atmospheres.

For molecules or atoms with more than one electron, the ionized state behaves just as if it were a different entity altogether. When the temperature becomes high enough to ionize calcium atoms, for example, the absorption lines of the original neutral calcium become weaker. At the same time the lines of ionized calcium appear in the spectrum. These lines then become stronger as the lines of the neutral element become weaker, as Figure 10.6 clearly shows.

With the recognition of a limited number of types of spectra for stars and the arrangement of these spectra in a logical sequence in terms of temperature, speculation arose as to the significance of the sequence as it might be related to the ages of stars. Initially it was thought simply that stars should cool off as they became older.* On this basis the hot, blue stars were thought to be young and the cool, red stars were thought to be older. Consequently, the terms **early** and **late** came into use by astronomers. The stars of early spectral type were the blue ones of classes O, B, and A, and the later spectral types were those at the other end of the sequence. Some years afterward the theory of the evolutionary processes of

*Later when nuclear processes became known as the main sources of stellar energy, it was thought that hotter stars ran out of hydrogen fuel sooner. Hence blue stars were considered young, i.e., early.

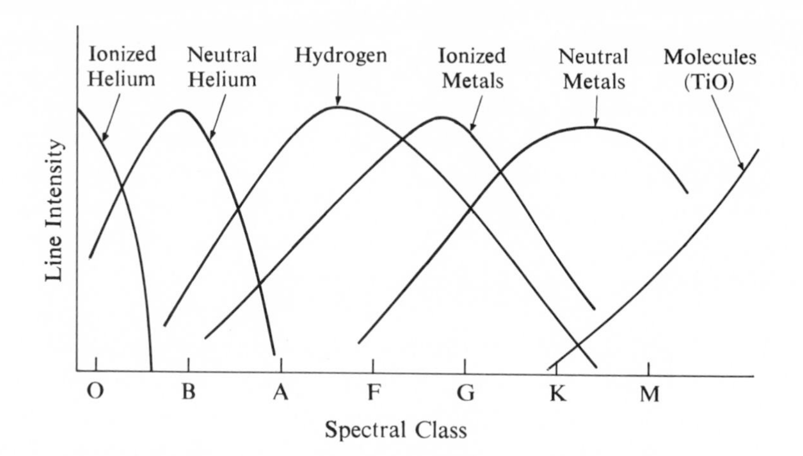

Figure 10.6. Increase in temperature results first in an increase in excitation, then in an increase in ionization.

stars was refined and given a much more meaningful theoretical basis. Even though we now know that the red stars are not necessarily extremely old, the terms "early" and "late" remain in use to indicate temperature directions on the spectral sequence. The student should understand the current usage of these terms and should not be confused by the implied reference to age.

As was to be expected, some of the astronomers actively engaged in large-scale programs of classification of spectra became quite skillful and were able to recognize differences on which they could base further subdivision of the seven principal spectral types. These subtypes are numbered from 0 to 9; therefore, the designation of the spectral type of an individual star is usually a letter followed by a numeral. The sun, for example, has a type G2 spectrum. Sirius has an A0 spectrum, and Barnard's Star has an M5 spectrum. It was later recognized that stars of the same spectral type were often very different in absolute magnitude. For example, Arcturus has an absolute magnitude of -0.3 and a type K2 spectrum while σ Draconis, barely visible to the naked eye, has an absolute magnitude of 5.92 and a type K0 spectrum. Astronomers are now able to detect and understand the subtle spectral differences which are the key to such differences in luminosity. (See Chapter 11.)

10.6 Classification Criteria

It is customary to list the spectral classes in the order O B A F G K M stated earlier. Nevertheless, in describing the important features used for classification let us begin with the cooler stars and work toward the hot ones. This is the most suitable approach since this is the direction in which the temperature causes an increase in the effects of excitation and ionization. Examples of each of the seven spectral types have been reproduced in Figure 10.4. Most of the significant absorption lines used for classification of stellar spectra lie in the violet and blue regions in the wavelength range from about 3500 Å to 5000 Å. In a color photograph the background continuous spectra similar to those in the figure would be violet at one end, green at the other.

Type M – Red stars with surface temperatures of about 3500°K. The dominant features are the numerous lines of neutral metals. The broad features at λ4585, λ4762, and λ4954 are the absorption bands of TiO molecules. The strong line at λ4227 is due to neutral calcium (CaI).

Type K – Orange stars with temperatures of about 4500°K. From the M stars through the subclasses of K stars the λ4227 line becomes weaker and the Fraunhofer lines H and K due to ionized calcium (CaII) become stronger with the rise in temperature. Only in the later K stars can the TiO bands be seen. The Balmer lines of hydrogen are weak but detectable.

Type G – Yellow stars with temperatures of about 6000°K. With this increase in temperature from the K stars the lines of ionized calcium and the Balmer lines become stronger. At the same time the λ4227 line of neutral calcium is becoming weaker as more of the neutral atoms are being ionized. The same applies to the large numbers of lines of other neutral metals which were conspicuous in the M stars.

Type F – Yellow-white stars with temperatures of about 7500°K. Within the temperature range of the F stars, calcium atoms begin to lose a second electron and the lines of CaII diminish in strength. With the rise in temperature the Balmer lines increase in strength. At F0 the once numerous lines of neutral metals have nearly disappeared.

Type A – White stars with temperatures of about 10,000°K. In the later A and early F stars, discrimination of the proper subclass is made on the relative strengths of the Balmer lines and the H and K lines of CaII. The H line at λ3968 and the fifth line of the Balmer series are nearly coincident and run together to make a conspicuous feature. The Balmer lines are actually strongest at A2. The H and K lines diminish rapidly through the temperature range of the type A stars.

Type B—Blue-white stars with temperatures of about 25,000°K. The Balmer lines weaken with the progressive rise in temperature through the stars from A0 to B0. As the hydrogen atoms become ionized, the helium atoms at last begin to become excited. Helium lines appear at $\lambda 4026$ and $\lambda 4472$. The line at $\lambda 4649$ is due to ionized oxygen. Other high excitation lines are present in type B stars.

Type O—Blue stars with temperatures of 50,000°K or higher. At such high temperatures the Balmer lines are very weak. In addition to the helium lines, lines of ionized helium may be seen.

With the above criteria as a guide, the student may examine the numbered spectra in Figure 10.1 and attempt to classify them. The correct spectral types for these stars are listed at the end of this chapter, preceding the Questions.

10.7 Radial Velocities

In order to classify stellar spectra according to the above criteria, it is usually enough to identify lines simply from the pattern seen in the complete spectrum. The distinctive patterns for various classes may be readily identified after only a little practice, and this is the manner in which spectra made with objective-prism spectrographs are normally studied. When actual determinations of wavelength are to be made, however, a reference system in the form of comparison lines of known wavelength is photographed along with the star's spectrum. This must be done with a slit-type spectrograph. (See Figure 10.7.) Measured wavelengths of readily identifiable lines, such as those of hydrogen and calcium, may then be compared with the known laboratory wavelengths of these lines, and the radial velocity may be computed from the difference. This procedure has been discussed at length in Chapter 4 and in connection with the planets and the sun. With reasonable care the precision of the measurements is such that radial velocities are among the most accurately determinable quantities in astronomy.

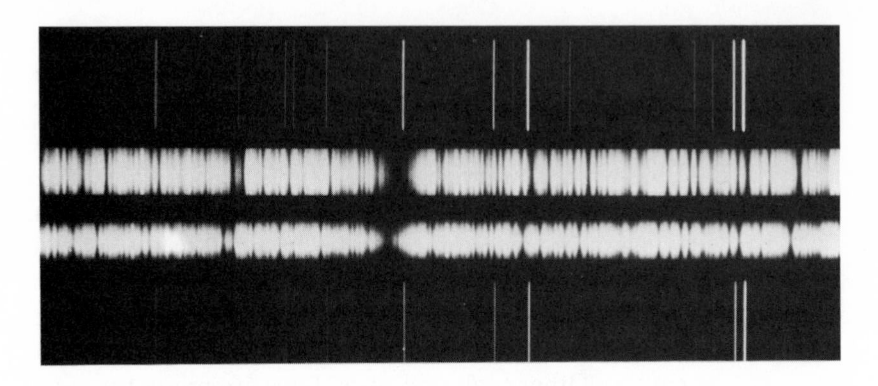

Figure 10.7. Spectra of Arcturus made six months apart. The radial velocity shifts are due to the earth's orbital velocity. (Hale Observatories photographs.)

Radial velocities for thousands of stars are now known. The directions of their spectrum shifts indicate approximately equal numbers approaching the sun and going away from it. For most of the nearby stars the radial velocities are less than 50 km/sec in either the positive or negative direction. (The radial velocity is positive when a star is moving away from us and negative when a star is moving toward us.) A few radial velocities greater than 100 km/sec have been measured. Extreme radial velocities for stars near the sun are about 250 km/sec. Finally, for faint galaxies and for quasars extremely large spectrum shifts toward the red have been observed. Such large shifts mean radial velocities which are a large fraction of the velocity of light.

The key to labeling Figure 10.1 is as follows: (1) M3III, (2) B3, (3) K0, (4) M8III, (5) G8III, (6) A0.

QUESTIONS

1. With reference to the basic formula which defines the magnitude system, explain why it is impossible to observe a single star and determine its magnitude.
2. Explain how a star's color index can be computed from observations made in two distinct systems.
3. How is the color index related to the actual color of a star?
4. Define absolute magnitude and state in your own words the significance of absolute magnitudes.
5. State the formula by which the absolute magnitude can be computed if the apparent magnitude and distance in parsecs are known.

6. List in order from blue stars to red stars the sequence of classes of stellar spectra. Which class represents the hottest stars and which class represents the cooler ones?
7. Explain why the lines of hydrogen are strongest in the stars of spectral type A and weaker in the spectra of other classes.
8. Explain the effects of excitation and ionization on the appearance of stellar spectra.
9. How are the radial velocities of stars measured from their spectra?
10. What is the range in radial velocity observed for typical stars?

Color Plate 1. The emission spectra of some common chemical elements and the solar spectrum showing some of the principal absorption lines. (Neon spectrum courtesy of Bausch & Lomb; mercury, sodium, helium, and solar spectra courtesy of Sargent-Welch Scientific Co.)

Color Plate 2. The Crab Nebula, M1, in Taurus. Compare this photograph with Figure 13.15. The network of red filaments indicates the presence of hydrogen emitting a bright-line spectrum. The yellowish-white mass embedded within the filaments shows a continuous spectrum with the characteristics of synchrotron radiation. The energy being radiated by this cloud is believed to come from the rapidly rotating pulsar which is the remnant of the star which exploded as the supernova of 1054 A.D. (Photograph courtesy of the Hale Observatories.)

Color Plate 3. The Pleiades, a galactic or open cluster in Taurus. The nebulosity surrounding the bright stars is an indication of the presence of dust particles throughout the cluster. These particles scatter light and make the dust cloud luminous. Since blue light is scattered more than red light, this reflection nebula has a definite blue color. (Photograph courtesy of the Hale Observatories.)

Color Plate 4. The Orion Nebula. This is another example of an emission nebula. The gases in this vast cloud are excited by radiation from the four hot stars of the "trapezium," easily seen near the center in this photograph. By comparing this photograph with Figure 15.1, one may note that the appearance of this spectacular object depends upon the manner is which it is photographed. Long exposures reveal the full extent of the cloud, but short exposures show the trapezium and the details near the center of the cloud. The Orion Nebula may be seen even in a small telescope by looking toward the star θ Orionis. The three bright stars in the upper right are not the stars of Orion's belt. (Lick Observatory photograph.)

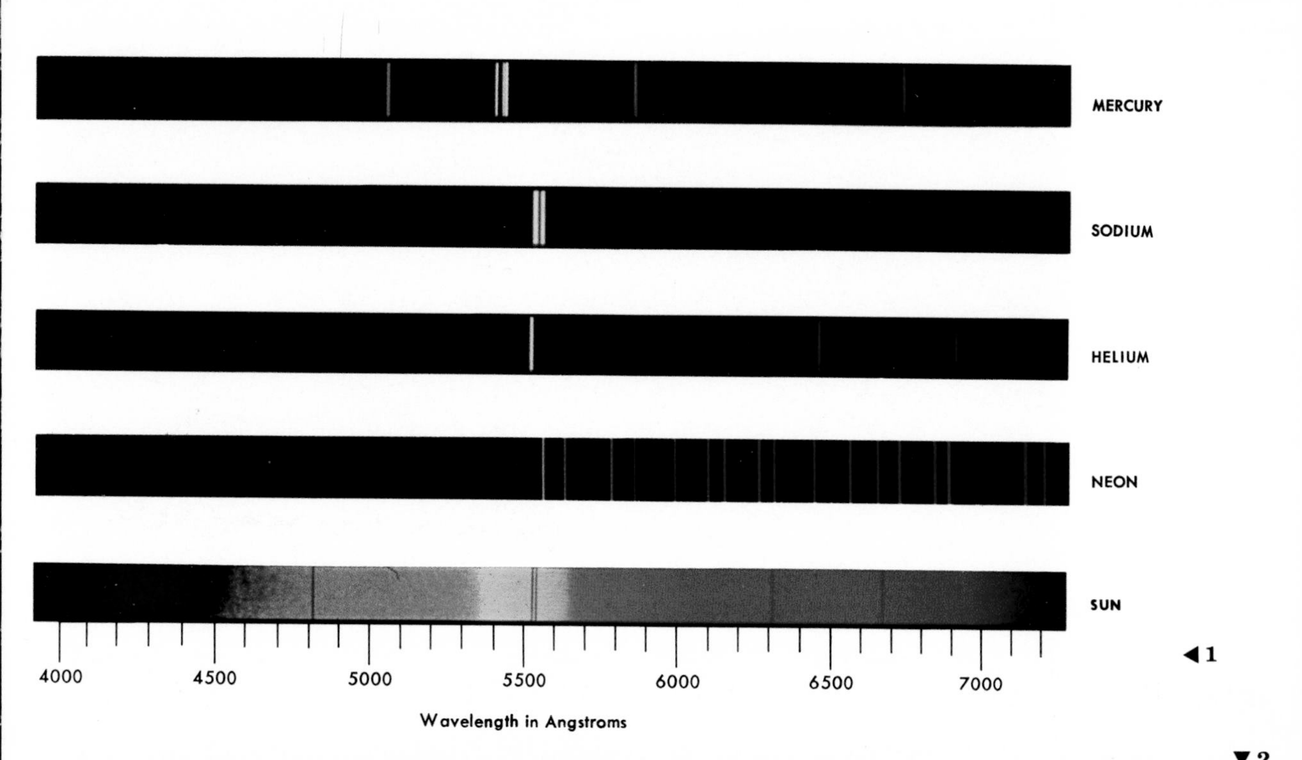
MERCURY
SODIUM
HELIUM
NEON
SUN
4000
4500
5000
5500
6000
6500
7000
Wavelength in Angstroms

◀ 1

▼ 2

3▲

4▶

Opposite page: ▲5

6▶

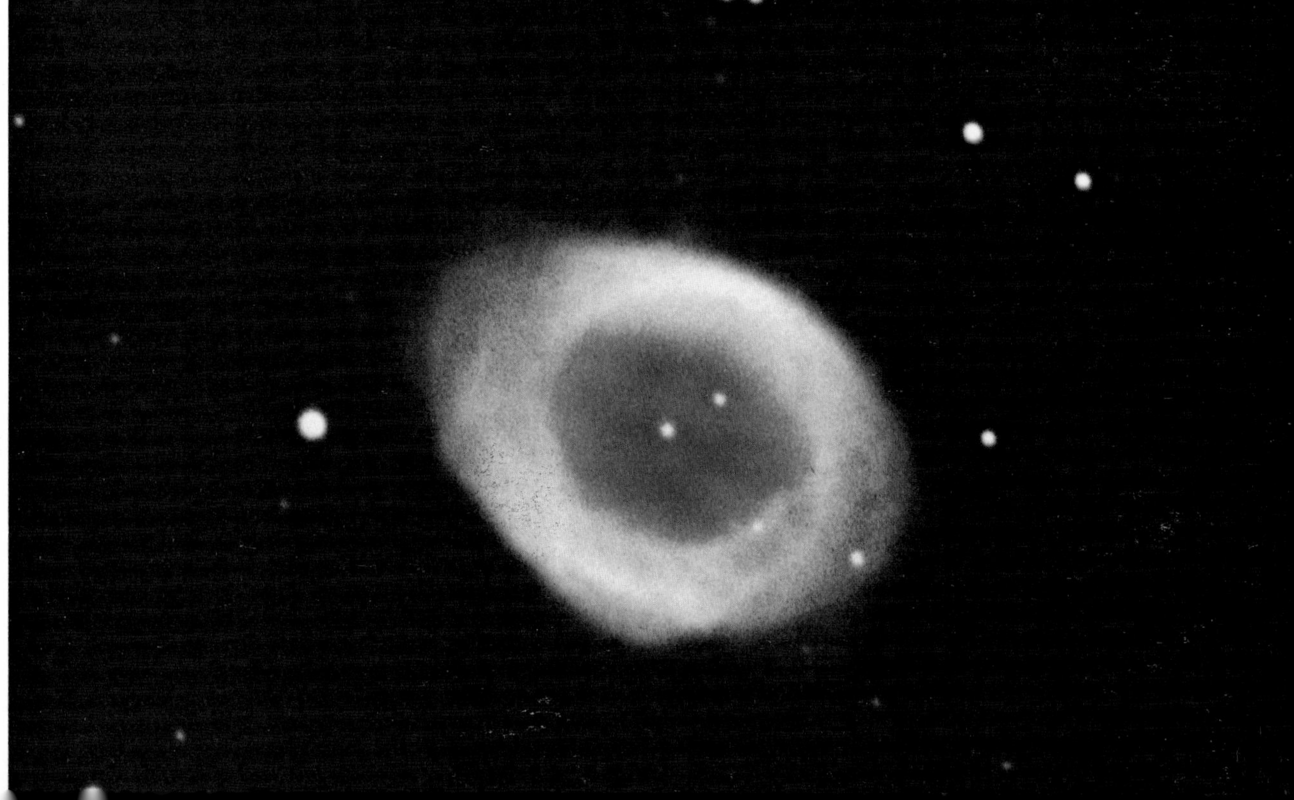

7▲

8▶

Color Plate 5. The Rosette Nebula in the constellation Monoceros. This is an example of an emission nebula, and this glowing gas shows a bright-line spectrum. The gas is excited by the group of hot stars in the center of the cloud. The low-density region around these stars may indicate that the radiation from the central stars has caused the gas to move outward. The small dark condensations in the upper right are believed to be areas in which star formation may soon begin. The angular diameter of the luminous gas is 80 arcminutes. At a distance of about 760 pc, this indicates a linear diameter of 17 pc (55.4 light-years). (Photograph courtesy of the Hale Observatories.)

Color Plate 6. The Ring Nebula in Lyra. This expanding shell of gas is believed to have been ejected by one of the stars seen within the ring. The differences in color are indicative of the differing degrees of excitation of the gases at differing distances from the exciting star. (Photograph courtesy of the Hale Observatories.)

Color Plate 7. A planetary nebula in Aquarius, NGC 7293. This is one of the largest known planetaries both in terms of angular and linear diameters. At a distance of 140 pc (450 light-years) it is also the closest planetary nebula to us. As in the case of the Ring Nebula, the color indicates the degree of excitation of the gas. (Photograph courtesy of the Hale Observatories.)

Color Plate 8. The Andromeda Galaxy, M31. This large spiral galaxy is barely visible to the unaided eye as a hazy patch of light. The outer spiral portions show up only on long-exposure photographs. The blue color of the spiral arms indicates that in spiral arms the brightest stars are young blue giants. (Photograph courtesy of the Hale Observatories.)

Figure 11.1. Constellations from an old star atlas.

11

A First Analysis of Stellar Data

Many of the fundamental astronomical data have been acquired by the basic observational methods described in the previous two chapters. The detailed methods and techniques at the telescope are in a continual state of change to keep pace with new advances in physics, electronics, and engineering. Broad new fields such as infrared and x-ray astronomy have been made possible by technological progress. Whatever the techniques, however, the accumulation of observational data is a fundamental part of the astronomer's task. The procedures by which the astronomer subsequently handles his data are similar to those used by all scientists. Observations are followed by classification and analysis. Analysis usually results in a theory or hypothesis which often goes beyond the existing data and suggests new experiments or observations. Let us now examine some of the analyses which became possible when the body of basic astronomical data had become large enough.

11.1 The Basic Nature of the Sun

Newton felt reasonably certain that the sun was a star even though in his time no parallax measurements had yet been made. Many years before, Greek astronomers had suspected the same thing. Only with the determination of the distance to one star was the nature of the sun as a star (or the stars as suns) clearly demonstrated. With the recognition of the vast distances to even the nearer stars, it was obvious that the sun, if removed to a comparable distance, would have a nominal brightness. This fact is so well known today that we mention it only because it is an example of a case in which a theory had to wait for critical observation before it could be confirmed.

11.2 Solar Motion from Proper Motions of Stars

Extensive programs in positional astronomy have resulted in accurate proper motions for thousands of stars. The examination and analysis of this mass of data reveal some interesting and significant facts concerning the motion of the sun through space.

Figure 11.2 indicates a segment of sky between two hour circles. In this segment several stars are shown with their proper motions indicated by arrows of appropriate length and direction. Each of these proper motions may be resolved into a component in right ascension (east or west) and a component in declination (north or south). Let us consider now only the proper motion components east or west. In the zone sketched in Figure 11.2, proper motions of five stars are indicated. Three of these have components of motion toward the east and two have components toward the west. If we

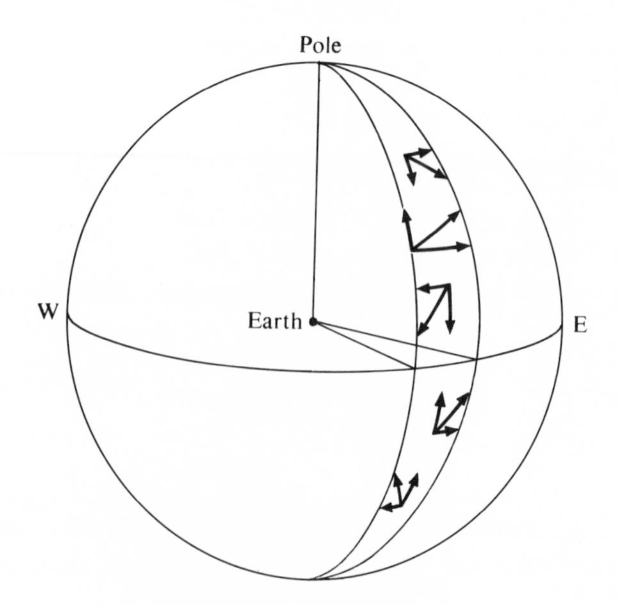

Figure 11.2. Every proper motion has a component in right ascension (east or west) and one in declination (north or south).

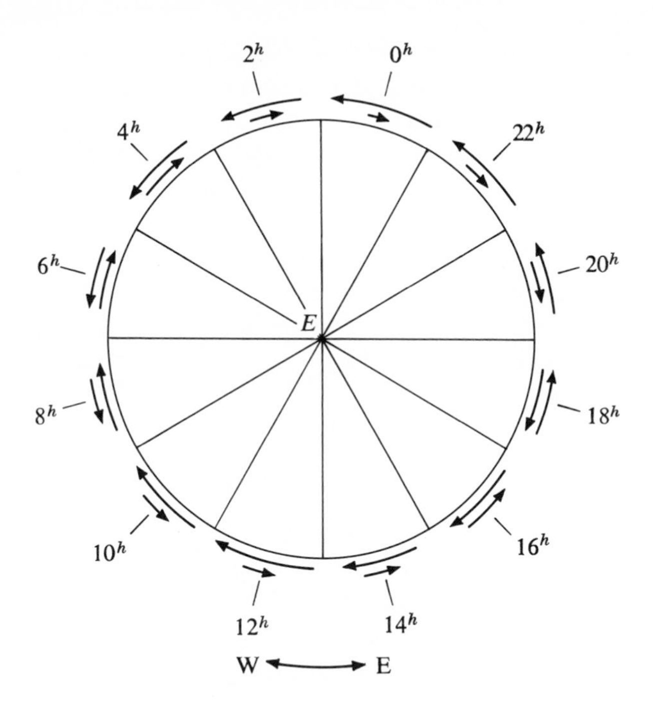

Figure 11.3. The numbers of eastward and westward proper motions in zones 30° wide around the celestial sphere. The earth is at the center.

divide the entire sky into similar zones 2 hours (30°) wide, we shall have 12 such zones. From the assembled proper motion data for a large number of stars we may then count the number of stars in each zone having proper motion components toward the east and the number having components toward the west. The numbers for the various zones are indicated in Figure 11.3 by the lengths of the arrows placed outside the circle. In zones centered on 18 hours right ascension and 6 hours right ascension the arrows are equal in length, indicating the same number of eastward as westward components. In zones at 12 and 24 hours, the numbers in the two directions are quite unequal. Some systematic effect is clearly present here to cause preferred motions in certain parts of the sky. If the stars moved about the sky in a completely random fashion, and if we consider motions of a large number of stars, we should count equal numbers of stars moving eastward and westward in each zone. From this, two possibilities are evident. Either there is a general drift of stars in the direction of the 6-hour zone or the sun is moving through the stars in the direction of a point somewhere in the zone at 18 hours. From other data we conclude that the effect is due to the sun's motion, and we may locate a point, the **apex** of the sun's motion, by making a similar analysis using only the proper motion components in declination. The actual apex is in the constellation Hercules at R.A. 18^h and Decl. +30°, not far from the bright star Vega. The point on the celestial sphere directly opposite the apex is referred to as the **antapex.**

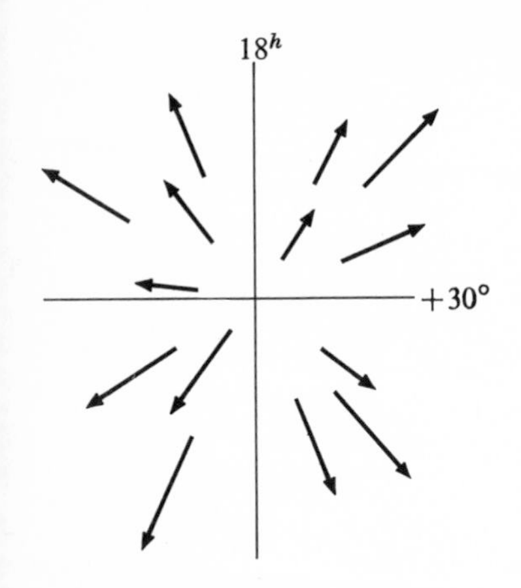

Figure 11.4. In the vicinity of the apex of the sun's motion, the proper motions are directed radially away from the apex.

Figure 11.4 shows the typical pattern of proper motions in the 18-hour zone in the vicinity of the apex. From the proper motions the stars appear to be moving radially outward from the apex. If we keep in mind that proper motions are actually angular motions, it becomes clear that the apparent radial pattern of Figure 11.4 is to be expected as the sun approaches the nearer stars in this region of the sky (Figure 11.5).*

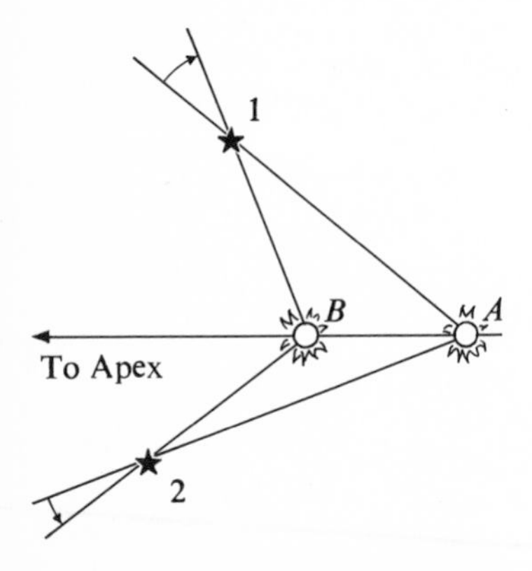

Figure 11.5. As a result of the sun's motion from *A* to *B*, the stars at 1 and 2 appear to move away from the direction of the apex.

11.3 Solar Motion from Radial Velocities

An analysis of a large number of radial velocities in many areas of the sky also leads to an indication of the direction of the sun's motion. The analysis is similar to that used for the proper motions. In the case of the radial velocities, however, the sun's actual velocity in space may be determined directly.

Again, zones 2 hours wide in right ascension may be considered. From catalogues listing thousands of radial velocities, the number of stars with positive radial velocities (i.e., motions away from us) and the number with negative radial velocities may be counted. In Figure 11.6 the lengths of the arrows represent the numbers of stars in each zone moving toward us and away from us. In the 18^h zone more stars are moving toward us than away from us. In the 6^h zone the reverse is true. In the zones at 12^h and 24^h the numbers moving toward us and away are just about equal. Again, we conclude that the sun is moving through a field of randomly moving stars. In the direction of the apex the average radial velocity is approximately −20 km/sec. In the direction of the antapex in Columba, the average radial velocity is approximately +20 km/sec. In the two zones at right angles to the direction of the sun's motion the average observed radial velocity is very close to zero. It is concluded from this, then, that the velocity of the sun must be approximately 20 km/sec toward the apex. It is not surprising to find that the coordinates of the apex determined from radial velocities agree very well with the coordinates of the apex determined from proper motions.

*It is interesting to note that in 1783 William Herschel obtained very reasonable coordinates for the solar apex using proper motions of only 13 stars.

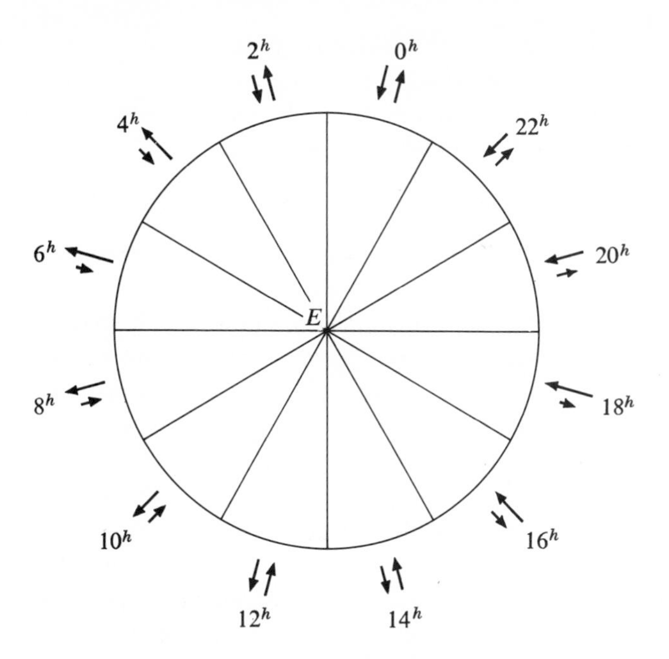

Figure 11.6. Analysis of radial velocities of stars in all parts of the sky leads to the determination of the direction of the apex as well as the velocity of the solar motion.

Proper-motion studies are limited to stars relatively close to the sun due to the fact that for increased distances a given velocity results in a smaller proper motion. (This point was discussed in Chapter 9.) Radial velocities, on the other hand, are measureable to large distances with remarkable accuracy.

11.4 Space Motions

The proper motion actually observed is an angular velocity observed as a change in the coordinates of the star with time. Depending upon the distance to the star, this proper motion can represent any true linear velocity through space. In Figure 11.7, stars A and B have the same proper motion. Being farther away from the earth, star A must move a greater distance than star B in any given time. The velocity of a star at right angles to the line of sight is referred to as its **tangential velocity,** and it may be computed if the distance (i.e., parallax) and proper motion are known. The expression for this is

$$v_T = 4.74 \frac{\mu}{p}$$

v_T is the tangential velocity in kilometers per second, μ is the proper motion in arcseconds per year, and p is the parallax in arcse-

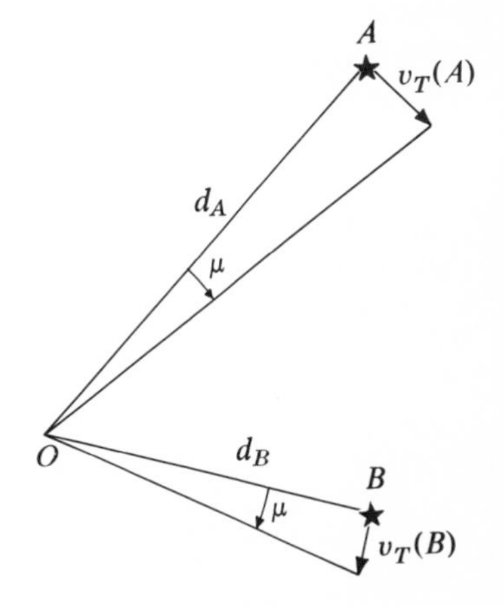

Figure 11.7. Relation between proper motion μ and tangential velocity v_T.

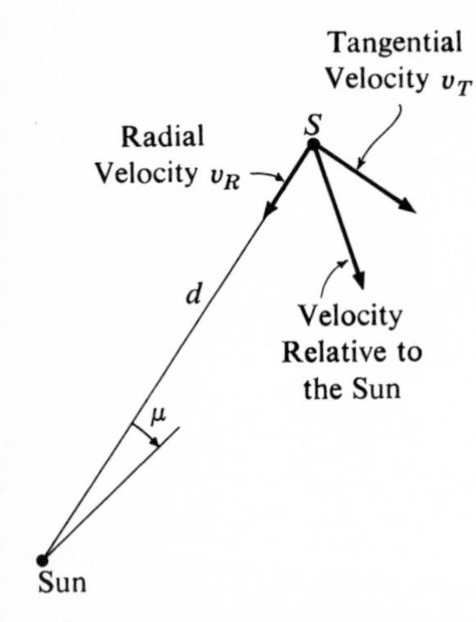

Figure 11.8. The velocity of a star relative to the sun may be calculated after the radial velocity has been measured and the tangential velocity has been computed.

conds. The constant 4.74 results from the manner in which the various units are combined. The computed tangential velocity may be combined with the observed radial velocity to give the **space velocity** of the star.

$$\text{space velocity} = \sqrt{v_T^2 + v_R^2}$$

where v_T is the tangential velocity and v_R is the radial velocity (Figure 11.8).

It has been shown that the sun's motion through space results in a systematic pattern of proper motions away from the solar apex and toward the antapex. Individual stars, however, show proper motions which may depart considerably from this pattern, indicating that the stars generally move in a more or less random fashion. Once the solar motion has been detected and understood, it becomes possible to remove the effects of solar motion from the observed space velocities of the stars and study their **peculiar velocities.** Thus, the radial velocity of a star near the apex may be corrected by adding 20 km/sec to the star's observed radial velocity. At right angles to the direction to the apex there is, of course, no correction and in the direction of the antapex, the 20 km/sec is subtracted from the observed radial velocity of the star. In specific terms the peculiar radial velocity v_R is given by the following formula:

$$v_R = v_{R(\text{obs})} + v_S \cos \lambda$$

where λ is the angle between the direction to the star and the direction to the apex, v_S is the sun's velocity toward the apex, and $v_{R(\text{obs})}$ is the star's observed radial velocity.

Correction of a star's observed proper motion for the effects of solar motion is a much more complex problem, and the student should satisfy himself that this correction requires a prior knowledge of the distance to the star. When the peculiar velocities of a large number of stars in the sun's vicinity have been determined by correcting for the sun's motion, it is seen that the stars move in what are essentially random directions within this relatively small volume of space.

11.5 The H-R Diagram

In the preceding chapter the magnitude and the spectral classification systems of stars were described. It was shown that there was a range of absolute magnitudes calculated for stars with accurate trigonometric parallaxes. Normal stars are rarely found brighter than $M = -5$ or fainter than $M = +18$. Spectral classifications from type O to type M were shown to indicate a range of temperatures from roughly 50,000°K to 3000°K. As data of these two types began to accumulate, it was inevitable that someone should question what combinations of absolute magnitude and spectral class could exist and which occurred most frequently. In about 1913 two men, Hertzsprung in Holland and Russell in the United States, simultaneously discovered the relation shown in Figure 11.9. When absolute magnitude was plotted against spectral class for stars with known parallaxes, it was seen that only certain combinations were possible. This plot came to be known as the **Hertzsprung-Russell diagram** (or more simply the H-R diagram), and astronomers at

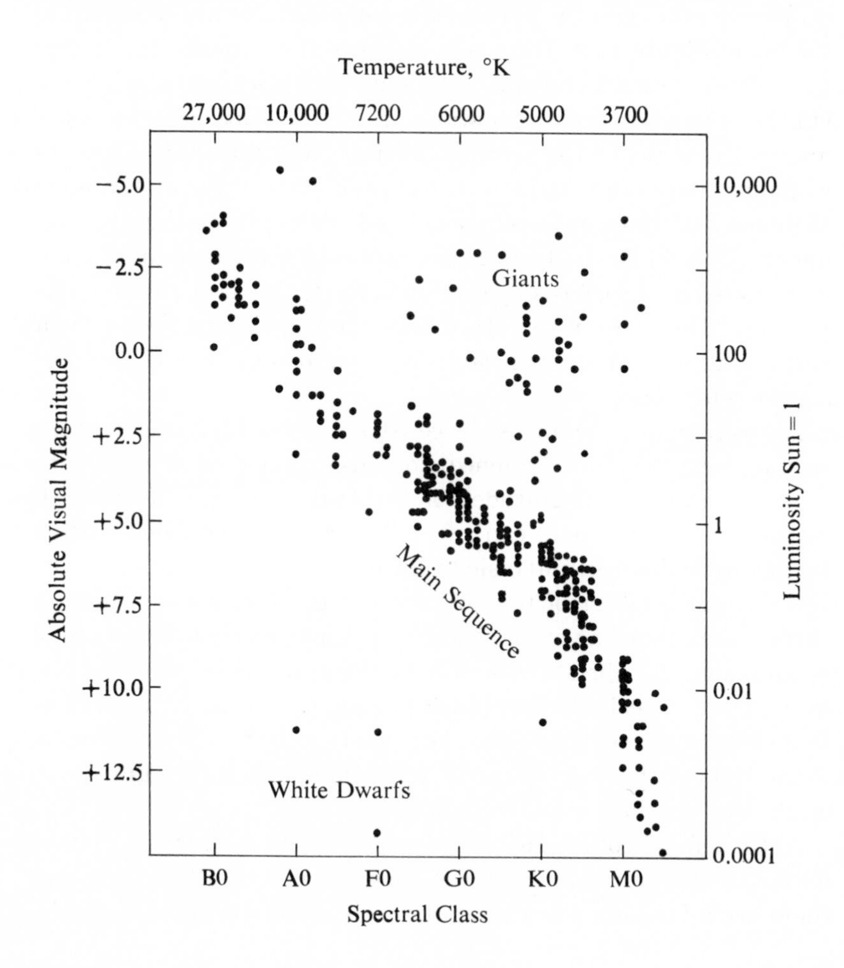

Figure 11.9. The Hertzsprung-Russell diagram, which shows important relationships between absolute magnitude and spectral class.

once realized that the diagram must be highly significant in the evolutionary development, or life cycle, of stars. In spite of the fact that radical changes in the interpretation of the H-R diagram have been made in the light of ever more complete information, it remains one of the most crucial and significant discoveries in astronomy. The conventional names for certain areas are indicated in Figure 11.9. The great majority of all stars, including the sun, lie on a diagonal belt which has been appropriately termed the **main sequence.** Stars in other regions have been called giants, supergiants, and white dwarfs.

11.6 Stellar Diameters from the H-R Diagram

The sequence of spectral classes from O to M is basically a temperature sequence, and the appropriate temperatures are indicated at the top of Figure 11.9. The main sequence shows quite clearly that the hottest stars are the ones which are intrinsically the brightest. On the diagram there are, however, a large number of stars above and to the right of the main sequence. At spectral type K0, for example, there are two distinct groups of stars: those on the main sequence and those in a poorly defined region above the main sequence. The Stefan-Boltzmann law states that the amount of energy radiated by a surface depends upon the fourth power of the temperature. Therefore, if two stars have the same temperature, their surfaces must radiate the same number of ergs per second per square centimeter.

Let us compare a main-sequence K0 star with a K0 star from the region above the main sequence. For this example we shall assume that the absolute magnitudes of the two stars are +6 and +1, respectively. How, then, can two K0 stars with the same surface temperature differ in intrinsic brightness by five magnitudes? Obviously, the brighter of the two stars must have a much greater surface area than the other star. Since a magnitude difference of 5 is equivalent to a brightness ratio of 100:1, it follows that the surface area of the brighter star must be roughly 100 times as large as that of the main-sequence star. For this reason the stars above and to the right of the main sequence have come to be called **giants** or, on the basis of their color, **red giants.**

At the left side of the H-R diagram the opposite situation may be noted. In the late B and early A-type spectra it will be seen that there are a few rare stars which lie well below the main sequence.

These stars must have surface temperatures higher than 10,000°K, but they are intrinsically fainter by 10 magnitudes or more than the main sequence stars of the same spectral class. By the reasoning advanced for the red giants, these stars must be relatively small and have come to be called **white dwarfs.**

Across the top of the H-R diagram are found a few more stars of extremely high luminosity, which have been given the appropriate name of **supergiants.**

The red supergiants and the white dwarfs represent the extremes of a broad range in actual diameters of stars. We may calibrate this diameter scale since we know with great accuracy the temperature, luminosity, and diameter of one star—the sun. In absolute units, such as ergs, the Stefan-Boltzmann law ($E=\sigma T^4$) relates the radiant energy per unit surface area to the temperature. Hence the total energy in all wavelengths radiated over the entire surface of a star may be computed from

$$L=4\pi R^2\sigma T^4$$

where R is the star's radius and σ is a constant. All of the quantities in this expression are well known for the sun. By comparing any star with the sun we may then say

$$L/L_S = R^2T^4/R_S^2T_S^4$$

When the absolute magnitude of a star is known, the ratio of the star's luminosity to the sun's is also known. A scale of such luminosity ratios has been drawn on the right side of Figure 11.9. The ratio of the temperature of a star to that of the sun may be computed from the knowledge of spectral classes. Finally, the desired ratio of the radius of a star to that of the sun may be calculated from these simple data. The method should not be expected to give exact results since the stars are not perfect radiators and can be affected by limb darkening. In addition, the absolute magnitude must be corrected for the fact that the magnitude systems measure energy over only a limited range of wavelengths. Nevertheless, a realistic idea of the range in stellar diameters may be obtained in this manner. Diameters along the main sequence range from 7 or 8 solar diameters for B-type stars to 1/10 solar diameter for late M-type stars. Red giants may be 25 times as large as the sun, but M-type supergiants such as Betelgeuse may be almost 400 times as large as the sun. At the extreme small end of the scale are the white dwarfs with diameters as small as the diameter of the earth.

11.7 Luminosity Criteria for Spectra

With the stars sorted out as they are on the H-R diagram, subtle differences of certain lines in their spectra were found to have particular significance. It was noticed, for example, that in the spectra of main-sequence stars the absorption lines are broader and more diffuse than lines in giants of the same spectral class. This difference may be clearly seen in the three spectra compared in Figure 11.10 and has been interpreted as a result of a great difference between the pressure of the gas in the atmosphere of giants and main-sequence stars. Because of the large diameters of the giants, the force of gravity at the photosphere will be relatively low and the gas pressure will be low. The lower pressure means, in turn, that within a given volume in the star's atmosphere there are relatively fewer atomic collisions to smear out the spectrum lines. The resulting lines in the spectrum of a giant are therefore narrow.

Giants may also be distinguished from main-sequence stars by the effect of pressure revealed in the relative strengths of the spectrum lines of ionized atoms and neutral atoms. In a gas of some high temperature there will be a mixture of atoms, ions, and electrons. If the pressure is high, it becomes easier for the ions to capture electrons and at a given time there will be fewer ions. If the pressure is low, on the other hand, there will be more ionized atoms. The result is that, at a given temperature, in the atmosphere of a star the strength of the lines of certain ionized atoms depends upon the gas pressure. Since the lines of certain neutral

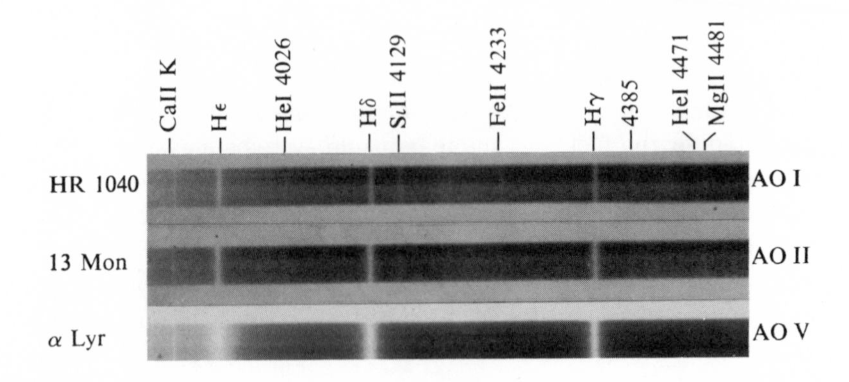

Figure 11.10. Differences in the appearance of lines in the spectra of giants (I and II) and main-sequence stars (V). In addition to narrower hydrogen lines, the giants also show stronger lines of certain ionized atoms. (Yerkes Observatory photographs.)

atoms are much less affected by pressure, a comparison between the strengths of the lines of ionized and neutral atoms can be used as an indication of pressure. These indications of pressure can then be correlated to absolute magnitude and luminosity. There are today a number of well-established criteria of this sort by which a series of luminosity classes have been established. These classes are designated by Roman numerals I to V as follows:

Ia – supergiants

Ib – less luminous supergiants

II – luminous giants

III – normal giants

IV – subgiants

V – main sequence stars

11.8 Spectroscopic Parallax

A thorough analysis of spectra by the methods outlined depends upon a luminosity calibration in terms of the stars for which trigonometric parallaxes have already been determined. Only for such stars are absolute magnitudes known, so that these were the only ones that could be used to construct the H-R diagram. When the H-R diagram was well understood, however, it became a valuable tool for determining the distances to stars beyond the reach of the parallax method. If the spectrum of a distant star can be photographed, the major features and the fine details tell us the luminosity class as well as the spectral type. The location of the star on the H-R diagram is then known, and the absolute magnitude may be read from the scale. The absolute magnitude determined in this indirect manner may then be used with the observed apparent magnitude in the expression

$$M = m + 5 - 5 \log r$$

to determine the distance in parsecs, r. Distances determined in this manner may be converted to parallaxes; hence the use of term **spectroscopic parallaxes.** In the case of bright stars, distances as large as 4000 pc may be determined.

In Chapter 15 we shall discuss the evidence for the presence of interstellar dust, which dims the light of distant stars and makes them seem farther away than they really are. In the practical use of the methods outlined here, it is necessary to take into account the effects of this interstellar material.

11.9 Rotation

Sometimes the detailed study of stellar spectra shows lines broadened in a manner different from the broadening due to high pressure in a star's atmosphere. The photographic density changes slowly when measured across the line, and the density profile would be described as "shallow." This type of broadening is attributed to the rotation of the star since rotation would cause the light from various parts of a star's surface to be Doppler shifted by varying amounts. The effect would be most noticeable for those cases in which the star's axis of rotation is at right angles to the line of sight. On the other hand, if the star's axis of rotation is directed towards us, there will be no line broadening of this sort. For axial orientations between these two extremes, we should expect an intermediate broadening.

This type of rotational broadening is quite commonly found in stellar spectra, and in the most extreme cases equatorial surface velocities as high as 250 km/sec are indicated. It is quite interesting also to note a relationship between spectral type and rotation. The O, B, and A stars are the most rapid rotaters, and the F stars rotate more slowly. No rotational line broadening is evident in the spectra of stars later than spectral type G.

The results obtained from rotation studies give another good example of the manner in which analysis and theory suggest more detailed observations. The new observations lead, in turn, to a new level of analysis. In this case a particular type of line broadening led to the discovery of rotation. When a large number of stars were studied, the relation between spectral type and rotation became apparent. This relationship must then become part of the complex theoretical studies of the formation and evolution of stars.

QUESTIONS

1. What sort of pattern will be seen in the proper motions of stars in the general direction of the apex of the sun's motion? Explain the reason for this effect.
2. How can the velocity of the sun's motion be determined from the analysis of a large number of stellar radial velocities?

3. What is the difference between proper motion and tangential velocity? Which of these quantities is computed and which is directly observed?
4. Define space velocity and peculiar velocity.
5. Sketch the H-R diagram and label the scales properly. Indicate the position of the sun on the diagram. Label the main sequence and the regions occupied by the giants and supergiants.
6. Explain the reasoning by which we have come to recognize that some stars with the same surface temperature as the sun must be many times larger than the sun.
7. Why are the lines in the spectra of giants narrower than the lines in the spectra of main sequence stars of the same class?
8. What is meant by the term *spectroscopic parallax?*

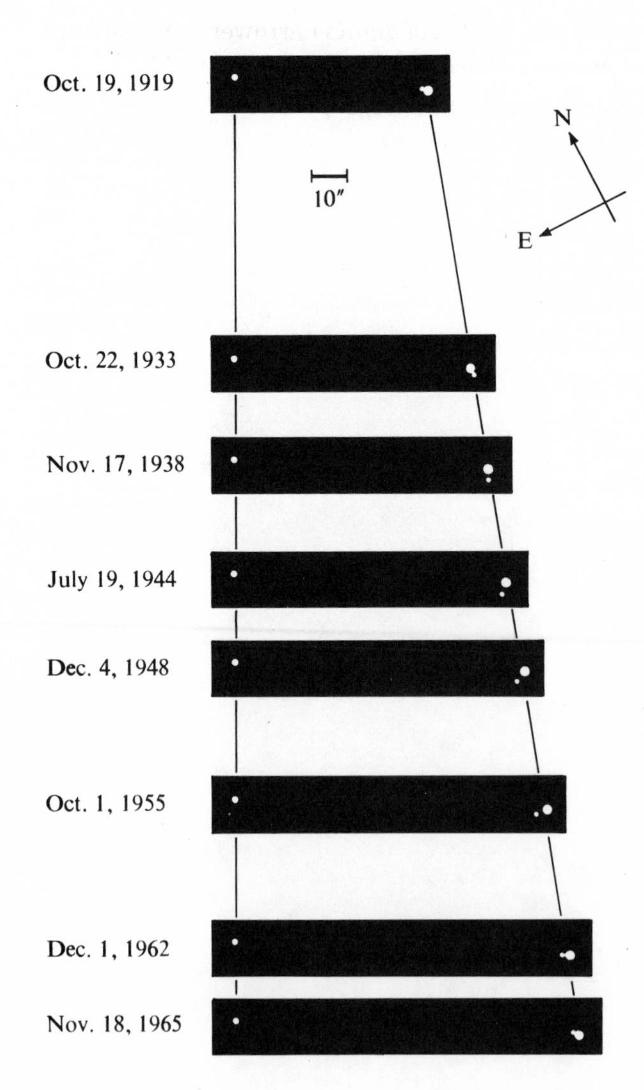

Figure 12.1. The pair of stars at the right comprise the visual binary Kruger 60. This diagram is adapted from a series of photographs which show both the orbital motion of the binary and the proper motion with respect to the star at the left. The original data are from the Sproul Observatory and the Leander McCormick Observatory.

12

Binary Stars

12.1 Visual Binaries

Along with many of his predecessors, William Herschel was anxious to observe stellar parallax as a proof of the earth's orbital motion. Herschel reasoned that the parallactic motion of a nearby star might show up best if the star happened to lie very close to the line of sight to a very distant star. The distant star should remain quite stationary and show no annual motion. Herschel searched the sky for pairs of such stars and compiled a list on which to concentrate his search for parallax. Instead of finding a parallactic motion, Herschel found, much to his surprise, that in a number of cases one star showed progressive motion around the other. He realized that he was seeing real binary star systems and not the chance combination of a near and far star that he had expected to find. These two-star systems are the **visual binary** stars, hundreds of which are now known and well studied. (See, for example, Figure 12.1.)

Observations of binary stars are usually carried out with a special measuring eyepiece mounted on the telescope. This eyepiece is rotated until a crosswire lies along the line joining the centers of the two stars. The direction of north in the field of view is known, and so the **position angle** between north and the line between the centers of the two stars is measured. Movable wires at right angles

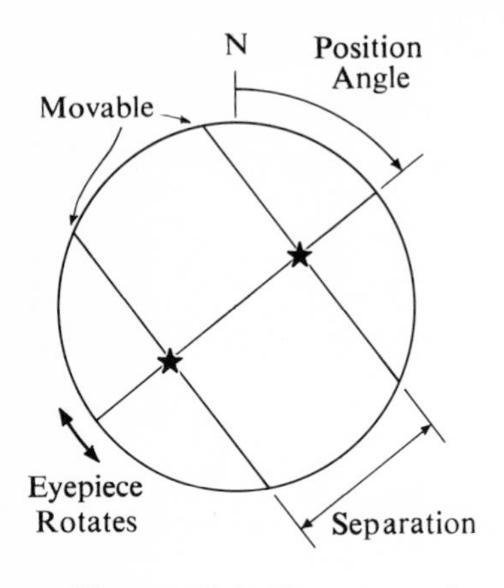

Figure 12.2. By means of a special telescope eyepiece the position angle and separation may be measured for the components of a visual binary system.

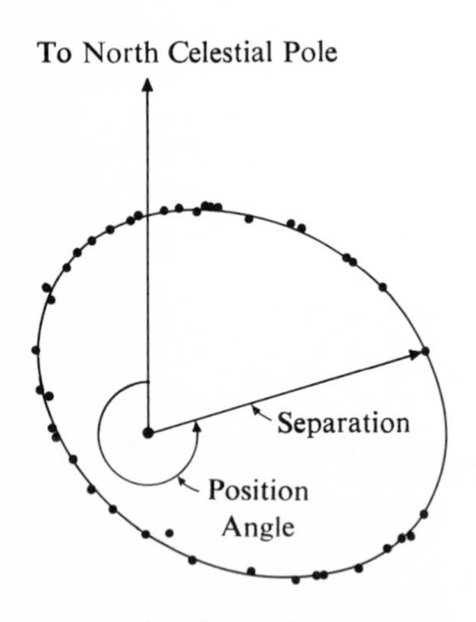

Figure 12.3. The apparent orbit of a visual binary may be plotted from a large number of observations of position angle and separation.

to the first wire are then placed through the star images in the telescope's focal plane. The measured distance between the star centers is then converted to an angular **separation** in the sky. Figure 12.2 shows the nature of these measures. When successive observations of position angle and separation over a long period of time are plotted, the orbital motion of one star around the other is clearly noted (Figure 12.3). In the case of very long periods, the efforts of a succession of observers are required to obtain enough points to derive the orbit.

The two stars in a binary system must obey Kepler's laws; therefore, the orbit should be elliptical. The star which is the brighter is referred to as the **primary;** the other is the **companion.** Only if the orbital plane is at right angles to the line of sight will the true shape of the orbit be seen. In most cases the orbit will have some specific **inclination** to the surface of the celestial sphere and we will see the projected or **apparent orbit.** This orbit will always have the shape of an ellipse, since an ellipse always projects as an ellipse. By careful examination of the apparent ellipse, moreover, some important information about the true orbit may be derived. The details of the complete analysis are somewhat complex, but one or two related points are quite interesting. The major axis of the apparent orbit does not coincide with the major axis of the true orbit. As indicated in Figure 12.4, however, the center of the apparent orbit does coincide with the center of the true orbit. Moreover, we know that the major axis of the true orbit must pass through the primary star at F. Thus, a line through the center and the primary star represents the projection of the true major axis into the plane of the sky. The true point of **periastron** (comparable to the perigee) is at point P in the figure. The orbital eccentricity is the ratio of the distance FC to the distance PC, and the ratio of these two line segments is the same in the projection as in the true orbit. The eccentricity is thus determined from the projected orbit. Some of the orbital elements, periods for example, are not affected by the inclination of the orbit.

Periods of revolution for known visual binary systems range from about two years up to several hundred years. There are simple reasons for these observational limitations. If the period were much less than two years, except in very nearby systems the two stars would appear so close together that the angular separation would be too small for us to resolve the two components with our telescopes. In the other cases in which the period is on the order of

100 years, the stars are widely separated and the orbital motion is very slow. From year to year, little or no change in the relative positions of the stars is observed. Some widely spaced pairs are suspected of being binaries simply because of their common proper motions.

So far we have discussed the orbit of one star around the other, considering the primary to be at one focus of the elliptical orbit of the secondary. In reality, both stars are moving around the center of gravity of the system. Each member is in its own elliptical orbit, and the center of gravity is at a focus of each of the orbits. The two stars must always be on opposite sides of this point. It is easily seen that the periods of the two stars must be equal and that the eccentricities of the two orbits must be the same. Furthermore, when the period of the secondary around the primary is considered, it is the same as the periods around the center of gravity. In addi-

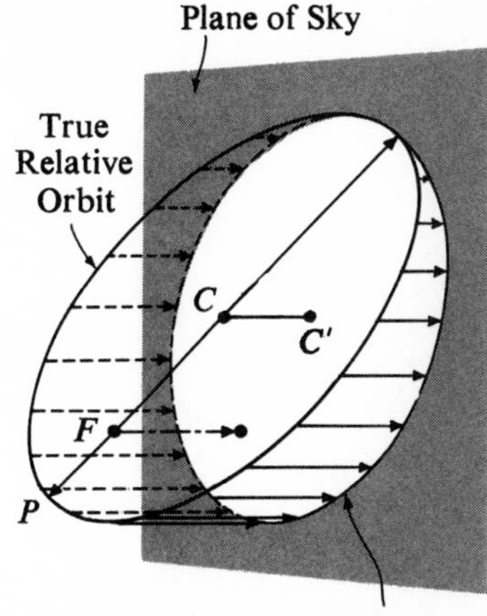

Figure 12.4. The relationships between the apparent orbit and the true orbit of a binary system. As seen from our position in space, C and C' coincide.

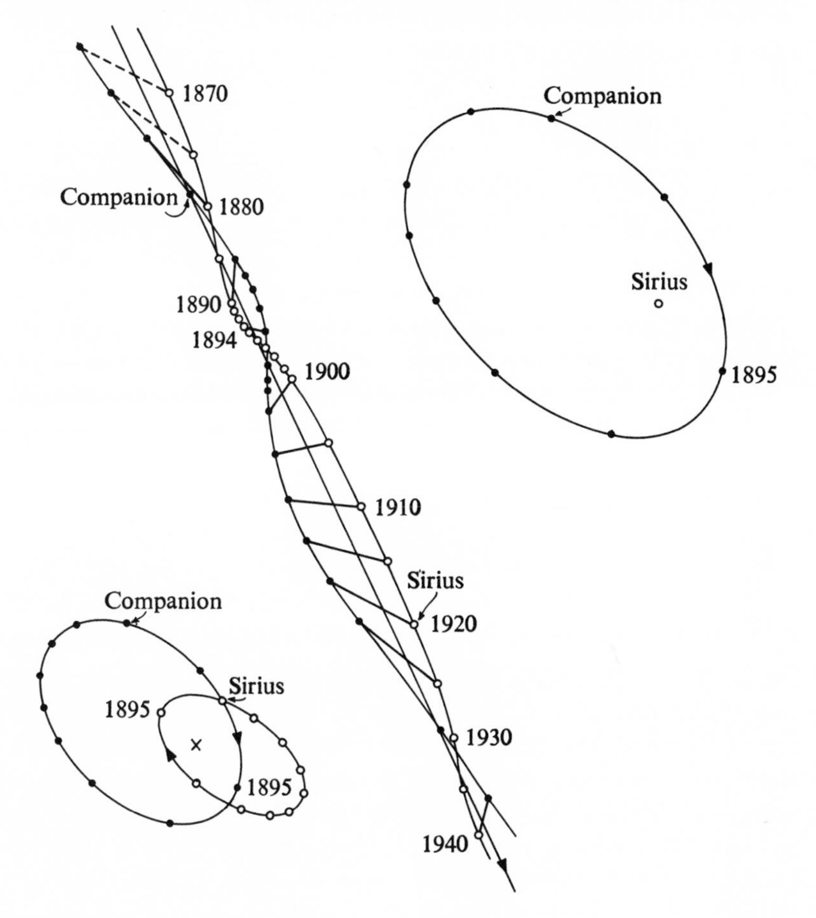

Figure 12.5. The motions of Sirius and its white dwarf companion. In the lower diagram the orbits of both stars around the center of gravity are shown. Above this, the motions of the two stars across the sky are shown. (From C. Payne-Gaposchkin and K. Haramundis, *Introduction to Astronomy*, 2nd ed., p. 336. © Copyright 1970 by Prentice-Hall, Inc.)

tion, the eccentricity of the relative orbit of the secondary with respect to the primary is the same as that of the orbits of the two individuals. The relative orbit is much easier to determine than the actual orbits of the two stars and, as we have said, gives the true period and shape of the orbit. The details of the motion with respect to the center of gravity can be determined only when the two stars can be referred to a fixed background of stars (Figure 12.5).

In a number of cases the secondary or companion star is actually invisible. The presence of the unseen companion is inferred from a periodic variation in the proper motion of the primary. This was the case for Sirius, which was noted to move along a wavy path rather than a straight one. It was recognized that the center of gravity of a binary system was moving along a straight path and that Sirius was moving back and forth around this center of gravity. Eventually the companion star was discovered when a large enough telescope became usable.

12.2 Stellar Masses

The two quantities observed for visual binaries are position angle and angular separation. When the parallax has been measured and the distance is known, the angular separation may be converted to a linear separation in astronomical units. Recalling the method by which the mass of a planet was determined in Chapter 5, we saw how Kepler's third law was used to find the sum of the masses of a planet and satellite when the period of the satellite and the distance between the two are known. Exactly the same analysis may be applied to binary star systems. Expressing the period P in years and the separation D in astronomical units, we write

$$P^2(m_1 + m_2) = D^3$$

Since P and D are known, we solve for $(m_1 + m_2)$, the sum of the masses of the two stars. Here the mass of the sun is taken as the unit of mass. A sum of masses determined in this way gives no information about the masses of the individual stars. For instance, if the sum of the masses happened to be seven times the mass of the sun, the individual masses could be 6 and 1, 5 and 2, 4 and 3, and so forth. More data are needed in order to find the individual masses.

In cases in which the motions of both stars are detected and the center of gravity has been located, the ratio of the two masses may be found. If r_1 and r_2 are respectively the distances of the two stars from the center of gravity, then the masses m_1 and m_2 are related by

$$m_1 r_1 = m_2 r_2$$

or

$$m_1/m_2 = r_2/r_1$$

By combining this mass ratio with the sum of the masses, the mass of each of the stars in the system may be determined. It should be noted that there really are not many binary star systems which lend themselves to this sort of analysis, since the system must be close enough for a reliable trigonometric parallax. There is also a great deal of effort required to locate the center of gravity and find the mass ratio from an extensive series of photographs.

Because of the great labor involved, reliable masses have been calculated in this manner for only 41 visual binary stars. Along the main sequence, masses range from 2.14 solar masses for Sirius, spectral type A1, to 0.08 solar mass for Ross 614B, a dwarf M star of spectral type M3. The largest reliable mass determined in this manner is that of Capella, a G-type giant, which is about 2.4 solar masses.

Stars with well-known masses are necessarily close enough to have accurate absolute magnitudes as well. Even from the rather meager data discussed here, it is apparent that there is a fairly direct relationship between mass and absolute magnitude or luminosity for main-sequence stars. It is not really surprising to learn that the intrinsically brighter stars are the more massive ones. The mass-luminosity relation based on accurate data from visual binaries is sketched in Figure 12.6. This relationship may be used to estimate the masses of single stars whose absolute magnitudes have been determined previously. When extrapolated to the very brightest stars, masses 40 or 50 times the sun's mass are indicated by this diagram.

Later in this chapter it will be seen that a few more points on the mass-luminosity relation may be obtained from a combination of photometric and spectroscopic data for certain unresolved binary stars of relatively short periods.

Stellar masses may be combined with the determination of the diameters of the stars as described in Chapter 11 to give the aver-

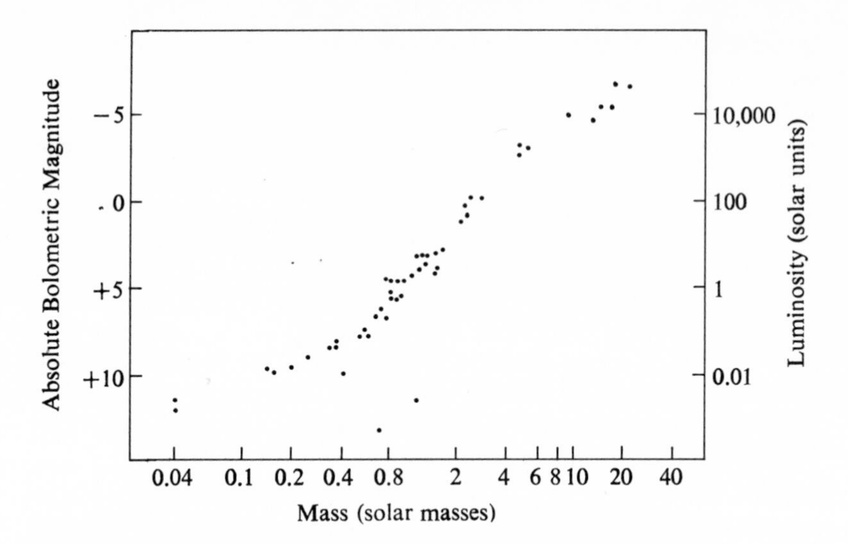

Figure 12.6. The mass-luminosity relations showing the relationship between absolute magnitude and mass. The masses less than 4M. were obtained from visual binaries. Masses greater than 4M. were obtained from spectroscopic-eclipsing binaries. (See Section 12.6.)

age densities of the stars. The results of such density measures are rather surprising. For the red giant K and M stars the mean density is so low that we would consider it the measure of a pretty good laboratory vacuum! For the white dwarfs a mass comparable to that of the sun is squeezed into a sphere only a few hundredths as large in diameter as the sun. Densities as high as one million times the density of water must exist in these stars. A liter of such material would have to weigh 500 kg if placed on the surface of the earth. Truly this is no ordinary matter. The atoms are almost completely ionized, and the electrons are moving very rapidly. Under these conditions the electrons are said to constitute a **degenerate gas.**

12.3 Spectroscopic Binaries

It was mentioned that when the revolution period of a binary system is less than about 2 years, the stars are so close together that we cannot resolve the individual components at their great distances from us. For such systems we see only a single dimensionless point of light. Nevertheless, the binary nature of many close pairs may be ascertained by spectroscopic or photometric means or by both.

The inclination of the orbit was defined as the angle between the plane of the orbit and the "plane of the sky." In reality this is a very small area of the celestial sphere in the vicinity of the binary.

In the cases in which the orbit is seen in its true shape, the inclination is zero. When the orbit is seen edge on, the angle may be as large as 90°. Clearly, when the inclination is anything other than zero, each star in a binary system must have a component of motion toward us during half of the period and away from us during the other half. If both the inclination and the orbital velocity are large enough, then a variable radial velocity should be detectable. A great many stars are known to have variable radial velocities resulting from this effect, and are called **spectroscopic binaries.**

The most commonly observed spectroscopic binaries are those in which the lines in the spectrum show a simple periodic motion first toward the red and then toward the blue. This is likely to be the case when one star is somewhat brighter than the other. The spectrum lines are visible only for the brighter star. When the two components are nearly equal in brightness, as in Figure 12.7, the lines will be periodically single, then double. The lines will appear single when the two stars are moving across the line of sight. A bit later, one star will be moving away from us and the other will be moving toward us. We then see the lines of one star shifted toward the blue and the lines of the other shifted toward the red.

Such double-line binaries are of great interest to astronomers because in a straightforward way they tell us something about the masses of the two stars. If the two stars are equal in mass, then both stars will be the same distance from the center of gravity. Both stars will be moving at the same velocity in their orbits, and so the shift in position of the spectrum lines will be the same for each of the two stars. When the mass of one star is smaller than the mass of the other, the smaller star will move in a larger orbit. Its velocity will therefore be larger also, and the shift in its spec-

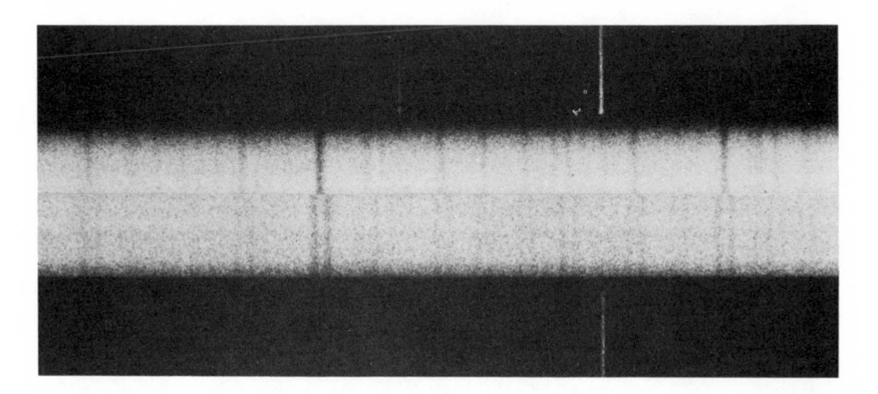

Figure 12.7. Spectra of the spectroscopic binary Mizar. This star is of spectral type A2, and the period is 20.5 days. The lower photograph was made almost one-quarter of a period after the first one. The separation of the two components in the lower spectrum represents a velocity difference of 140 km/sec. (Hale Observatories photographs.)

trum lines will be greater. For double-line binaries, the mass ratio is simply the inverse of the ratio of the maximum orbital velocities. This is true regardless of the inclination of the orbit.

The period is determined simply by measuring the time interval from zero radial velocity to maximum negative to maximum positive radial velocity and back to zero. The periods actually observed are usually fairly short, that is, about 10 days or less. If the period is very long, the variations in radial velocity are difficult to detect since the orbital velocities are slow. For this reason, more spectroscopic binaries are known with short periods than with long ones.

It is customary to show the radial velocity variations graphically by plotting a radial velocity curve such as those in Figure 12.8. The first curve would result from a star moving in a circular orbit. For many spectroscopic binaries the orbit is elliptical rather than circular. In the center and right-hand parts of Figure 12.8 it is interesting to notice how the orientation of this ellipse relative to the line of sight affects the shape of the radial velocity curve. The orbits are sketched below the velocity curves, and in each case the observer is assumed to be looking from the bottom of the page into the plane of the orbit. In each case, points marked *a* and *c* are the points at which the radial velocity is zero, and points *b* and *d* are the points of maximum positive and negative radial velocity,

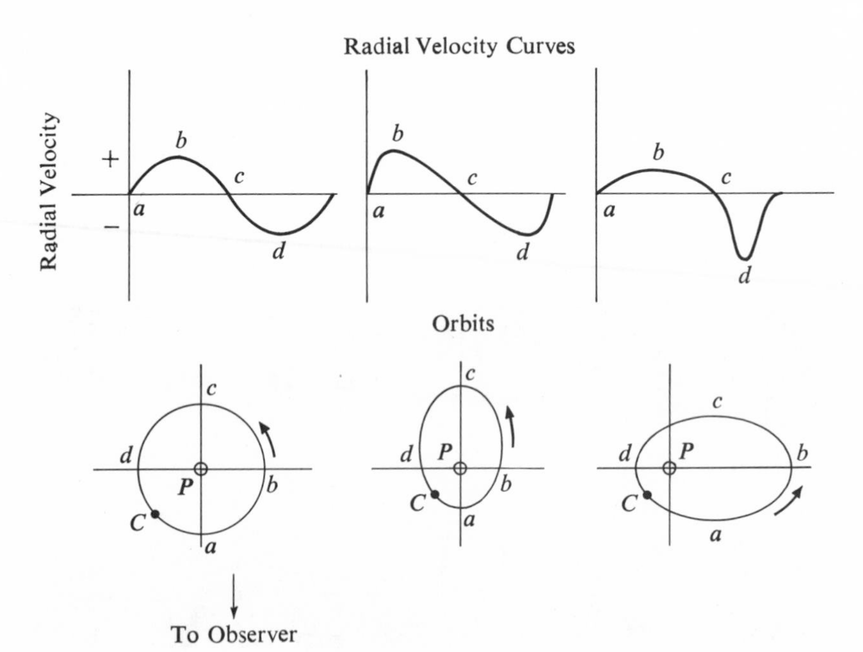

Figure 12.8. The shape of the radial velocity curve of a spectroscopic binary depends upon the shape of the orbit and its orientation with respect to the line of sight.

respectively. Remembering that in the relative orbits sketched here the companion is moving most rapidly when closest to the primary (periastron), one may note the times required to go from a to b and from b to c, and so forth. In the center diagram the companion moves from a to b in less time than from b to c. The maximum positive and negative radial velocities at b and d are equal. In the third diagram, however, the times from a to b and b to c are equal but are long compared with times from c to d and d to a. Furthermore, the maximum positive radial velocity occurs at apastron, when the companion is moving slowly. The maximum negative radial velocity occurs at periastron, when the companion is moving very rapidly. It would be rare for an astronomer to find a case precisely like one of those pictured here, but the curves which are observed do lend themselves to this type of analysis.

The spectra of some binaries are extremely complex and show a number of varying features which are very difficult to explain. Such a star is β Lyrae, whose series of spectra may be seen in Figure 12.9. In this series a narrow dark line originating in an interstellar gas cloud has been used as a reference to align the successive spectra. This line is stationary, and the shifts due to orbital

Figure 12.9. A series of spectra of the spectroscopic binary β Lyrae showing changes throughout the complete cycle. (*Sky and Telescope* photographs, Vol. XVI [July, 1957], p. 418.)

motion are clearly seen for some of the stronger lines. The behavior of some weaker lines and the presence of emission lines, however, indicate some unusual effects. The β Lyrae system consists basically of a B9 star and a star thought to be of spectral type F. The period of revolution is 13 days, and the two stars are close enough to each other to be significantly distorted from their normal spherical shapes. Their close proximity also means there can be an actual streaming of gas away from each of the stars. There may be some exchange of gas between stars, but there also seems to be an expanding ring of gas around the whole system because most of the streaming gas spirals outward. All of these details have been inferred from the manner in which certain of the absorption and emission lines vary in displacement and in strength. According to modern interpretations, certain highly displaced lines come from the gas streams seen in front of the stars, while the emission lines come from the expanding spiral of gas. Such a loss of mass is bound to have serious effects on the components and on the system as a whole. In the case of the β Lyrae system, the mass loss results in an observable increase in the period. Given enough time the physical characteristics of the components must also be changed.

12.4 Eclipsing Binaries

It is the inclination of the binary star orbit that makes the system detectable as a spectroscopic binary. When the inclination is 90° (or very nearly 90°) another effect can lead to discovery of binary systems. With a 90° inclination, our line of sight to the star lies in the plane of the orbit. There must, therefore, be times when the companion star passes either in front of or behind the primary. At these moments the total light of the system becomes less. For such systems we detect a periodic and systematic variation in the light intensity.

Figure 12.10a indicates a series of phases in the revolution of a hypothetical eclipsing binary system. We shall assume here, as is often the case, that the primary is a large cool star and that the companion is actually smaller and brighter. Figure 12.10b is a representation of the **light curve,** which shows the variations with time in the observed light of the system. Points on the light curve numbered 1 through 9 correspond to the phases sketched in Figure 12.10a and described here.

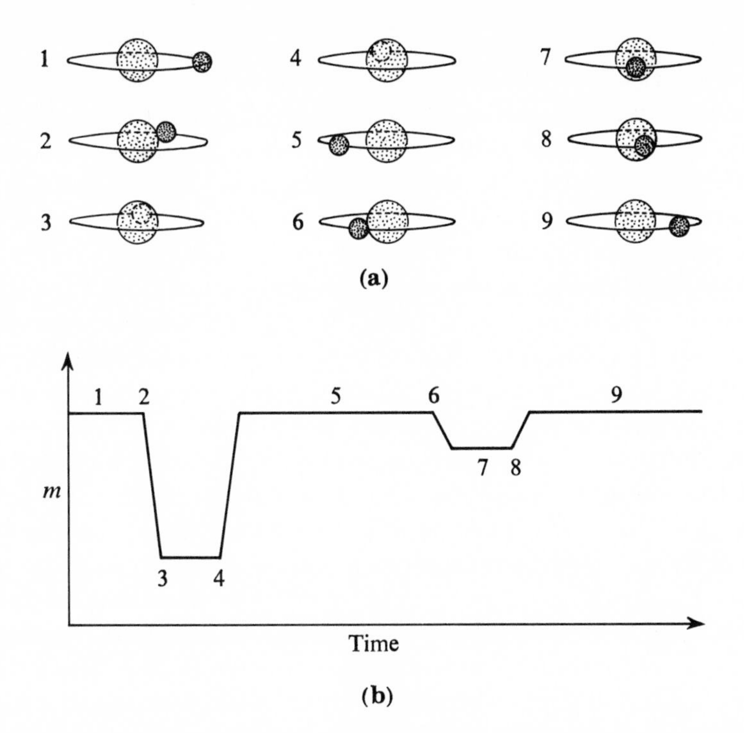

Figure 12.10. The geometry of an eclipsing binary system and the resulting light curve.

1. Both stars are exposed to view. We measure the total light of the system.
2. The leading limb of the bright star is about to go behind the dimmer one.
3. The bright star is completely eclipsed. We measure only the light of the dimmer member.
4. The leading limb of the bright star begins to emerge.
5. The full light of both stars is again seen.
6. The leading limb of the bright star begins to cover the faint one.
7. The bright star is completely in front of the dim one but does not completely cover its disc.
8. The leading limb of the bright star is at the limb of the dim one.
9. Both stars are again completely visible. The cycle is ready to repeat itself.

Light curves similar to this idealized one are quite common and are interpreted as described above. It is to be emphasized, however, that we actually view a single dimensionless point of light. Variations in the intensity of this light give rise to the light curve, and the model of an eclipsing binary system is constructed from the pattern of the light curve. There is little doubt that this interpretation is a correct one, because it is well supported by spectroscopic observations. Any eclipsing binary should also be observable as a spectroscopic binary unless the period is rather long and orbital velocity is low. Maximum radial velocities should be observed at times about midway between eclipses, that is, near time 1 and time 5 on the light curve. This is, in fact, the case for those binaries that can be observed both photometrically and spectroscopically.

Whereas there are many light curves which resemble the idealized curve quite closely, there are many others which are somewhat different. The factors causing such differences are related either to the shape and orientation of the orbit or to special characteristics of the individual stars. One should not expect the orbits of all eclipsing systems to be inclined exactly 90°. The fact that we see as many eclipsing systems as we do must mean that the total number of binary stars is very large, and most of these numerous systems will have their orbital planes oriented in such a way that no eclipses will occur. There must also be many systems for which the inclination is large, but not quite large enough to cause total eclipses. The naked-eye star Algol has a light curve which indicates only partial eclipses. In Figure 12.11a the primary never completely disappears behind the companion, and the whole disc of

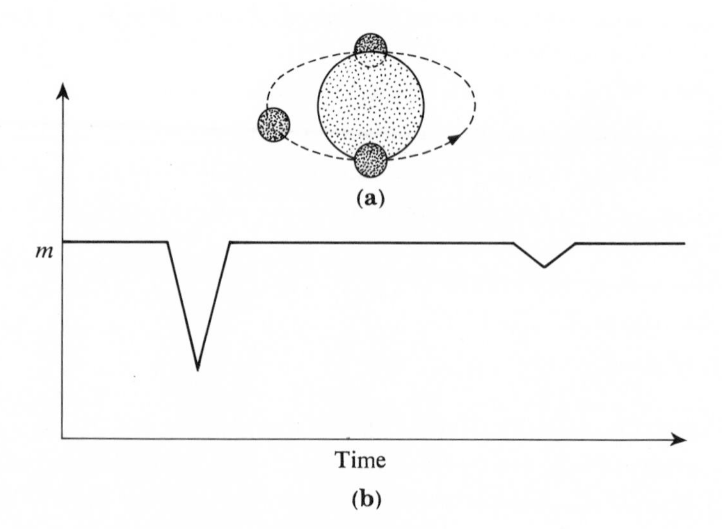

Figure 12.11. Light curve of an Algol-type eclipsing binary, showing the effect of eclipses which are only partial because of the inclination of the orbit.

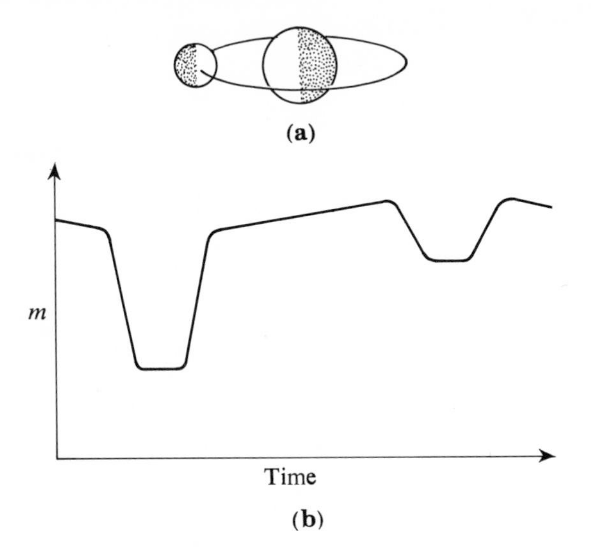

Figure 12.12. The reflection effect in close binary sytems causes departures from the idealized light curve of Figure 12.10.

the primary is never projected onto the secondary. Thus the eclipses are never complete, and the minima on the light curve (Figure 12.11b) always are pointed rather than flat.

Eccentricity is another orbital characteristic which can sometimes be recognized in the light curve. If the orbit is circular, the interval from primary minimum to secondary minimum ought to be the same as the interval from secondary minimum back to primary minimum. If the orbit is elliptical and the major axis is at 90° to the line of sight, then these two intervals will be unequal. There can, of course, be orbits in which the eccentricity may exist but does not show up in the light curve at all.

When the period of an eclipsing binary is less than about 10 days, the two stars are apt to be fairly close together. In some extreme cases the primary and companion may be nearly in contact. The effects of such close proximity of the two stars appear in the light curve in two ways. First, if the stars are large and close together, each one should block a considerable amount of the light from the other. Much of this intercepted light is actually reflected back toward its source with the result that both stars are hotter and brighter on one hemisphere than on the other (Figure 12.12a). As the two stars revolve, the bright face of each star will periodically be toward us and away from us. The total light from each component is variable and the light curve takes on the appearance of Figure 12.12b with rounded "corners."

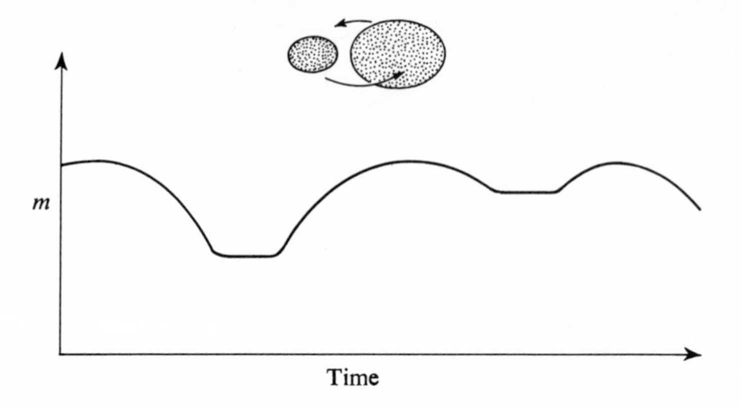

Figure 12.13. Tidal distortion in close binary systems means that greater surface area is toward the observer between eclipses, and the light is therefore brightest midway between eclipses.

The second proximity effect arises from the fact that two large gaseous bodies close to each other will cause mutual tidal distortions. Thus instead of two revolving spherical bodies there will be two revolving ellipsoidal bodies. Throughout the period each of these will present a larger surface area between eclipses than during eclipses. As indicated in Figure 12.13, the ellipticity effect results in a nonuniform brightness during the portions of the period between eclipses. These two proximity effects usually go together and present a challenging problem to the double-star astronomers who must try to analyze the light curve.

As mentioned in Chapter 8 (page 174), we observe the effect of limb darkening in the sun. The apparent disc of the sun is less bright near the edges than at the center. We should expect the same effect in other stars as well, and we have evidence for this in the light curves of some eclipsing binaries. When the period is more than about 10 days, the two stars will be far enough apart that the proximity effects mentioned previously will be minimal. In the presence of limb darkening a nonuniform disc is gradually obscured as one star passes behind the other. This means that the decline in light intensity begins slowly rather than abruptly, and changes gradually as minimum light is approached. On the light curve, then, we find rounded corners as the eclipse begins and as the light level reaches minimum.

12.5 Stellar Diameters from Eclipsing Binaries

It should be clear that an eclipsing binary should also be observable as a spectroscopic binary unless it is one of the rare cases with a very long period. Furthermore, if the eclipses are total, the inclination of the orbit must be 90° or very nearly 90°. The maximum observed radial velocity must therefore be very nearly the true orbital speed at the time of the observation. When the orbit is circular, this velocity is constant; when the orbit is elliptical, the velocity varies. In either case the velocity at other points in the orbit is predictable, and the velocity at the time of an eclipse may be computed from the combined spectroscopic and light-curve data.

Referring again to the light curve in Figure 12.10, we see that the interval from 2 to 3 represents the time required for the smaller star to pass completely behind the larger one. This time interval multiplied by the orbital speed at the time of the eclipse gives the diameter of the smaller star. Similarly, the interval from 2 to 4 represents the time required for the leading limb of the smaller star to travel a distance equal to the diameter of the larger star. This time interval multiplied by the same orbital speed gives the diameter of the larger star. Thus, in a relatively simple manner the diameters of both members of a binary system may be determined. Although the method outlined here is straightforward, the observations and analysis are actually difficult and require considerable care.

12.6 Masses from Spectroscopic-Eclipsing Binaries

We know that from spectroscopic observations of double-line binaries the radial velocities of both stars may be found. Furthermore, the ratio of the masses of the two stars is the inverse of the ratio of the two radial velocities. If the sum of the masses could also be found, we would be able to supplement the meager data on stellar masses. Obtaining the sum of the masses, however, requires knowledge of the radius of the orbit. If the orbital velocity and the period are known, the orbital circumference and radius may be calculated. Radial velocities can be observed, but because of the inclination of the orbit to the line of sight, the observed radial velocity may not be the true orbital speed. There are, fortunately, a few cases in which double-line spectroscopic binaries are also eclipsing binaries. In these cases, we must be looking into (or almost into) the edge of the orbit. The observed radial velocity is now very nearly the true orbital speed, and the sum of the masses may be found.

Reliable masses have been determined in this manner for only

14 stars. Some of these stars, however, have very large masses, and have made it possible for the mass-luminosity relation to be extended to the system γ Cygni, in which the two components each have masses about 17 times that of the sun. Less reliable, because it is not an eclipsing binary, are the masses of the components of a double-lined spectroscopic binary known as "Plaskett's Star." The components of this system are both O type stars and probably have masses about 70 times that of the sun. These are among the most massive stars known.

12.7 Statistics and Formation of Binary Stars

With a little thought and study of Table 12.1, the student should realize that the presence of any eclipsing binaries is enough to imply the actual existence of large numbers of binary stars. We detect as eclipsing binaries only those cases with orbital inclinations of nearly 90°. Since there are so many other possible orientations for orbits, there must be a great many undetectable cases. Considering all types of binary stars, the statistics suggest that multiplicity is very common, and this is borne out by the stars close to the sun for which the sample may be considered to be reasonably complete. Among the 30 stars (or stellar systems) *nearest* the sun there are 13 multiple systems. The multiple systems are made up of a total of 29 individual stars. Thus, more than half of the stars in this sample are members of multiple systems. For the 30 *brightest* stars the statistics offer similar support for the prevalence of binaries. Fifteen members of this group provide a total of 41 stars. There is no doubt that multiplicity is very common among stars, perhaps being the rule rather than the exception.

In 1965 a long study of Barnard's Star was completed by P. van de Kamp of Swarthmore College. This study revealed very small periodic irregularities in the proper motion, which were interpreted as evidence of an invisible companion with a mass between that of a large planet and a small star. Van de Kamp's analysis gives a mass of 0.0015 solar mass for the companion—twice the mass of Jupiter. There are a number of other stars known to have variable proper motions, and in all probability they are also accompanied by small unseen companions.

Thus, reliable observations show that multiple star systems are quite common and that some stars are accompanied by companions

Table 12.1. Summary of binary star properties.

	Visual	Spectroscopic	Eclipsing
Periods	two years to 200 years	mostly less than 10 days	hours to hundreds of days
	If period is shorter than this, the stars are too close together to be resolved. If period is longer than this, orbital motion is too slow to be detected.	If period is longer than this, the orbital velocity is too slow to cause measurable Doppler shifts in spectrum lines.	Shorter periods are more frequent because they are more likely to be discovered. Inclination of orbit is very critical for longer periods.
Masses	0.8 to 2.4 $M_{\odot}$	For systems observable both as eclipsing and spectroscopic binaries masses have been determined from 0.64 to 17 $M_{\odot}$	

of relatively small mass. Are these small bodies best described as planets or stars? Certain factors place both upper and lower limits on the possible masses of stars, but the question remains difficult to answer. Clearly, the formation of multi-body systems is common in space, and perhaps many such systems do consist of a central star and a system of planets. Astronomers would like to understand the factors that favor the formation of a planetary system over a double- or triple-star system.

Today astronomers favor the idea that stars form in the large clouds of gas and dust which are known to exist in space. Conditions in space often cause the formation of clusters in which the space density of stars is relatively high. Within a cluster the random motions may sometimes result in a situation in which three moving stars pass near each other. At such a time the kinetic energy of the three bodies can be redistributed. One star will acquire a large amount of energy and will leave the system at a high velocity. Having lost some of their kinetic energy, the other two stars "capture" each other and form a true binary system. The mathematical theory underlying the capturing process favors the creation of well-separated pairs with relatively long periods.

Early in this century it was suggested that a rapidly rotating body could become unstable and split itself into two small bodies. Proposed as the origin of the earth-moon system, the process, however, failed to account properly for the angular momentum we now know exists in the earth-moon system. Again, the mathematical theory for the division of rotating bodies was well worked out and seemed to be quite feasible for binary stars of short periods. It is now thought that short-period binaries may have been formed in this manner, whereas the longer-period binaries resulted from captures.

A third possibility is that in the actual process of star formation more than one body is formed. Thus, there may be times when a pair of stars would form while in other situations one star and a series of planets might form.

QUESTIONS

1. Sketch the orbit of a visual binary and indicate the positions of the primary and secondary. Indicate the separation and position angle.
2. Describe fully the observations and calculations required for the determination of the masses of stars.
3. Why are there upper and lower limits to the observed periods of visual binaries?
4. What is the mass-luminosity relation?
5. Describe the ranges in mass and density for stars.
6. Why is it unlikely that spectroscopic binaries might have periods as long as those observed for visual binaries?
7. Why is it likely that most binaries with short periods are undetected as binaries?
8. How is the mass ratio found for spectroscopic binaries in which the lines of both components are visible?
9. Sketch the light curve of an eclipsing binary. Label the two scales and indicate the eclipses on the light curve.
10. How will the shape of the light curve of an eclipsing binary be affected if the stars are not spherical but are elongated; if the orbit is elliptical; if the angle between the line of sight and the orbit plane is relatively large?

11. Explain how the diameters of the components of an eclipsing system can be determined.
12. Outline the evidence and reasoning in support of the idea that binary stars represent a very large percentage of all the stars.

Figure 13.1 Two overlapping photographs of the variable star WW Cygni showing the change in brightness from maximum to minimum. (Hale Observatories photographs.)

13

Variable Stars

Observations continued over some time show regular variations in brightness for a great many stars. For a few of these stars the variations are interpreted in terms of eclipsing binary systems. The other stars actually do change their intrinsic brightness. The total output of energy of these **variable stars** changes over a considerable range in a regular manner, and the light curve may repeat itself over and over with the same form. The brightness of such a star is thus predictable for any time in the future once an accurate light curve has been established. From such factors as the length of the period and the shape of the light curve, the variable stars fall into several well-defined groups. Each of these groups represents an important problem to the astronomer interested in stellar evolution, for he must try to decide whether or not variability is a stage in the evolution of every star or is due to some peculiar properties found only in certain stars. (See Figure 13.1.)

With the ever-increasing refinement of photometric methods, large numbers of variable stars have been discovered in all constellations. Variability of the brighter stars was naturally detected fairly early, and these stars already had names or designations from one of the earlier star catalogues. The astronomer Bayer, for example, assigned Greek letters to the bright stars in each constellation in approximate order of their brightness. Thus α Lyrae is the brightest star in the constellation Lyra, the Harp; β Lyrae is the next brightest, and so forth. δ Cephei is an example of a star from Bayer's catalogue that was later found to be variable. In a similar manner, Flamsteed assigned numbers to stars in each constellation, moving across each constellation in the direction of increasing right ascension. As the fainter variables were recognized in increasing numbers, no one was able to predict the number of variables that would eventually be known in each constellation. A scheme for designating variable stars was needed, and at first the letters R through Z were used (U Geminorum, for example). When even more variables were found the letters RR through RZ were used, then SS through SZ, and so forth. This naming scheme was not suited to the task, but it was too widely used to be changed by the time its faults became known. Eventually this scheme reached its limits, and variables found since have been given designations which consist simply of the letter V followed by a number and the name of the constellation. V381 Cygni is an example. The student need not be perplexed by this scheme, nor should he attach any particular significance to the designations.

13.1 Light Curves and Color Curves

The basic observational datum for an individual variable star is its light curve. As mentioned in the earlier discussion of eclipsing binaries, the light curve represents a graph of many magnitude determinations plotted against time. Figure 13.2 shows a light curve for a rather typical variable star. For some classes of stars the light curve has exactly the same shape through cycle after cycle. For these the **period,** or interval between maxima, may be determined with great precision. Obviously, the longer the series of observations, the more accurately determined will be the period. If a period has been determined from only two successive maxima, then a rather small error may result in a large difference between predicted and observed times of maxima after 100 additional cycles. When an approximate period is known, the number of cycles between two widely spaced maxima may be computed. The total time interval divided by the number of cycles must give a much more accurate period. Stars showing regular and repeating variations have measured periods which may range from a few hours to 100 days. Stars showing less regular variations may have periods of several hundred days.

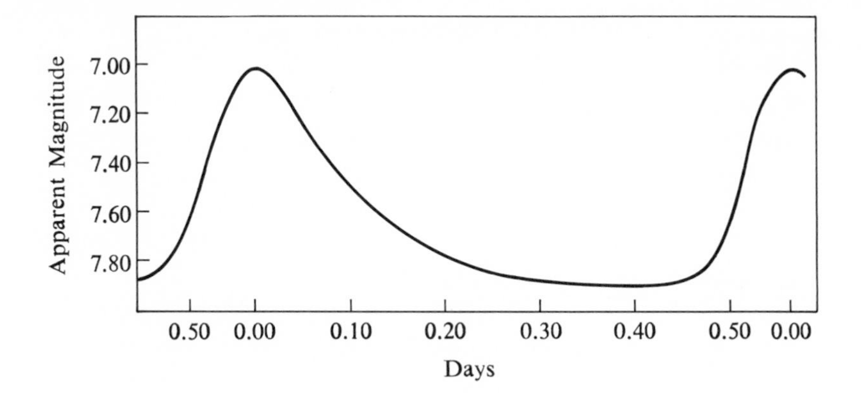

Figure 13.2. The light curve of the variable star RR Lyrae. (After A. J. Wesselink, *Bulletin of the Astronomical Institutes of the Netherlands.*)

The total range from maximum light to minimum light defines the **amplitude** of the light curve and is usually expressed in magnitudes. Thus for the star RR Lyrae in Figure 13.2, the amplitude is 0.85 magnitudes. When the period is regular, the amplitude is usually regular or constant. For some of the very short-period variables the amplitude may be only a few hundredths of a magnitude. For others the change can amount to several magnitudes.

Variable stars may be observed in any of the color systems described in Chapter 10. When observed in two colors, B and V for example, the variations of the color index with time may be noted. A plot of this is referred to as a **color curve** and can yield some interesting information very quickly. When the color index is small or negative, the star must be hot. Then as the color index becomes larger, the star must have a lower temperature.

13.2 RR Lyrae Variables

It has been customary for astronomers to name classes of variable stars for the prototype star of the class. Thus when a number of stars were found to have characteristics similar to those of the star RR Lyrae, the entire group of stars became known as RR Lyrae variables. Similarly, stars whose light curves resemble that of the star δ Cephei are known as Cepheid variables.

Referring again to Figure 13.2, note that the light curve rises rather steeply from minimum light to maximum light, falling off steeply at first, then more slowly as it approaches minimum again. For RR Lyrae the period from maximum to maximum is roughly 0.6 day, and the amplitude is 0.85 in magnitude. Other stars of this class have light curves very similar to this one, and their periods are all less than one day.

A most significant quantity is the absolute magnitude of the RR Lyrae stars. Unfortunately, none of these stars is close enough for the determination of a parallax, but statistical methods applied to the motions of some of the brighter ones have led to distances and absolute magnitudes for this group of variables. The surprising fact was that all of the RR Lyrae stars seemed to have nearly the same average absolute magnitude, namely $M = 0.5$. Thus, once a variable star has been observed and classified from its light curve as an RR Lyrae variable, its distance can readily be computed. The quantity $M = 0.5$ is simply combined with the observed average apparent magnitude for the star, and the distance is computed by the same means as outlined in Chapter 11 for spectroscopic parallax. With the knowledge that a star whose absolute magnitude is near zero is really very bright, it is clear that RR Lyrae variables may be studied from extremely great distances. It is for these reasons that they have come to be called **distance indicators.** In earlier times they were commonly referred to as **cluster-type variables** because of their common occurrence in the globular clusters. Indentification of RR Lyrae variables in such a cluster gave first the distance to the variables and then, more importantly, the distance to the entire cluster. This deduction is easily justified since the distance to the nearest such cluster is many times greater than the dimensions of the cluster itself.

The spectra of RR Lyrae variables would usually be classified as type A or F, but the relative strengths of some of the lines of the metals and of hydrogen indicate that metals are not as abundant as in the sun. The spectrum varies slightly as the light varies and would be given an earlier classification at maximum light than at minimum. Very few of the variables of this type are bright enough for precise radial velocity studies to have been made. From the meager data that exist, it is clear that the radial velocity varies with the same period as the light. The radial velocity variations for RR Lyrae have been plotted in Figure 13.3, resulting in a situation very similar to that for the Cepheid variables, to be discussed below. It is sufficient here to mention simply that the radial velocity variations are taken to indicate that the surfaces of these stars pulsate, becoming larger and smaller as the light varies. For those cases in which the mean radial velocity can be measured, the RR Lyrae variables seem to have rather high velocities with respect to the sun.

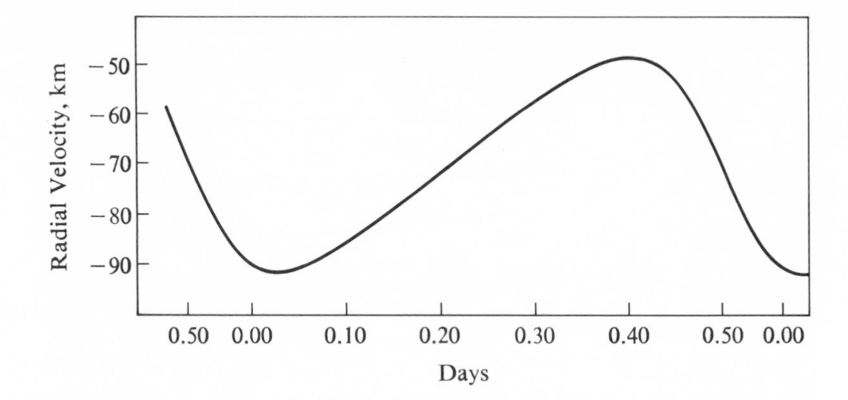

Figure 13.3. Radial velocity variations measured from spectra of RR Lyrae. (After C. C. Kiess, *Lick Observatory Bulletins.*)

Among all stars over the sky, RR Lyrae variables seem to be quite common. They are by no means uniformly distributed, however. We have mentioned already that these stars are quite common in globular clusters. They also occur in large numbers in the direction of the constellation Sagittarius, where the center of our entire system of stars lies. Other RR Lyrae variables are found in the thinly populated areas at great distances from the plane defined by the Milky Way.

13.3 Cepheid Variables

The star δ Cephei is of magnitude 4.5 and easily seen with the naked eye. It has been marked on the appropriate star chart in Chapter 1. Near it in the sky are two stars—one brighter and one fainter—to which the magnitude of δ Cephei may be compared. The observer who carefully observes δ Cephei from night to night over a period of a week or so can readily watch the star grow brighter and then dimmer. The complete period from bright to dim and back to bright is about $5\frac{1}{3}$ days, and the range in brightness is about one magnitude, or nearly the difference in magnitude between the two convenient comparison stars. The results of naked-eye observations pursued with care often show surprising consistency. The light curve begins to take shape after only a few weeks of effort.

A great many stars show light curves similar to that of δ Cephei, as pictured in Figure 13.4, and the class of stars known as Cepheid variables is rather large. Periods for these stars range from about one day up to about 45 days, with periods between six and seven days occurring more frequently than either the longer or the very short periods. The distribution in space of the Cepheids is sufficient

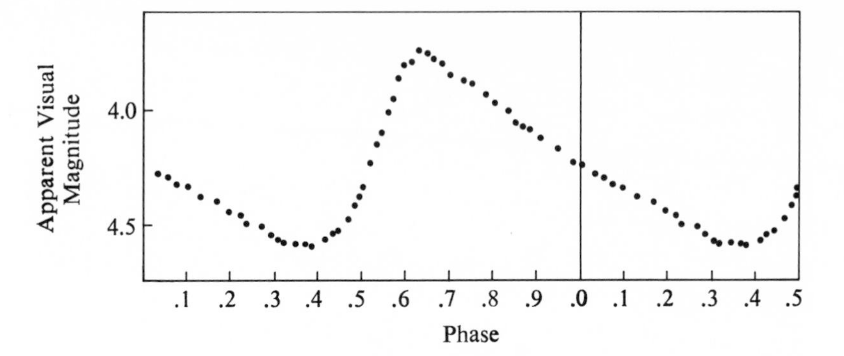

Figure 13.4. Light curve of δ Cephei constructed from a large number of individual observations. (After A. J. Wesselink, *Bulletin of the Astronomical Institutes of the Netherlands.*)

grounds for a clear distinction between them and the RR Lyrae variables. The Cepheids are found almost exclusively near the plane of the Milky Way, a region avoided by the RR Lyrae variables.

With their earliest spectral type about F2, the Cepheids are of somewhat later spectral type than the RR Lyrae variables. There are some rather definite changes in spectral type throughout the period, and the spectral class is always earliest at maximum light. The changes in spectral class are rather small for the shorter-period Cepheids, but become much more significant for the longer-period stars. For example, for a particular Cepheid with a period of 2.24 days, the spectral type changes from F6 at maximum light to F9 near minimum light. By contrast, a Cepheid with a period of 43.7 days varies from F8 at maximum light to K2 near minimum.

The radial velocities of Cepheids are variable just as they are for the RR Lyrae variables. It was from attempts to analyze these radial velocity variations that the concept of pulsating stars first arose. For a long time it was thought that the light and radial velocity variations for Cepheids must arise from some double-star

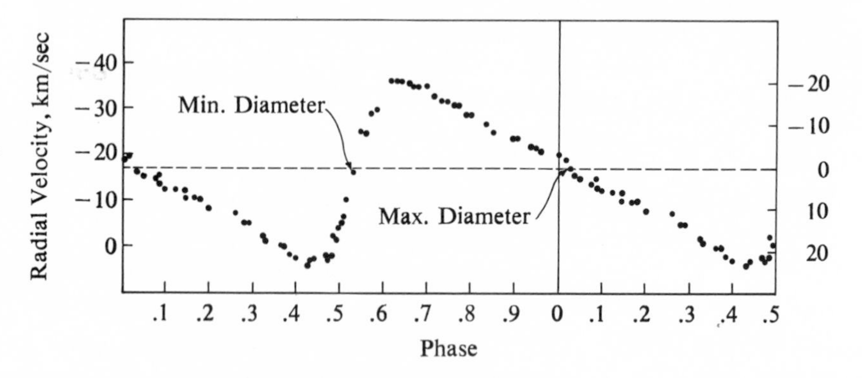

Figure 13.5. Radial velocity variations measured from the spectra of δ Cephei. The star is brightest when the rate of expansion is greatest rather than when the star is actually at its greatest diameter. (After J. H. Moore and T. S. Jacobson.)

situation such as those described in Chapter 12. Finally, however, Harlow Shapley was able to show that for stars as massive and voluminous as the Cepheids must be, a double-star interpretation of the shortest periods could hold only if one star revolved inside of the other. Today, pulsation as the cause of the variability has been unquestionably accepted. The observational analysis is supported, moreover, by theoretical studies of stellar evolution which indicate that variability like that of the Cepheids is to be expected at a certain stage in the life of massive stars.

Study of the velocity curve of Figure 13.5 along with the light curve of Figure 13.4 is helpful in understanding the changes that take place as the star varies.

13.4 Period-Luminosity Law

The astronomer trying to measure the parallaxes of Cepheids to determine their absolute magnitudes encounters the same problem found for the RR Lyrae variables. None is close enough to have a measurable parallax, so other, less reliable methods were initially invoked. The history of the attempts to find accurate absolute magnitudes for Cepheids is long and complex.

Certainly the most significant single fact relating to the Cepheids is the relationship between absolute magnitude and length of period. The Cepheids with the longest periods are the ones which are actually the brightest. This interesting phenomenon could never have been guessed at from studies of Milky Way Cepheids alone, because astronomers had no good means of finding the absolute magnitudes of any of them. In the years just prior to 1920, however, Shapley and one of his co-workers, Henrietta Leavitt, had been making extensive studies of the variable stars in the two Magellanic Clouds visible in the southern hemisphere sky. These bodies have the appearance of detached sections of the Milky Way, but they are actually small independent stellar systems some 150,000 light-years distant from us. Leavitt noticed that in the Small Magellanic Cloud the Cepheids with the longest periods were the ones with the brightest apparent magnitudes. The Magellanic Clouds were known to be very distant, and so it was safe to assume that distances within the clouds were very small compared with the total distance from the sun to the clouds themselves. Leavitt's findings clearly meant that absolute magnitude must depend on period for the Cepheids. Her original plot of apparent magnitude versus the logarithm of the period in days is shown in Figure 13.6.

It was soon realized that this relationship between period and luminosity could be extremely valuable if the absolute magnitude scale could once be established. All that was needed was the parallax of one Cepheid of known period. This would be enough to establish the "zero point" of the period-luminosity relation which was so clearly indicated by the Cepheids in the Small Magellanic Cloud. As stated earlier, however, no Cepheid was close enough for an accurate parallax determination. Astronomers were therefore forced to make an assumption which later proved to be significantly in error. We have stated that the periods for the RR Lyrae variables are all less than one day and that periods for Cepheids are all longer than one day. It was assumed (quite naturally, on the basis of information then on hand) that the Cepheids were closely related to the RR Lyrae variables, and that the absolute magnitudes of short-period Cepheids must be similar to the absolute magnitudes of the RR Lyrae variables. By means of some earlier statistical arguments based on the motions of the RR Lyrae variables, a value for the absolute magnitude of these stars had been established. Therefore, on an M-versus-log P graph, such as

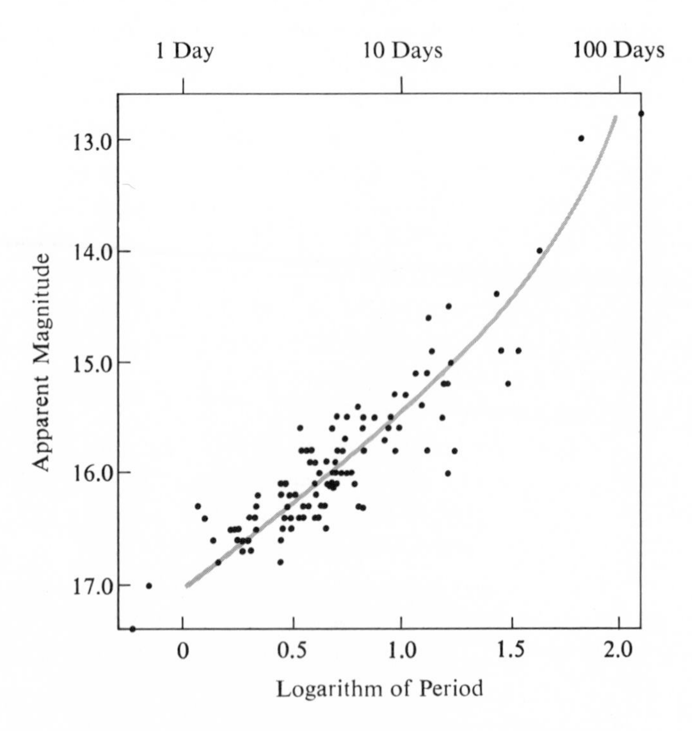

Figure 13.6. The period-luminosity law as determined by H. Shapley and H. Leavitt from the Cepheids in the Small Magellanic Cloud.

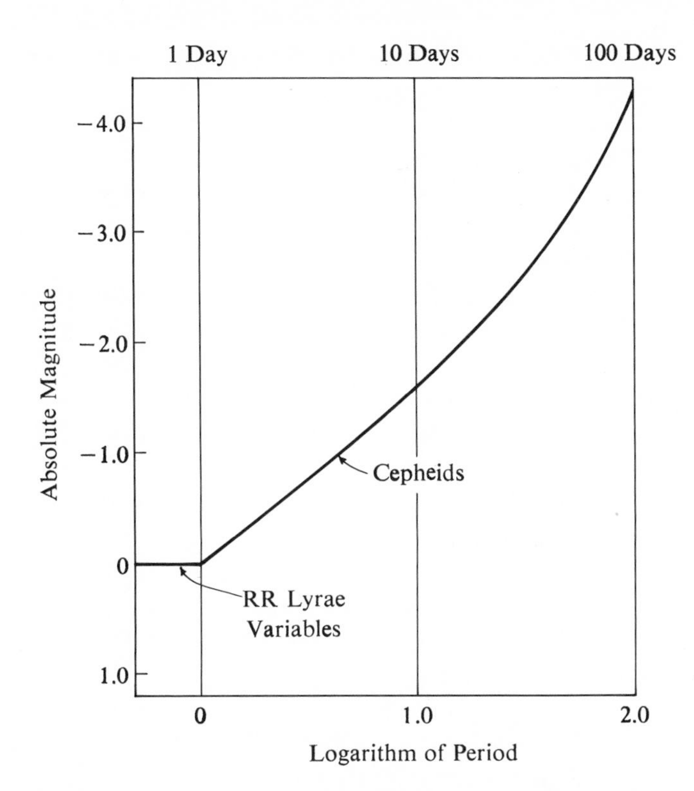

Figure 13.7. A preliminary period-luminosity relation in which the Cepheids of short period have been assumed equal in absolute magnitude to the RR Lyrae variables.

that in Figure 13.7, a line representing the period-luminosity relation for Cepheids could now be drawn. The slope for the line was taken from Leavitt's data from the Small Magellanic Cloud and the line was made to pass through the point ($M = 0$, $\log P = 0$). In effect, this tied the Cepheids to the RR Lyrae variables.

This first version of the period-luminosity relation for Cepheids proved to be quite a useful tool and made possible some of the first estimates of the distances to galaxies at vast distances from us. Whenever a variable star was detected, its light curve could be plotted. From the light curve the period could be determined. The logarithm of the period could then be used to find the absolute magnitude using the period-luminosity relation. The absolute magnitude is then used in the expression

$$M = m + 5 - 5 \log r$$

to solve for the distance r in parsecs as described previously. Since the Cepheid variables are really very bright stars, they can be identified at great distances, and they thus afford a means by which very great distances may be measured. The first estimate

(750,000 light-years) of the distance to the great spiral galaxy M31 in Andromeda was based on the Cepheids observed there.

Many Cepheid variables in the nearer galaxies had been studied with the 100-inch telescope at Mount Wilson. The distances to these galaxies had been based on the Cepheids observed, and these distances had provided the basis upon which distances to very faint galaxies had been estimated. The Cepheid variables were the real key to our whole concept of the size of the universe. By an interesting coincidence, the RR Lyrae variables in M31 were expected to be just about two magnitudes too faint to be detected in the 100-inch telescope. When the Mount Palomar 200-inch telescope went into operation in about 1950, it was thought that the RR Lyrae variables would be just barely within the limiting magnitude of that new instrument. One of the first programs undertaken with the 200-inch was, therefore, a search for RR Lyrae variables in M31. The search proved to be fruitless, however, and it was soon clear that there had been an error in the zero point of the period-luminosity relation for the Cepheid variables. This difference seemed to be about 1.5 magnitudes, so the zero point of the period-luminosity law was appropriately revised. The absolute magnitudes of RR Lyrae variables were known with reasonable accuracy, and so the Cepheids must really be 1.5 magnitudes brighter than previously imagined. This revision by astronomer Walter Baade was subsequently confirmed and accepted. Its consequences were of overwhelming significance.

If the Cepheids in the Andromeda galaxy were intrinsically brighter than originally estimated, it was clear that they must also be farther away. It is interesting to see what happens to a distance estimate when the absolute magnitude is made 1.5 magnitudes brighter. Let us assume that a Cepheid is observed to have an apparent magnitude of 12.0. If we assume that the absolute magnitude of the same star is −3.0, then by using the formula $M = m + 5 - 5 \log r$ we may compute the distance as follows:

$$-3.0 = 12.0 + 5 - 5 \log r$$

or

$$\log r = \frac{20}{5} = 4.0$$

Therefore,

$$r = 10{,}000 \text{ pc}$$

Now if we make the absolute magnitude brighter by 1.5 magnitudes we have

$$-4.5 = 12.0 + 5 - 5 \log r$$

$$\log r = \frac{21.5}{5} = 4.3$$

and

$$r = 20{,}000 \text{ pc}$$

Thus the change of 1.5 in M results in a distance twice as large as first computed. The Cepheids in the Andromeda galaxy were, therefore, at a distance of 1,500,000 light-years rather than 750,-000 light-years. The distances to all other galaxies were related to this distance and they also had to be doubled. The universe was suddenly twice as large as previously thought, all because of an

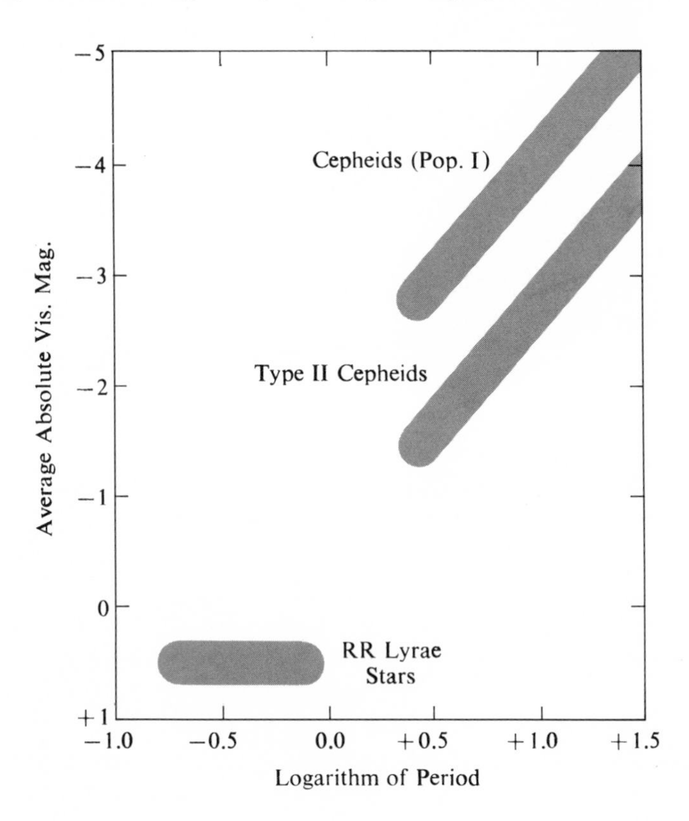

Figure 13.8. The period-luminosity relation after recognition of two types of Cepheids. (Wyatt, p. 446.)

adjustment of 1.5 in the absolute magnitudes of the Cepheid variables.

In the years since Baade's work there have been other revisions in the period-luminosity relation for Cepheids and RR Lyrae variables. For the RR Lyrae stars the absolute magnitude is now thought to be about 0.5. For the Cepheids of shortest periods the accepted absolute magnitude is a little bit brighter than −2.0. It is also recognized that there is some natural scatter along the line. At a particular period there is actually a range of possible absolute magnitudes for Cepheids. It is also known that the original class of Cepheids is really composed of two distinct groups for which the period-luminosity relationships are different. These two groups are shown in Figure 13.8. Stars which are represented by δ Cephei are indicated here as Cepheids (Pop. I). Members of the second group, which really seem related to the RR Lyrae variables after all, are known as W Virginis stars. The light curves of these stars are often characterized by a slight bump on the downward branch of the light curve, and their distribution in space is more like that of the RR Lyrae variables than that of the Cepheids.

13.5 Long-Period Variables

A large class of late-type stars are best described simply as **long-period variables.** A number are often referred to also as **Mira-type** variables after the star Mira, in the constellation Cetus, the Whale. This interesting star had been known from ancient times to be one which varied in brightness over an unusually large range. Its name, which literally means *wonderful*, was appropriate to a star which became invisible during much of its 400-day period.

A few general statements may be made about the long-period variables as a class. The periods which we here define as *long* are all more than 200 days, and may be as long as 1000 days. The magnitude changes through a range of at least 2.5 magnitudes, with a range of 5 or 6 magnitudes being typical. Spectral types are all late. A few K-type spectra are found, but most are M-type. Emission lines are usually seen, also. Absolute magnitudes have been found for some long-period variables and indicate that these are giant stars rather than main-sequence ones. The mean absolute magnitude is nearly zero for cases in which the period is near 200 days. For the stars with periods at the other extreme, i.e., about 1000 days, the mean absolute magnitudes are slightly fainter, $M=+1$. The long-period variables are found almost exclusively in the Milky Way, just as are the Cepheids.

An interesting explanation of the spectacular change in luminosity is derived from the fact that the long-period variables pulsate in much the same manner as the Cepheids and RR Lyrae variables. Changes in temperature accompany the pulsations, and the stars are brightest when they are hottest. The actual temperature change is relatively small, however, and it can be difficult to explain the large changes in luminosity in terms of such small temperature changes. By contrast, for Cepheids the temperature change is greater and the amplitude of the light curve is less than for the long-period variables. The explanation for the seeming anomaly lies in the low surface temperatures of the long-period variables. At temperatures of 3400°K the maximum of the Planckian energy distribution is in the infrared. Only the short-wavelength portion of the curve is in the visible region, as indicated in Figure 13.9. A similar curve has also been drawn for the star at its maximum temperature. The total area under the upper curve is only a few percent greater than the total area under the lower curve. The amount of energy within the visible wavelength range, however, is considerably greater and can account for much of the great range in luminosity for these stars.

A second factor affecting the luminosity in the visible wavelength region is related to the broad absorption bands of certain molecules, particularly titanium oxide. The titanium oxide (TiO) bands appear as broad absorption features in the visual and infrared regions. Very sensitive to temperature, the bands are quite pronounced in the spectra of cooler stars but disappear in the hotter ones. Such broad absorption features must block a significant

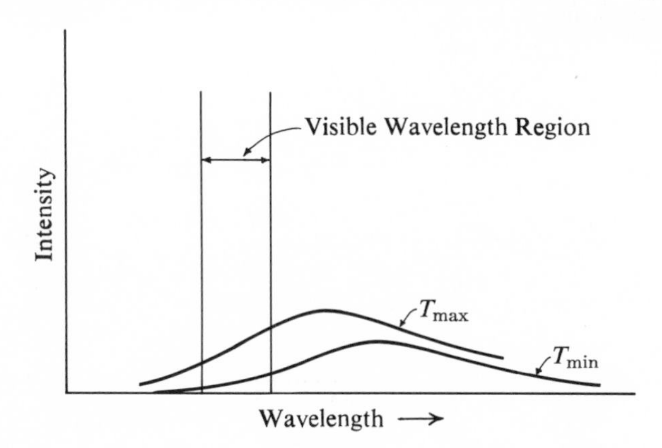

Figure 13.9. The energy in the visible wavelengths increases substantially when there is a relatively small increase in temperature.

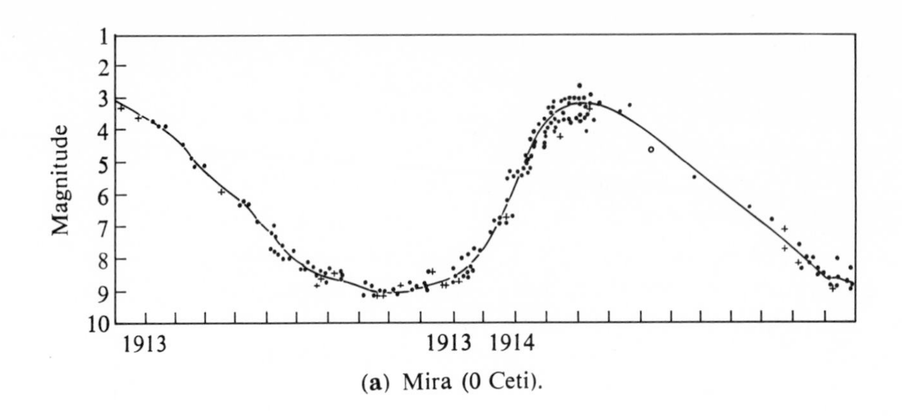

(a) Mira (0 Ceti).

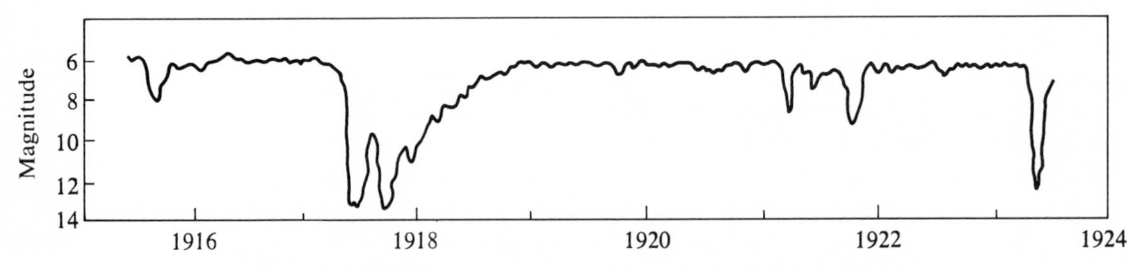

(b) R Corona Borealis. Constant maximum with irregular decreases.

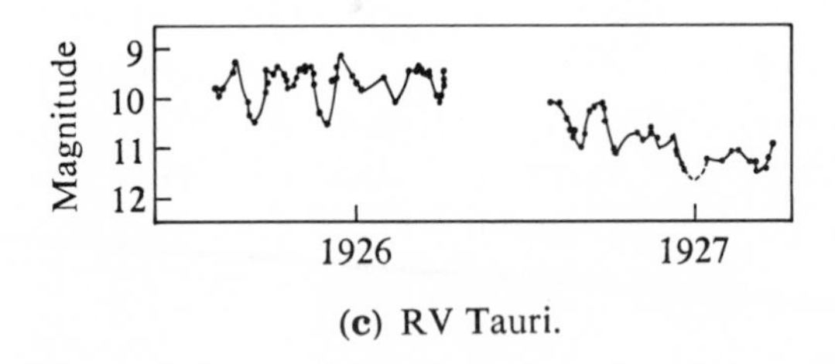

(c) RV Tauri.

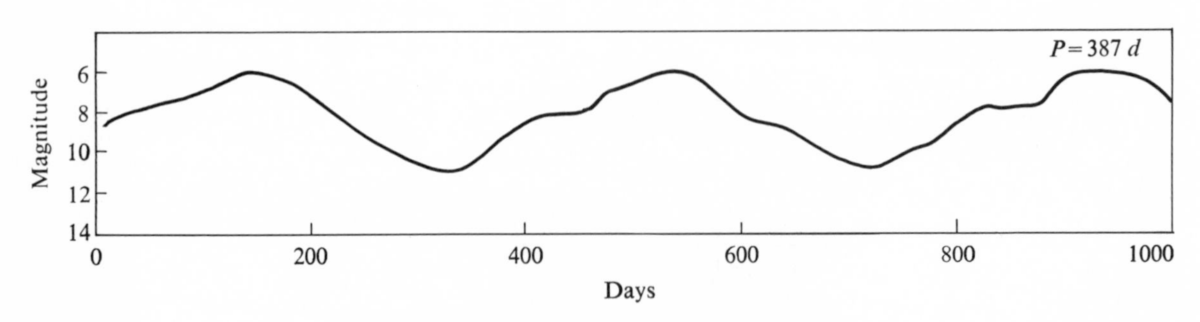

(d) The Long-period variable T Cephei.

Figure 13.10. Light curves for a number of other types of variable stars.

amount of the total radiant energy of the star. Therefore, variations in the strength of the TiO bands occur as a result of variations in temperature. As the temperature becomes lower, the bands become stronger. The visual energy, in turn, varies with variations in the strength of the bands. As the bands become weaker, more light in the visual wavelengths is radiated. Here, then, is a second mechanism which contributes to the large observed variations in the magnitude of the long-period variables. Changes in temperature affect the TiO bands and these in turn affect the amount of visible radiation. Combined with the changes in Planck radiation, the molecular absorption permits satisfactory explanation of the very great magnitude variations.

13.6 Other Types of Variable Stars

We have described in some detail three of the more populous classes of variable stars. There are many other classes with fewer known members. Light curves of typical members of a few of these are shown in Figure 13.10, and pertinent data for these classes is summarized in Table 13.1. In Figure 13.11 the locations of the vari-

Table 13.1. Characteristics of some classes of variable stars.

Class	Range in Periods	Typical Variation in Magnitude	Spectral Type	Mean Absolute Magnitude
Cepheid	2 to 40 days	1.0[1]	F to G[2]	−4[3]
RR Lyrae	0.1 to 1.0	1.3[4]	A to F[2]	+0.5
Mira type	100 to 400	4.5	M4[5]	+2[6]
δ Scuti	0.08 to 0.19	0.2	F2	+1.8
W Virginis	1 to 50	0.8[1]	F to G[2]	−3.0[3]
β Canis Maj.	0.15 to 0.25	0.1	B	−4
RV Tauri	75[7]	1.3	G to K	−2
Semiregular	100[7]	1.6	G to M	−1
R Coronae Bor.	Sudden decrease	4	G to K	−5 ?
UV Ceti	Sudden flare	2	M3 to M6	+15
U Geminorum	60[7]	3.6	B, A	+8
T Tauri[8]	Irregular	3	F, G[5]	+5

[1]Greater for longer periods.
[2]Later at minimum light.
[3]Brighter for longer periods.
[4]Less for longer periods.
[5]Emission lines in spectrum.
[6]Fainter for longer periods.
[7]Typical.
[8]Usually associated with interstellar matter.

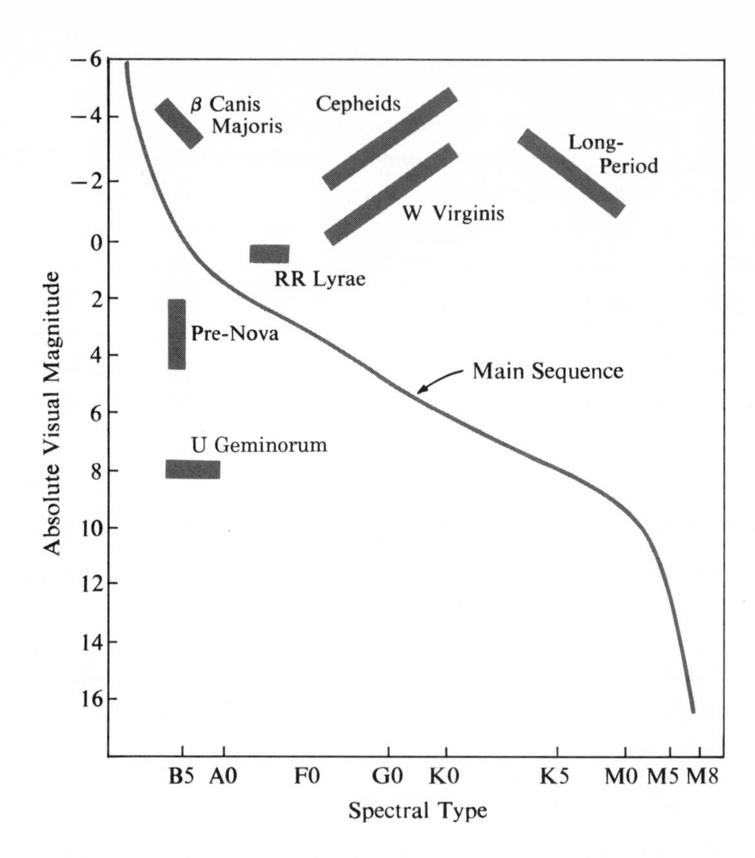

Figure 13.11. The general areas occupied by the variable stars on the H-R diagram.

ous groups on the H-R diagram are indicated. We have already suggested that the H-R diagram holds some clues to the evolution of stars, and Figure 13.11 suggests that the variable stars must enter the theory at significant points.

13.7 Novae

On rare occasions astronomers have found that, in places where no star had previously been known, a star suddenly has appeared, persisted for a few months, growing steadily dimmer, and then disappeared. These events came to be known as **novae** (singular: nova) or **new stars.** We know today that a nova is really an already existing star which undergoes an extreme and rapid increase in luminosity. Several novae are detected each year, and astronomers have been fortunate enough to have obtained fairly complete sets of observations for a few of them. For a number of the novae observed in the last 50 years, astronomers have been able to find earlier observations which give an idea of the nature of the star before its eruption. From the available data a reasonably complete description has been drawn.

The pre-nova star is usually one of early spectral type, B or early A, and has an absolute magnitude of about +5. Thus, the pre-nova star is considerably fainter than the normal main-sequence stars of the same spectral type. When the eruption occurs, the star increases in brightness quite rapidly, becoming 10 or 12 magnitudes brighter in a period of 10 days or so. The absolute magnitude may reach −8.0 or −9.0. Gradually the star fades, becoming fainter by 3 magnitudes in about a year. There are often fluctuations in brightness after the initial rapid decline. The light curve of Nova Aquilae 1918 may be seen in Figure 13.12. Eventually the star returns to very nearly its original brightness and is apparently only a little the worse for wear; that is, the amount of actual mass loss is only a fraction of the total mass of the star. No two novae behave quite the same in all respects, but this general pattern is seen in the behavior of most.

We discussed before the relationships between diameter, surface temperature, and luminosity for stars, and earlier in this chapter we showed that the light variations for Cepheids result chiefly from the changes in surface temperature which result from the pulsations. Similar lines of reasoning may be used to gain initial insight into the nova phenomenon. In the nova case, a star that is relatively hot to begin with suddenly increases in brightness by several hundred thousand times without any great change in surface temperature. The explanation must be that the star's radiat-

Figure 13.12. The light variations of Nova Aquilae 1918 from June to December, 1918.

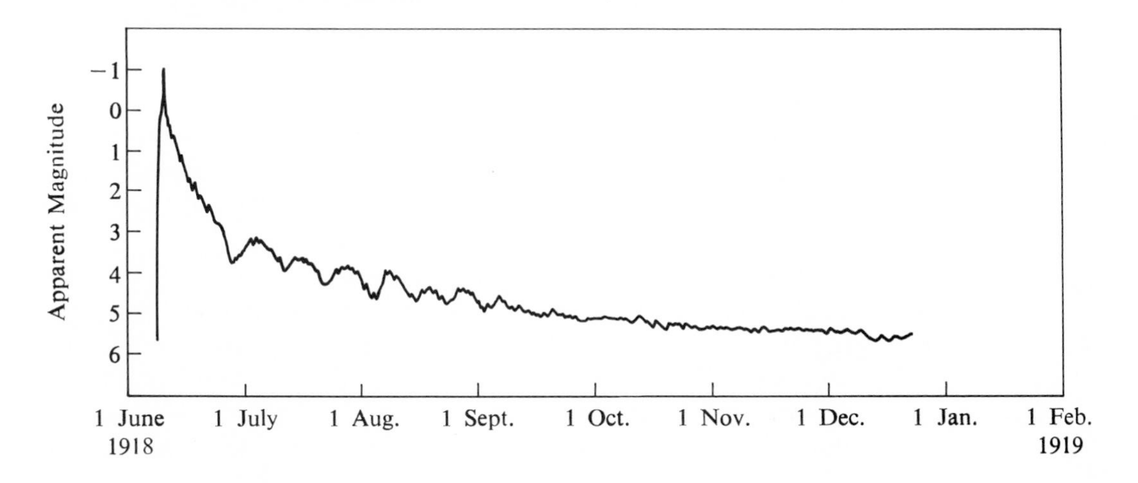

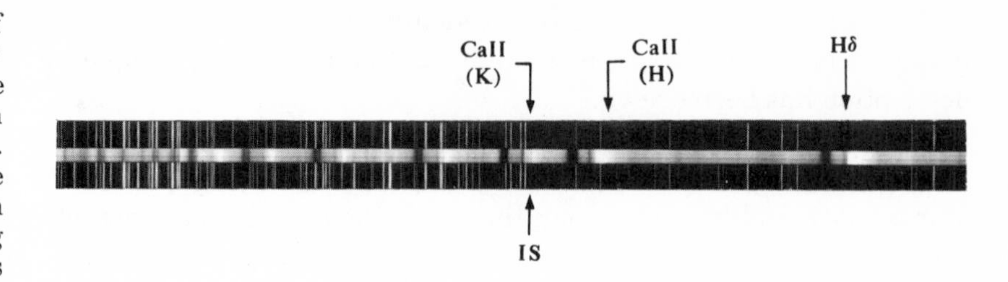

Figure 13.13. The spectrum of Nova Delphinius 1967. The Hδ and the K lines of CaII have been marked, and the emission features can be easily seen. Shifted toward the violet are the multiple absorption features, indicating expanding gas shells. At the wavelengths of the K line an interstellar line has been marked IS. (Lick Observatory photograph.)

ing surface area increases by several hundred thousand times. By quickly releasing incredible amounts of energy the star must expand rapidly without becoming cooler at the surface.

Our greatest insight into the nova phenomenon comes from the spectra that have been obtained during a number of nova outbursts. During the brief period of rapid brightening, the absorption lines in the spectrum of a nova become broadened, weak, and may disappear – only to reappear considerably shifted toward the violet. Again we see the effect of a rapid increase in radiating surface area. Sometimes there are multiple patterns of lines all strongly shifted toward the violet, suggesting that several concentric spheres or shells of gas account for the increase in light. Just after the nova reaches its maximum brightness, the spectrum changes radically with the appearance of emission lines. The centers of the emission lines are essentially unshifted, but the lines are very much broadened. Part of a typical nova spectrum seen in Figure 13.13 shows clearly the broad emission feature with a distinct absorption feature at its violet edge. The interpretation of these characteristic spectrum features is derived from the notion that shells of gas are expanding away from a central star.

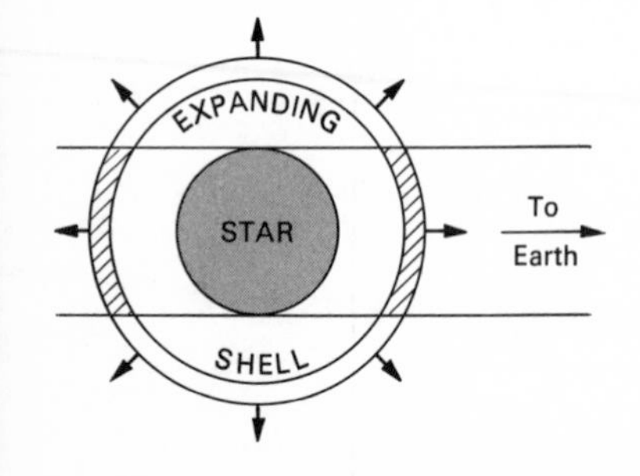

Figure 13.14. The expanding shell of gas around a nova accounts for the broad emission features and the violet-shifted absorption features seen in a typical nova spectrum. (Wyatt, p. 453.)

If a shell of gas is expanding, then from any direction the gas must be moving simultaneously in many directions with respect to the line of sight. If this gas has been excited by intense radiation from a central star, the expanding shell should show emission lines greatly broadened due to the motion of the gas. Part of the gas in the shell must lie between the observer and the central star. Being cooler than the central star, this gas causes absorption lines in the star's spectrum. Furthermore, because of the rapid expansion of the shell, the gas which produces the absorption lines is moving toward the observer at a high speed. The resulting absorption lines are, therefore, strongly shifted toward the violet. The expanding shell is shown schematically in Figure 13.14, and one may see the parts which account for the broad emission features and the violet-shifted absorption lines.

The spectra of novae are actually somewhat more complex than indicated above. The main characteristics serve to confirm the idea that the rapid increase in brightness is due to a suddenly expanded radiating surface. The surface expansion may be confirmed some years after the explosion when the star appears to have returned nearly to its initial condition. Since the ejection velocities are nearly constant and in excess of the escape velocity, the expanding shell must eventually become large enough to show a measurable angular diameter. When the angular diameter of the expanding shell is visible, we have a novel but rather accurate means by which the distance to the nova may be determined. Suppose for example, that the nova shell has become very large and is fairly even in all directions around the central star. The shift of the emission lines can tell us the actual velocity of the gas toward us and away from us. Changes in the angular size of the shell can give the angular velocity of the gas at right angles to the line of sight. We know from the observed radial velocity and the assumption of uniform expansion what the actual tangential velocity of the gas must be. From here it is a straightforward matter to compute the distance at which this tangential velocity would give rise to the observed angular velocity (see Chapter 11). This method, with some refinements, is actually quite a good one and has led to some of our best data on the absolute magnitudes of novae.

13.8 Supernovae

Once in a while the nature of a stellar explosion is such that the star reaches an absolute magnitude in excess of −16. The star has become some hundreds of millions of times brighter than our sun. Such outbursts as these are appropriately referred to as **supernovae.** One star is suddenly and briefly as bright as the combined light of all the stars in a small galaxy. Supernovae do not occur very frequently—perhaps only once in 50 or 100 years in any galaxy, but if they occur within 1000 pc it is not likely that they will be missed by astronomers. Tycho Brahe saw one in his time, and Kepler saw one less than 100 years later. Most surprisingly, an earlier supernova was missed completely by Europeans, but was seen and recorded by Chinese and Japanese astronomers in 1054 A.D. Their records quite accurately located it in the constellation

Taurus, and today we see in its place the strange irregular cloud pictured in Figure 13.15 and in Color Plate 2. The Crab Nebula is one of the most interesting of all celestial objects.

Proper motions or angular velocities of the expansion of the gas have been combined with radial velocity data as described above to give a distance of 1100 pc to the Crab Nebula. The long dimension of the cloud as we see it, nearly 2 pc, combined with the present rate of expansion, indicates that the cloud cannot have been expanding for more than about 1000 years. This is further evidence that the Crab Nebula resulted from the supernova of 1054 A.D. Quite obviously, the original star has been dramatically affected and a considerable fraction of the original star mass is contained in

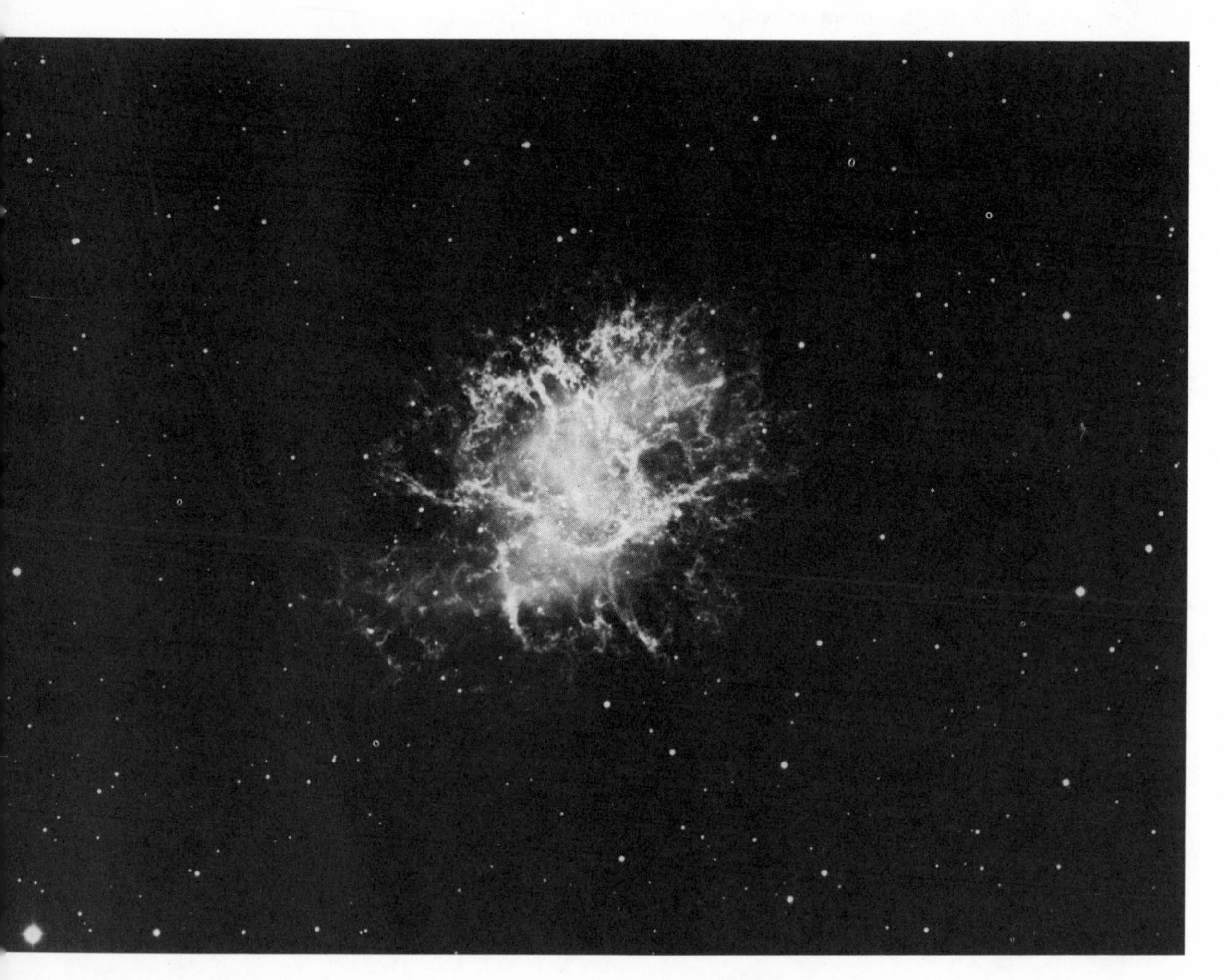

Figure 13.15. The Crab Nebula in the constellation Taurus, the remnant of the supernova explosion of 1054 A.D. (Hale Observatories photograph.)

the nebula, unlike a normal nova in which the post-nova star seems to be little changed. Buried in the gas of the Crab Nebula there is a faint star which has long been thought to be the remnant of the original star. Since 1969 this star has been seen to fluctuate very rapidly at radio, visible, and x-ray wavelengths. It is a member of a newly discovered class of variables known as pulsars. (See page 287.)

The complexity of the Crab Nebula can be vividly seen in the comparison of the observational data obtained from a number of independent studies. Special narrow-band filters can be used in photographing the Crab to record a large structureless mass emit-

Figure 13.16. The Veil Nebula in the constellation Cygnus. It is also a supernova remnant. (Hale Observatories photograph.)

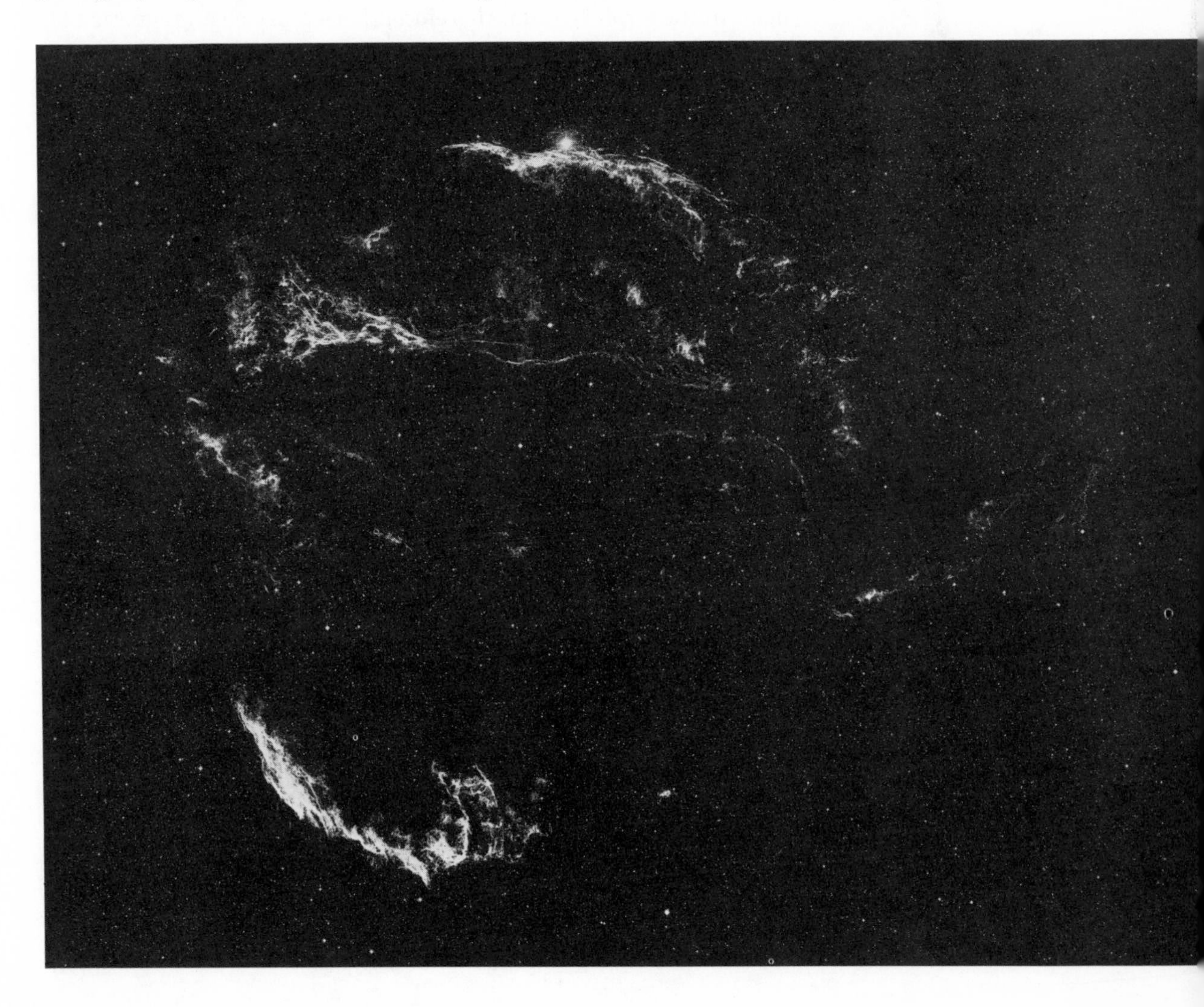

ting a continuous spectrum. Photographs made with other filters show that the luminous, excited hydrogen is distributed in a complex array of filaments. The light from the structureless mass is strongly polarized as well. Radio observations show continuous emission in the radio region of the spectrum, and here again the radiation is strongly polarized. Astronomers today believe that they are seeing **synchrotron** emission from the Crab. This name is derived from the similarity between the light emitted by the Crab and the light emitted from a particle accelerator called a synchrotron. The frequency distribution and polarization of the light is characteristic of electrons moving at high speeds in a magnetic field. In the Crab Nebula, therefore, there must be a magnetic field and there must be electrons moving at many different velocities to account for the observed continuous emission in both the optical and radio wavelengths. Rocket experiments have disclosed the Crab as the first known source of x rays in the sky. From almost all of the observational techniques which have been tried on the Crab, new questions have arisen before the old ones have been fully answered. Nevertheless, the studies have greatly added to our knowledge of supernovae and supernovae remnants. Radio sources and faint optical nebulae have recently been found near the locations at which Tycho and Kepler saw supernovae in 1572 and 1604, respectively. Some of the visible parts of another supernova remnant are shown in Figure 13.16.

13.9 Recurrent Novae

In a supernova explosion the star is so drastically altered that such a catastrophe can only occur once in the lifetime of a single star. On the other hand, the normal nova explosion seems to leave stars only slightly changed, and some stars which have been observed as novae have brightened a second time some years later. For example, the nova T Coronae Borealis was seen in 1866. Eighty years later, in 1946, the star became a nova again. T Pyxidis has been seen as a nova on four occasions since 1890. Other stars, such as U Geminorum, show large, sudden increases in brightness and then after a few days return almost as quickly to their original brightness. These outbursts may come as often as two months apart, but are not regular enough to be predictable. Neither is the brightness at maximum predictable, as is true for the Cepheids and the RR Lyrae variables. There does seem, however, to be a relationship

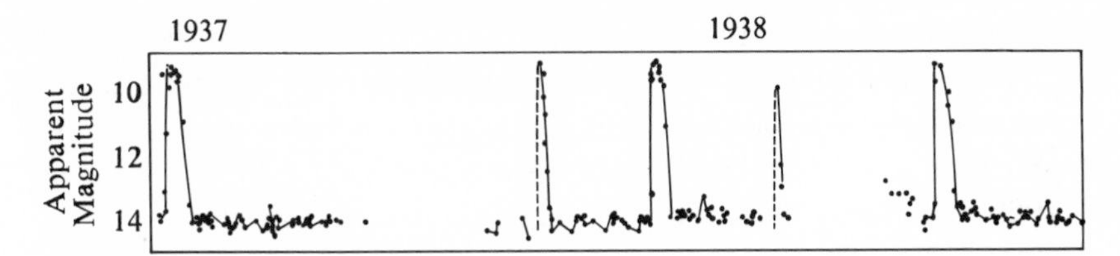

Figure 13.17. Light curve showing sudden brightening of U Geminorum.

between the intensity of an outburst and the interval between outbursts. The longer the interval, the brighter the star when the outburst occurs. The light curve of U Geminorum may be seen in Figure 13.17.

13.10 Novae in Binary Systems

After a star has become a nova, astronomers watch it closely for any further signs of change in brightness or spectra. Robert Kraft of Lick Observatory deduced from his own observations and those of other astronomers that many of the old novae and U Geminorum stars were members of binary systems. Several of the old novae were found to be spectroscopic binaries, and several recurrent novae were found to be eclipsing binaries. In all cases the periods were very short. For DQ Herculis 1934, for example, the period between eclipses was only 4 hours and 39 minutes. For the recurrent nova WZ Sagittae, the spectroscopically determined period was only 81.5 minutes. As data of this sort began to accumulate, some astronomers began to wonder if all novae might not be members of binary star systems. If this should prove to be the case, it should furnish some important clues to the mechanism which triggers the nova eruption. The critical factor may be the transfer of matter from one star to the other at a stage in which the inflow of new material results in serious imbalances in the rate of energy production. On the other hand, it may well be that the binary nature of novae is simply a coincidence, since most stars are members of multiple star systems anyway. In this case, the mechanism causing the nova could be such that any star could become a nova at some point in its evolution.

13.11 Pulsars

In the winter of 1968, radio astronomers at Cambridge, England announced the discovery of a new class of variable radio sources which they described as **pulsars.** These sources radiate short pulses of radio frequency energy at regular intervals of the order of one second. Four pulsars were discovered in the original survey, and four more had been found before the end of the summer of 1968. It is very likely that the number of detectable pulsars may be quite large. The adopted scheme for identifying pulsars uses the first letter of the name of the discoverer's observatory followed by the letter 'P' and a number which represents the right ascension of the source in hours and minutes. Thus, CP 1919 is a Cambridge pulsar at R.A. 19^h19^m.

Very soon after the original Cambridge announcement, other astronomers began to study the detailed form or structure of individual pulses. They found that while the period between pulses is extremely constant, there is considerable variation in shape from one pulse to the next. The pulsars were originally studied at a frequency of 81.5 megahertz (megacycles per second). Later it was noted that less energy was received at higher frequencies up to approximately 400 megahertz. A difference in the time of arrival of the various frequencies was also noted. The energy at the lower frequencies arrives slightly later than the energy at the higher frequencies. This delay may amount to several seconds and is an effect caused by the passage of the radio signals through ionized interstellar gas. It is partly from the measured delays in arrival times, coupled with certain assumptions about the density of electrons in space, that distances within our own galaxy have been estimated for the pulsars.

As soon as sufficiently accurate pulsar positions became known from the radio studies, optical astronomers began to look for stars which might be the source of the radio emissions. The optical identification of the pulsars was not simple, however. The principal problem was that conventional optical detectors such as photographic films and photoelectric photometers could not discriminate fluctuations in brightness as rapid as those of the pulsars. Very specialized techniques had to be devised before the optical variations could be detected. At first the search with high-speed detectors proved fruitless. None of the stars near the radio positions of the pulsars showed any optical variations, and none showed any strange colors or spectrum features to distinguish them from "normal" stars. The breakthrough came when a radio pulsar was found with the coordinates of the Crab Nebula, mentioned above as a supernova remnant. Two stars have long been visible in the Crab

Nebula, and one of these was soon found to show rapid changes in brightness at visual wavelengths. Not surprisingly, the optically determined period is the same as the radio period.

At about the same time, observations of the Crab Nebula pulsar were made by an x-ray detector carried aloft by a high-altitude rocket. Rapid fluctuations were again found, and again the periods were in agreement.

The picture that emerges, then, is that a pulsar is very likely to be a stellar fragment left after a supernova explosion has caused a drastic modification in the original star. Support for this lies in the fact that another pulsar has been found near the center of a supernova remnant in the constellation Vela. The Vela pulsar has not yet been detected at optical wavelengths, however. The association of pulsars and supernovae is interesting and significant, but the pulsars still present a very challenging problem to those astronomers who seek to understand the basic physical mechanism of the periodic signals. What kind of a star can vary in brightness at many wavelengths and with a period which may be as short as 0.033 sec?

Since pulsation provides a satisfactory explanation of the variations in light for stars such as Cepheids and RR Lyrae variables, the pulsation theory was applied to the pulsars. The result required that if the star had roughly the mass of the sun, it would have to be as small as a moderate-sized asteroid. The density of the matter in such a body would have to be thousands of times higher than that found in the white-dwarf stars, and this seems difficult to accept. A second plausible theory is that the pulsars are rapidly rotating stars which radiate strongly from certain discrete regions on their surfaces. Then the rate at which we receive pulses is equal to the rate of rotation of the star. The observed subpulses could be explained by the presence of more than one radiating spot. The analysis in this case suggests a star perhaps 1000 km in diameter if the mass is comparable to that of the sun. Again, the density would be extreme but not quite as high as in the case of the pulsating star. In both of these possible cases the small diameter means that the star must necessarily be intrinsically faint. This helps to explain the fact that only two optical pulsars have been discovered so far.

The extreme conditions suggested by the rotational hypothesis are in agreement with the concept of a neutron star, to be discussed in Chapter 16. When the nuclear energy sources in a star's interior have been expended, the star could be expected to collapse under its own gravitational attraction. In the process the outer layers might be blown off in a cataclysmic explosion. In the remaining core, however, the high density would cause the transformation of the various atomic nuclei and electrons into neutrons. Such a star could have the size and density implied in the rotational model of the pulsars. Thus, the current consensus is that pulsars are rapidly rotating neutron stars. The theory suggests that the rate of rotation should be decreasing and that the period between pulses should be lengthening. Such an increase in period has actually been observed for the pulsar in the Crab Nebula.

QUESTIONS

1. For what reasons is a star classified as an RR Lyrae variable?
2. How do radial velocity variations lead to the conclusion that RR Lyrae and Cepheid variables pulsate as they change from bright to dim?
3. What is the range in period for RR Lyrae variables and for Cepheid variables? What are the ranges in amplitude?
4. From the changes in spectrum during the period of a Cepheid variable, when is the star's surface temperature highest and lowest?
5. What is the period-luminosity relation for Cepheid variables and how is this relation used to find the distances to Cepheids?
6. What are some of the important characteristics of the long-period variables?
7. Describe briefly one of the two explanations for the very large fluctuations in visible radiation in the long-period variables.
8. What is the basis of the belief that the great increase in brightness of a nova is due to a very great increase in diameter rather than a rise in surface temperature?
9. What is the interpretation of the broad emission lines with violet-shifted absorption lines seen in the spectra of novae?
10. State the absolute magnitude at maximum light of a nova and a supernova. What are the approximate absolute magnitude and spectral type of a pre-nova star?

11. What significance can be attached to the discovery that most novae are members of binary star systems?

12. Why are the fluctuations of pulsars so difficult to detect at optical wavelengths?

14

Galactic and Globular Clusters

In several areas on the celestial sphere are groups of stars which seem to form small compact constellations. Of those groups bright enough to be seen easily with the naked eye, the most familiar are the Pleiades and the Hyades. Both are situated in the constellation of Taurus, and the Pleiades is shown in Figure 14.1 and in Color Plate 3. The telescope reveals many other similar star clusters varying over a wide range in numbers of member stars and in degree of concentration. Studied in detail, the clusters have been the source of vital parts of our knowledge of stellar evolution and of the structure of our galaxy.

There are two quite distinct types of clusters. The first are the **galactic clusters,** such as the Pleiades, the Hyades, and Praesepe (Figure 14.2). As the name suggests, galactic clusters are found almost exclusively in the direction of the Milky Way. The second type are the **globular clusters,** such as M13 in Hercules pictured in Figure 14.3. Globular clusters appear quite similar to one another and are never found in the Milky Way. Their distribution on the celestial sphere is quite uneven, since they are found principally in one half of the sky. In addition to the differences in the distribution of the galactic and globular clusters, there are differences in the types of stars to be found in each.

◄**Figure 14.1.** The Pleiades, a galactic cluster in the constellation Taurus. (Yerkes Observatory photograph.)

Figure 14.2. Praesepe, a galactic cluster in the constellation Cancer. (Yerkes Observatory photograph.)

Figure 14.3. The globular cluster M13 in Hercules. (Hale Observatories photograph.)

14.1 Galactic Clusters

There are nearly 800 known galactic clusters, and they show considerable range in actual dimensions and in their total numbers of stars. The Pleiades cluster has a diameter of 4 pc and contains 120 stars, while in the constellation Vela there is a cluster containing only 15 stars. The two largest clusters form the beautiful double cluster in Perseus. Each of these clusters is some 20 pc in diameter, and each contains more than 250 stars (Figure 14.4). Galactic clusters certainly vary greatly in appearance. They are often referred to, however, as **open clusters** because their overall appearance suggests the low density of the member stars.

Membership Criteria

The overall appearance of a galactic cluster can give an initial idea as to which stars are actually members and which are in the foreground or in the background. When we see even a few stars of similar brightness close to each other in the sky, we are fairly safe in assuming that they are actually near each other in space. If, however, we are to study only the cluster, then we must be able to de-

Figure 14.4. The double cluster in Perseus. (Lick Observatory photograph.)

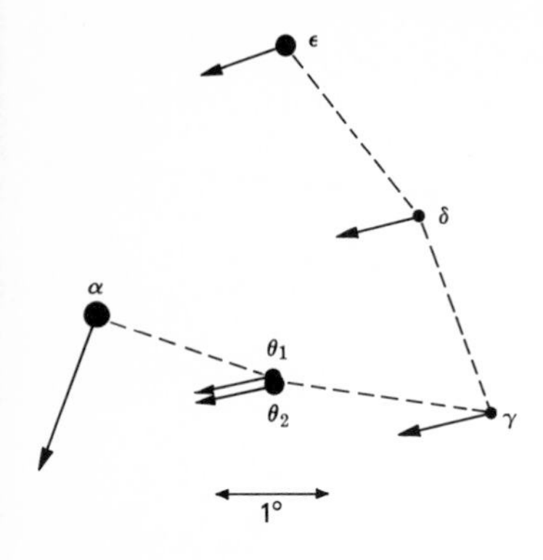

Figure 14.5. Proper motions of the brighter stars in the Hyades. Most of the motions appear to converge toward the left. The motion of Aldebaran is clearly unrelated. (Wyatt, p. 477.)

cide which stars really belong and which are simply in the same direction on the celestial sphere.

If a cluster is to maintain its identity over hundreds of millions of years, all of its stars must be moving through space in parallel paths and with the same space velocities. Some of this motion is revealed to us as proper motion, and for true cluster members the proper motions should all be the same or at least very nearly the same. We may mention the Hyades cluster again to see how proper motion can confirm cluster membership. The brightest star within the boundaries of the Hyades is the first magnitude star, Aldebaran. The proper motion of Aldebaran is quite different from that of any other star in the area (Figure 14.5), and so it is obvious that Aldebaran is not a cluster member but is a foreground star.

A further point may be noted regarding the proper motions of the Hyades stars. The proper motions seem to be converging toward a point well outside the boundaries of the cluster. This convergence is actually a perspective effect resulting from the general motion of the cluster away from us. In other cases an apparent divergence of proper motions indicates that clusters are coming toward us. By noting the angular distance between the convergent point and the cluster itself, the true direction of the cluster's motion may be calculated. For three or four clusters the proper motions and their convergent point may be combined with radial velocity data to find the distance to the cluster. The details of this so-called **moving-cluster parallax** will not be described here, but the method has been applied to the Hyades with significant results.

Again, because of their common space motions, the true members of galactic clusters should show common radial velocities. For the more distant clusters the proper motions become too small to be detected, and the radial velocities are useful in confirming cluster membership.

Color-Magnitude Diagrams

Cluster membership may also be confirmed by means of a very useful device known as a **color-magnitude diagram.** This is simply a diagram in which color index is plotted against apparent magnitude. The observations required for this are quite simple since all that is needed is an observation in each of two color systems for each star. Color index is dependent upon temperature (see Chapter 10) and is simply the numerical difference between the magnitudes in blue and visual light (B − V). The cooler stars are red and the color index is positive, reaching an extreme value of about +2.0. For stars of spectral type A, (B − V) = 0 and for hotter stars the color index is negative. Since the spectral type is also dependent upon a star's temperature, spectral type and color index actually mean the same thing. The color index, however, may be determined from very direct observations with fairly simple equipment. Being relatively close together in space, the members of a cluster are all very nearly the same distance from the sun. For this reason, the members which are really the brightest are the ones which will appear to be the brightest. In other words, there will be a direct relationship between the absolute and apparent magnitudes for the member stars in the cluster. This relationship may be likened to a group of people standing together some distance from an observer. Without knowing the actual height of any individual or the distance to the group, it is apparent which persons are taller and which are shorter.

Thus color index is similar to spectral class, and for the cluster members, apparent magnitude is an indication of absolute magnitude. When apparent magnitude is plotted against color index, therefore, the resulting color-magnitude diagram should resemble the H-R diagram described in Chapter 11. The color-magnitude diagram for the Hyades is shown in Figure 14.6. A well-populated main sequence is unmistakable. Aldebaran with an apparent magnitude of 0.80 and a (B − V) color index of +1.55 would be off the diagram and could not be a member of the cluster. We know that stars so far above the main sequence simply do not exist. The color-magnitude diagram has told us where the stars lie with respect to

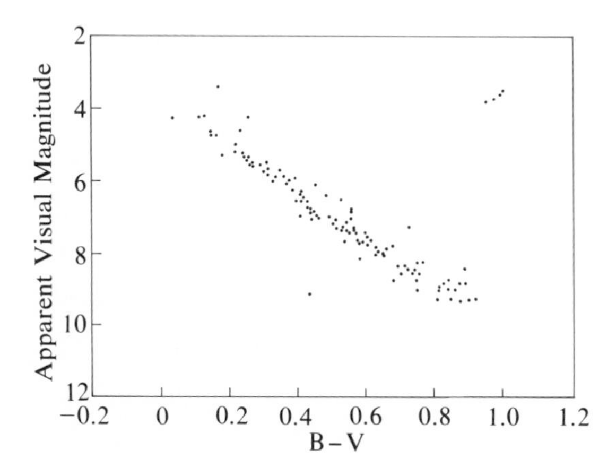

Figure 14.6. The color-magnitude diagram of the Hyades cluster. (Plotted at Leander McCormick Observatory from published data of H. L. Johnson and C. F. Knuckles.)

the standard H-R diagram. If a star suspected of cluster membership falls in an impossible portion of the color-magnitude diagram, that star is clearly *not* a member of the cluster.

The most significant aspect of the color-magnitude diagram is that from it the distance to the cluster as a whole may be determined with a high degree of accuracy. Knowing that the color-

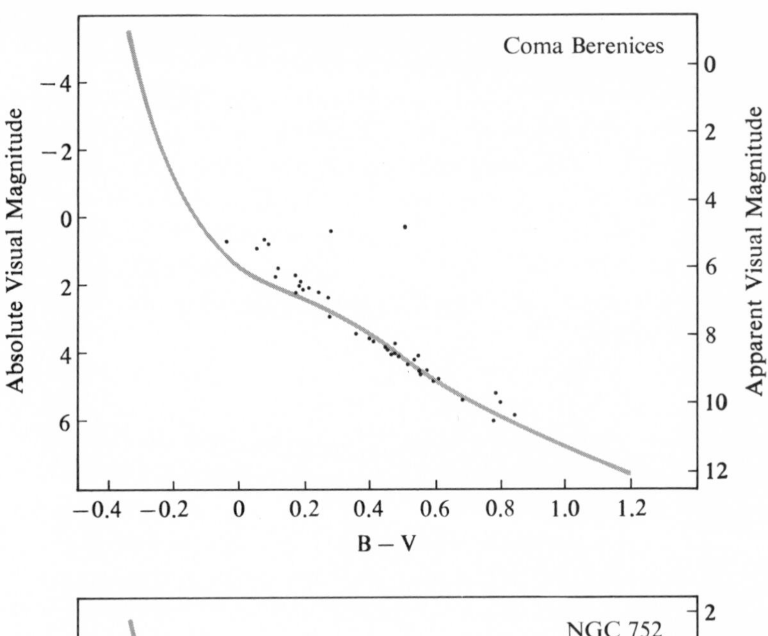

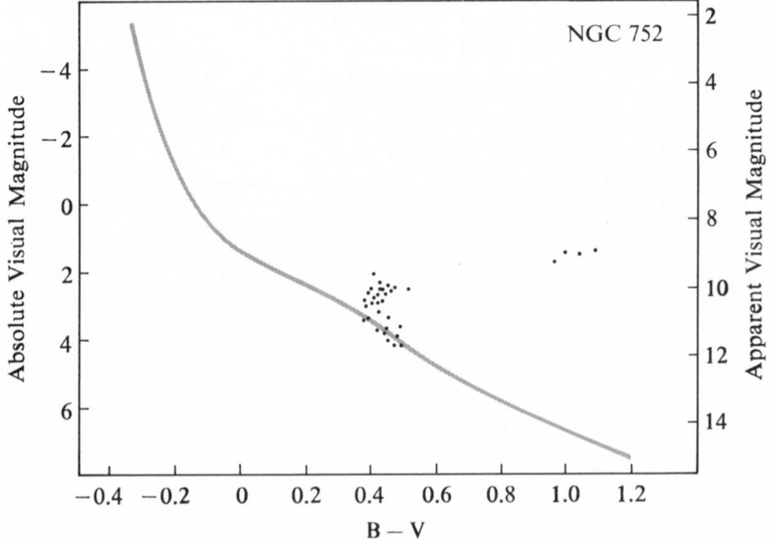

Figure 14.7. Color-magnitude diagrams for four galactic clusters. The solid line in each diagram is the "zero-age main sequence," which corresponds to the main sequence of the H-R diagram. If fainter stars had been included, it is likely that the actual main sequence in each diagram would be better defined. (Plotted at Leander McCormick Observatory from published data.)

magnitude diagram ought to resemble the H-R diagram, we may superimpose one diagram upon the other. Then assuming that nothing has affected the observed (B − V) colors of the stars, the color-magnitude diagram need only be adjusted vertically until its main sequence matches that of the H-R diagram. When this has been done the absolute magnitude of any individual star may be

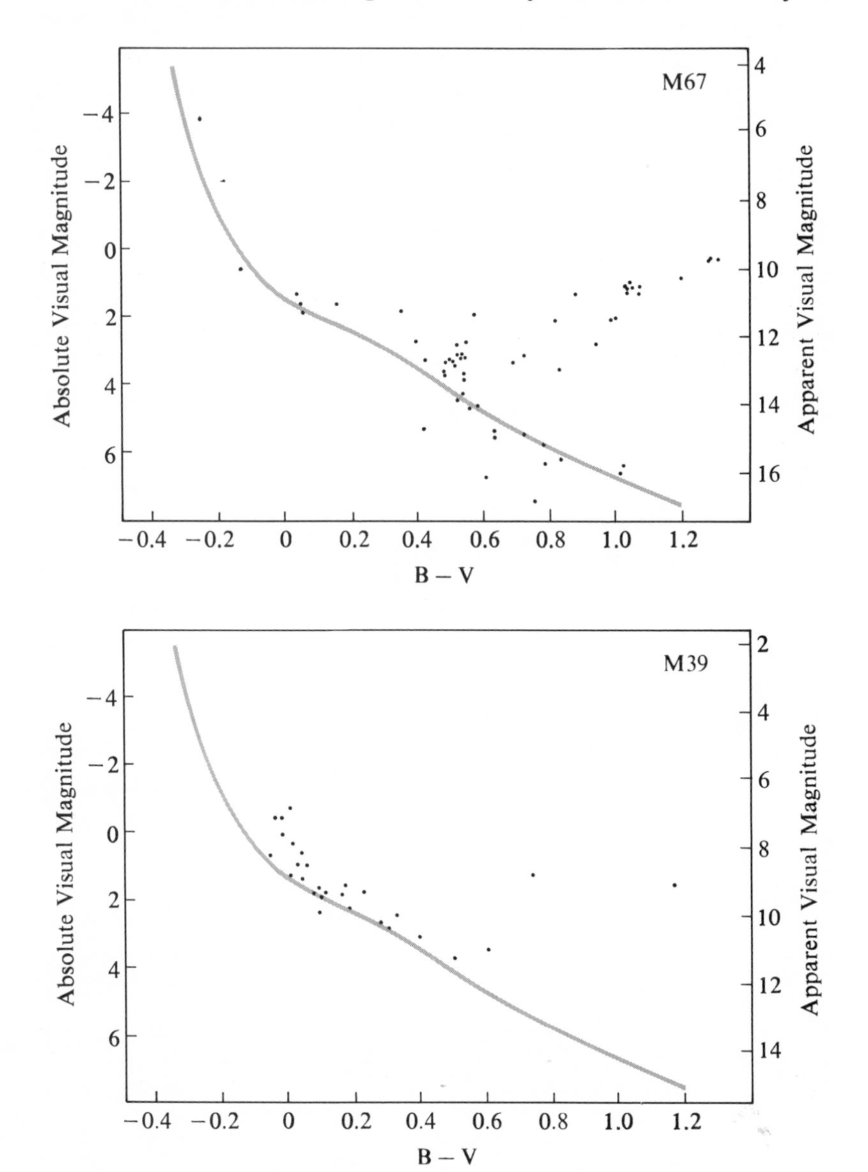

Figure 14.7. (continued).

read from the diagram. For each star, then, absolute and apparent magnitude may be substituted into the formula

$$M = m + 5 - 5 \log r$$

and the distance r may be calculated as in previous chapters.

The difference $m - M$ is referred to as the **distance modulus** and is a convenient quantity in common usage among astronomers. Clearly, for all members of a cluster, the distance modulus should be the same. Slight variations may arise due to the known width of the main sequence and to unavoidable observational errors.

Whereas all color-magnitude diagrams show a basic similarity to the H-R diagram, each is distinct from all the others. In Figure 14.7 the color-magnitude diagrams for a number of clusters have been reproduced. In each diagram the normal main sequence has been indicated. The clusters all show some stars on the main sequence, but there is considerable variety in the distribution of stars in the upper half of each diagram. Because of the great distances to these clusters, faint stars at the lower end of the main sequence are not detectable.

14.2 Associations

In addition to the galactic or open clusters, there are in the Milky Way certain less obvious and less dense groups of stars known as **associations.** About 80 such groups are recognized, and they contain anywhere from a few dozen to several hundred individual members. The associations have been very poorly defined, but we do know they have in common surprising numbers of stars of spectral types O and B. It was this characteristic which led to their first recognition as groups of related stars. In some cases a galactic cluster seems to exist as a kind of nucleus for an association. For example, the double cluster h and χ Persei is part of what has come to be known as the Zeta Persei Association. Another interesting property of the associations is that they seem to be expanding at surprisingly high rates, and a few stars seem to have been ejected from certain associations at truly large velocities. A few of the stars in the Orion Association may be seen in Figure 14.8 and in Color Plate 4.

From the manner in which the stars in associations are moving in relation to each other and from the expansion of the associations

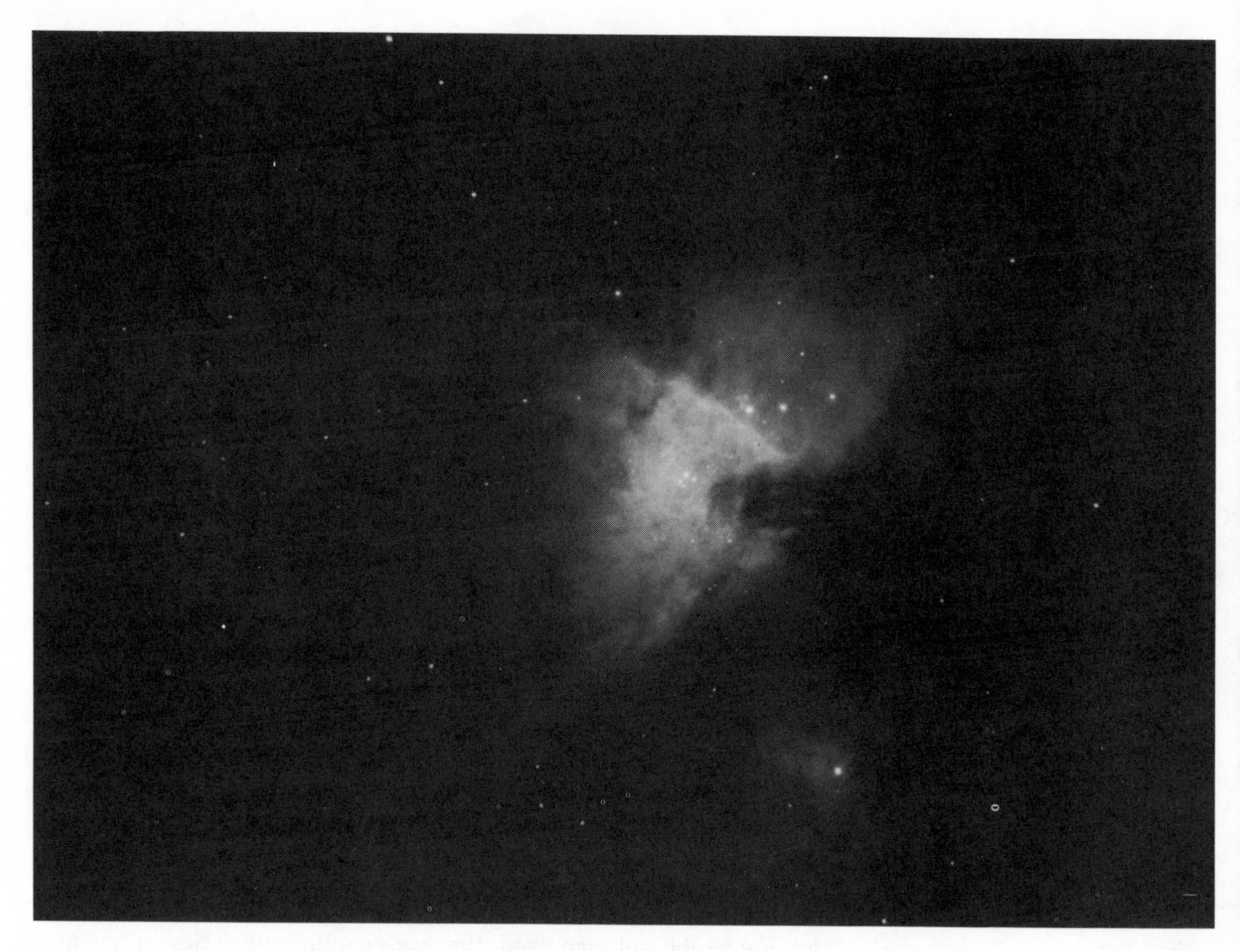

Figure 14.8. The bright stars embedded in the Orion Nebula form the central part of the Orion Association. (Photograph by J. C. Duncan.)

themselves, it is clear that all of the associations will lose their identities in a relatively short time, i.e., a few million years. The associations must therefore be recognized as young on the cosmic time scale. If we are to assume that their present appearance results from the formation of their member stars in the same region of space at more or less the same time in the past, then the member stars themselves are relatively young also.

Color-Magnitude Diagrams of Associations

A comparison of Figure 14.9 with Figures 14.6 and 14.7 reveals a rather striking difference between the color-magnitude diagrams for galactic clusters and those for associations. In the associations some stars are to be found with colors and magnitudes which locate

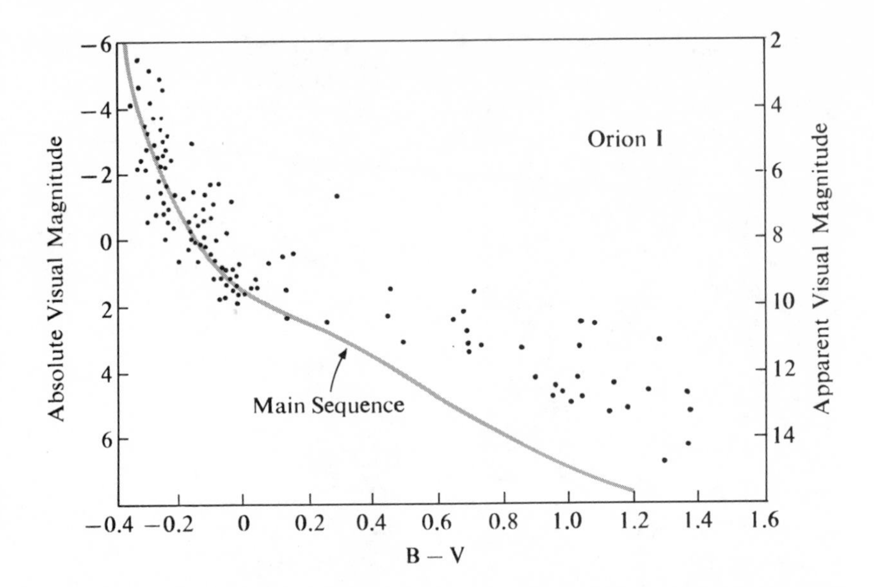

Figure 14.9. The color-magnitude diagram for the association Orion I. The cooler stars (those toward the right) are abnormally bright for their color indexes. This may mean that these stars are still in the process of contracting to a stable main-sequence configuration. (Plotted from published data of H. L. Johnson.)

them on the upper end of the main sequence. Down toward fainter, cooler stars, however, the stars appear to lie above the main sequence. This must mean that the stars are abnormally bright for their surface temperatures. It may be that excessive brightness is due to diameters which are larger than normal for stars of these surface temperatures. In Chapter 8 it was mentioned that contraction must be an important source of energy in the initial stages of a star's life. On the basis of this, the suggestion has been made that some of the young stars in associations are still contracting and have not yet acquired the stable characteristics of normal main-sequence stars.

14.3 Globular Clusters

The variety in appearance found among the galactic clusters is not found among the globular clusters. All globular clusters look similar to the cluster M13 in Figure 14.3. This cluster is in the constellation Hercules and may be seen with a small telescope during the summer months. The designation M13 indicates that it was the thirteenth in the catalogue of 105 faint, extended objects published by Charles Messier in 1781. In Figure 14.10, photographs of five globular clusters have been reproduced. All of these photographs are on the same scale, and the distance and actual diameter of each may be found in Table 14.1.

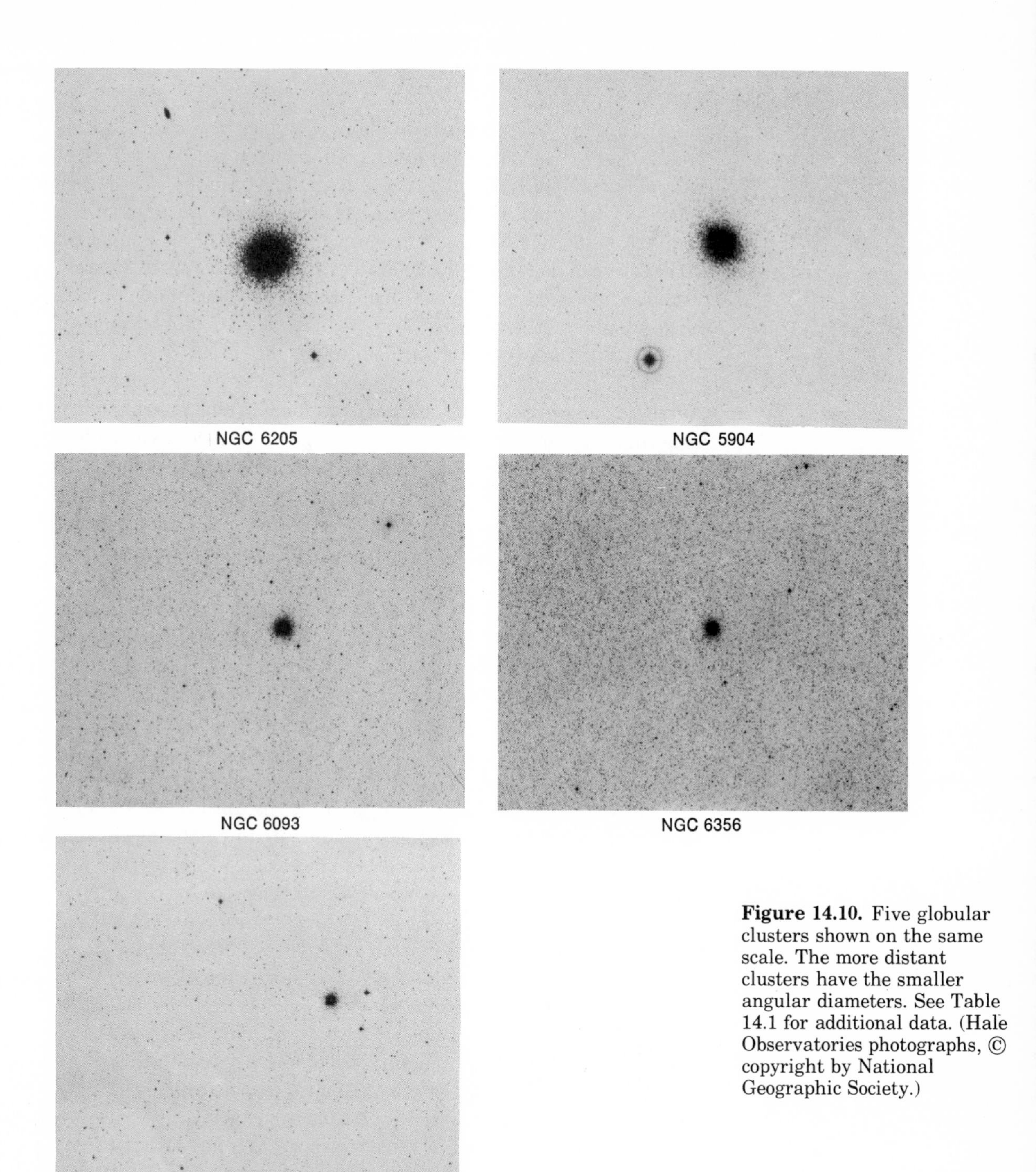

Figure 14.10. Five globular clusters shown on the same scale. The more distant clusters have the smaller angular diameters. See Table 14.1 for additional data. (Hale Observatories photographs, © copyright by National Geographic Society.)

Distances to Globular Clusters

As in so many astronomical problems, distance is a crucial factor that must be determined before the real characteristics of the globular clusters may be known. These huge aggregations of stars are much too far away for direct parallax measurements. Some of the indirect methods of Chapters 11 and 13 must therefore be invoked. Calculation of distances to the faintest globular clusters ultimately depends upon what is known about the closer ones. Fortunately, large numbers of RR Lyrae variables have been found in some of the nearer globulars and identified from light curves derived from a series of observations spaced in time (Chapter 12). So many RR Lyrae variables are known to exist in globular clusters that for a long time stars of this type were known as **cluster-type variables.** In the globular cluster Omega Centauri, for example, there are 167 variables. The distance to the clusters is so large compared with the dimensions of the cluster itself that the distance to any individual star, such as one of the variables, is reasonably assumed to be the distance to the entire cluster.

In practice, then, a series of observations of apparent magnitude m serves to identify a star as an RR Lyrae variable. The mean between the apparent magnitudes at maximum and minimum may then be used in the formula

$$M = m + 5 - 5 \log r$$

to find the distance, r. The mean absolute magnitude M is 0.5 for these stars, as noted in Chapter 13. Thus, in any globular clusters in which variable stars of this type have been identified the distance to the cluster may be determined with reasonable certainty.

Many globular clusters are so far away (see Figure 14.10 and Table 14.1) that the RR Lyrae variables are not detectable and one sees only the brightest individual stars in the outer parts of the cluster. From the nearer clusters it is known that the brightest stars are about two magnitudes brighter than the variables. Therefore, when the apparent magnitudes of the brightest stars have been established, a distance may be calculated using $M = -1.5$ for the brightest detectable stars in the cluster.

The most distant globular clusters appear simply as hazy spots on photographs, and no individual stars may be picked out. Distance estimates must now be based on the apparent magnitude of

Table 14.1 Selected globular clusters. These clusters have been selected because their linear diameters are all somewhat similar. (Data from G. Kron and N. Mayall, Astronomical Journal, Vol. 65, page 581, 1960.)

NGC No.*	Linear Diam. (pc)	Ang. Diam. (arcsec)	Distance (kpc)
6205 (M13)	34	12.9	9.1
5904 (M5)	30	10.7	9.5
6093	33	8.6	13.0
6356	31	6.3	17.0
6229	31	3.6	30.0

*NGC refers to the New General Catalogue of Nebulae and Clusters of Stars by J. L. E. Dreyer, published in 1888.

the cluster as a whole. For the nearer clusters whose distances are known, an absolute magnitude may be computed for the integrated light of the entire cluster. There is some variety in the absolute magnitudes of the nearby clusters, but from these it is possible to establish that they range from about $M=-5$ to $M=-9$. These values may then be used with the apparent magnitudes of the faint globular clusters to provide estimates of their distances. In a similar way the range in linear diameters of the clusters may be ascertained for the nearby clusters and used with the observed angular diameters of the faint clusters in order to estimate distances.

The Omega Centauri cluster is one of the largest globulars known and is also one of the closest, being about 4800 pc from the sun. The faintest globulars are of the order of 70,000 pc from the sun. The accuracy of the methods used to estimate these vast distances is admittedly not very great. Nevertheless, the distances are known with sufficient accuracy to allow the general distribution of the globular clusters to be established.

Color-Magnitude Diagrams for Globular Clusters

A color-magnitude diagram may be constructed for a globular cluster in exactly the same manner as for the galactic clusters. Magnitudes in two color systems are determined in order to provide a color index. Visual magnitude is then plotted against the color index. When such a diagram is constructed for a globular cluster, the surprising result is that it does not look like the H-R diagram at all. The kinds of stars found in the globular clusters seem to be rather different from those found in the solar neighborhood and in the galactic clusters. The color-magnitude diagram for the globular cluster M13 is shown in Figure 14.11, and the appearance is quite striking. If this diagram is superimposed on the H-R diagram, one can see that only the lower part of the main sequence is present. There do not seem to be any main-sequence stars brighter than $M=2.0$. The long extension upward and to the right must represent yellow and red subgiants and giants. Other very bright giant stars fall along what has come to be known as the **horizontal branch.** The empty space in the horizontal branch is noticeable only because the RR Lyrae variables have not been plotted in this diagram. If included, they would fall in this space.

It is somewhat difficult to construct the color-magnitude diagrams for globular clusters because of the difficulty of picking out the faint stars. If a photograph is to record the faint stars, the exposure time must be rather long and the central portion of the cluster will thus be overexposed. The faint stars on the diagram are necessarily from the outer parts of the cluster. The stars from the lower end of the main sequence are probably present in the globular clusters, but they are simply too faint to be detected at such great distances from the sun.

Distribution of the Globular Clusters

There are 119 globular clusters known today, and it has been shown statistically that there are probably very few that have not yet been discovered. Almost all of the known globulars lie in the half of the sky which is in the direction of the constellation Sagittarius. Of the total number, more than 30 lie in the direction of Sagittarius itself and fall within an area that represents only about 2% of the area of the celestial sphere. In the early years of this century, Harlow Shapley sought for the correct interpretation of this very uneven distribution of globular clusters and gave us our first insight as to the overall size and shape of the great system of stars of which our sun is a part.

By the methods described earlier in this chapter, Shapley worked from the nearest globular clusters to the most distant ones, estimating distances to a large number of them. Knowing their distances and their directions in space, he could show clearly that the globular clusters filled a roughly spherical volume of space.

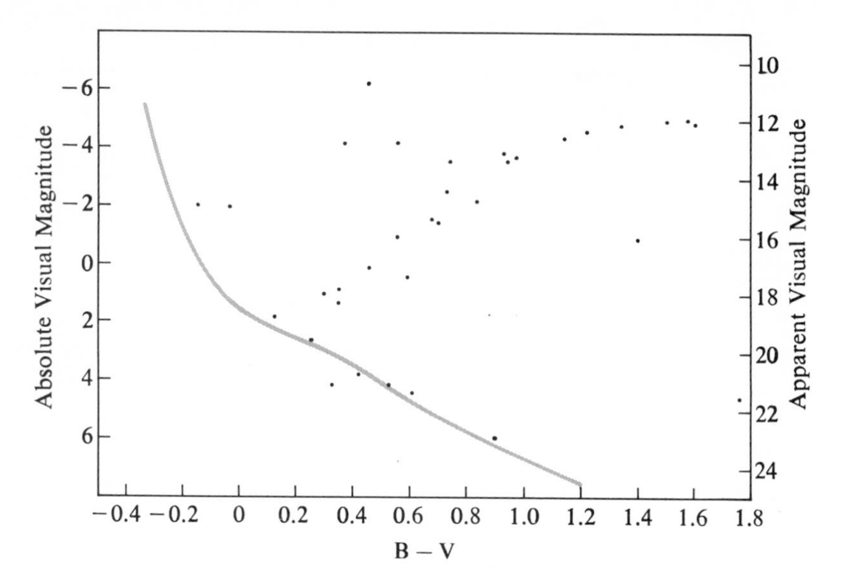

Figure 14.11. The color-magnitude diagram for the globular cluster M13. The data have been plotted in such a way that the cluster main sequence coincides with the zero-age main sequence. Note especially that the faintest stars are near the limit of the 200-inch telescope. (Plotted from published data of W. A. Baum.)

Most important was the fact that this globular cluster system was centered not at the sun, but at a point some 15,000 pc away from the sun in the direction of Sagittarius. Some notion of this concept may be obtained from Figure 14.12, which shows the positions of the globular clusters projected into the plane of the Milky Way. The constellations of the Milky Way lie in the plane of the page, and the true positions of the clusters should be either above or below the plane of the paper.

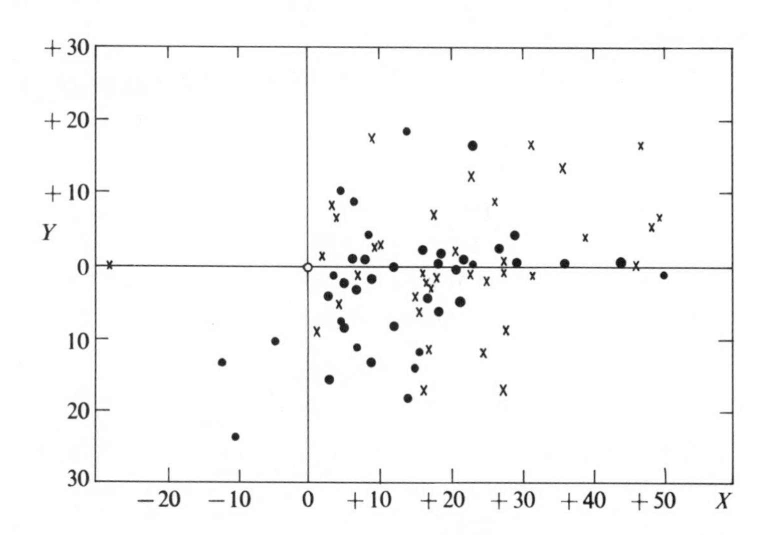

Figure 14.12. Locations of globular clusters projected into the plane of the Milky Way. The crosses indicate clusters north of the plane. The dots indicate clusters south of the plane. Diminishing size of the symbols indicates increasing distance from the plane. The sun is represented by the open circle.

Shapley went on to infer that the center of the vast globular cluster system was also the center of the system of stars. In his view, the sun was well out from the center of a fairly thin, disc-shaped mass of billions of stars around which the globular clusters formed a sort of a swarm or halo. The Milky Way, that conspicuous band of stars running around the entire celestial sphere, is simply the view of this flattened system seen from a point inside it. Earlier astronomers, including William Herschel and J. C. Kapteyn, had also looked upon the Milky Way as a flattened system of stars. It was Shapley, however, who first attempted to locate the center of this system and estimate its true size. Today we can find little fault with Shapley's reasoning or with his conclusions. His model of the distribution of the stars and globular clusters is essentially correct. The role of interstellar material in dimming the light of distant objects was not recognized at the time of these early studies and, as a result, Shapley's estimate of the distance to the center of this large system was somewhat larger than the currently accepted value of about 10,000 pc.

QUESTIONS

1. Describe the major differences between the galactic clusters and the globular clusters.
2. How can membership in a galactic cluster be confirmed observationally?
3. Sketch, with properly labeled scales, the color-magnitude diagram for a galactic cluster. How can such a diagram be used to estimate the distance to such a cluster?
4. What is an association and why are associations recognized as relatively young groups of stars?
5. What is the principal method by which the distances to the nearer globular clusters were determined?
6. Compare the color-magnitude diagram of a globular cluster with that of a galactic cluster.
7. Describe the general distribution of the globular clusters in the sky.

8. How did the distribution of globular clusters lead Shapley to the first reliable value for the distance from the sun to the center of our galaxy?

15

Interstellar Material

A telescopic view of the "sword" in the constellation Orion shows extended luminous clouds (pictured in Figure 15.1). Long-exposure photographs show that the fainter parts of these clouds cover an area of the sky comparable in size to the full moon. Similarly, photographs of the Pleiades show that the member stars of that cluster are also embedded in vast wispy clouds. High magnification with larger telescopes fails to resolve these clouds into individual stars as Galileo resolved the Milky Way. Such extended bright areas are true clouds of interstellar material. They are, however, only the most obvious of the many forms in which great quantities of matter are to be found in space. The term **nebula** (the Latin word for *cloud*) has been used to refer to many of the clouds in which interstellar matter is situated. Some of these are nonluminous, and so we speak of **dark nebulae** as well as bright nebulae. Much of the actual material between the stars is revealed only by its effect on the light of stars shining through it, and some of the most significant recent studies have made use of the fact that some of the clouds emit radio signals which can be detected with radio telescopes here on earth.

With the exception of some clouds detected by radio astronomers which do not radiate in the visible spectrum, all of the interstellar material within our own system of stars lies within 20° of the

◄**Figure 15.1.** The Orion Nebula, an example of an emission nebula. This immense cloud of gas is illuminated by the radiation from a group of very hot stars in the overexposed central region of this photograph. Compare this photograph with Figure 14.8. (Hale Observatories photograph.)

direction to the Milky Way. This is the same general region in which the galactic clusters, associations, and Cepheid variables are found.

15.1 Bright Nebulae

Astronomers have studied the light from bright nebulae with special spectrographs and, in doing so, have found that the bright nebulae fall into two dsitinct groups. Some, such as the Orion Nebua, show a bright-line spectrum and are known as **emission nebulae**. (See Figure 15.2 and Color Plates 4 and 5.) Others show a continuous spectrum and are called **reflection nebulae** since material in the cloud is actually reflecting the light of a nearby star. The nebulosity surrounding the Pleiades (Figure 15.3 and Color Plate 3) is an example of the latter type.

Emission Nebulae

Quite obviously, a bright nebula which shows an emission-line spectrum must be composed of gas, and the kinds of atoms composing the gas may be identified from the spectrum. The principal gas in emission nebulae is hydrogen, but there is evidence also for the presence of helium, oxygen, and nitrogen. Within or near an emission nebula, one or more very hot stars are always to be found. These stars are always spectral type B1 or hotter and are, therefore, strong emitters of ultraviolet radiation. This energy is absorbed and reradiated to produce all of the observed spectral features of the bright nebula.

The intense ultraviolet radiation from the star ionizes all of the hydrogen atoms in its neighborhood. The distance to which the star can ionize its surrounding gas depends upon the density of the gas and the temperature and luminosity of the star. A hot O6 star can, for example, ionize all of the hydrogen within a radius of 100 pc. A B0 star can ionize hydrogen out to about 40 pc and an A0 star is

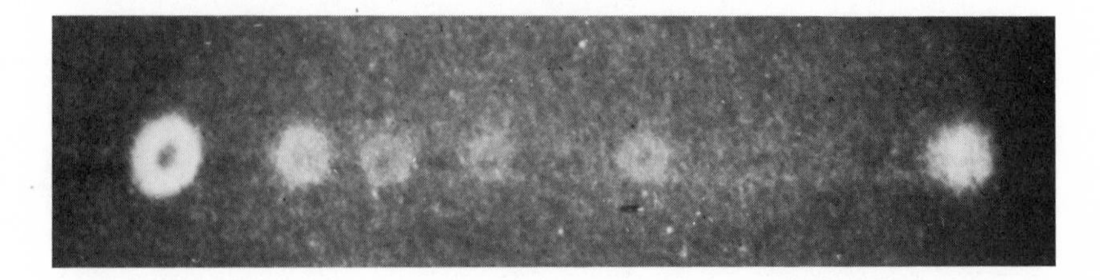

Figure 15.2. A portion of the spectrum of the Ring Nebula, a particular type of emission nebula known as a planetary nebula. Each image of the ring is at a different wavelength in this objective-prism spectrogram. (Yerkes Observatory photograph.)

Figure 15.3. Nebulosity surrounding stars in the Pleiades. An example of a reflection nebula in which light from the embedded stars is reflected from dust particles in the clouds. (Official U.S. Navy photograph.)

effective in ionizing hydrogen only within about 1 pc of itself. It is customary to refer to neutral hydrogen as HI and ionized hydrogen as HII. Similarly, ionized oxygen is OII, and doubly ionized oxygen is OIII. In this terminology the areas of ionized hydrogen around hot stars are referred to as **HII regions.** Within the HII regions there are always some atoms which have just recaptured an electron, and it is from these atoms that the bright-line spectrum of hydrogen comes. The Balmer lines are the only lines detectable from the earth's surface, but the lines of the ultraviolet Lyman series and the infrared Paschen series are present as well. They are absorbed in the earth's atmosphere. After an electron has been captured, the atom quickly loses energy by emitting light and returns to the ground state. The flux of ultraviolet energy from the nearby hot star is so great, however, that in a very short time the atom will be ionized again. Any individual atom will spend most of its time in an ionized state. Only occasionally will it capture an

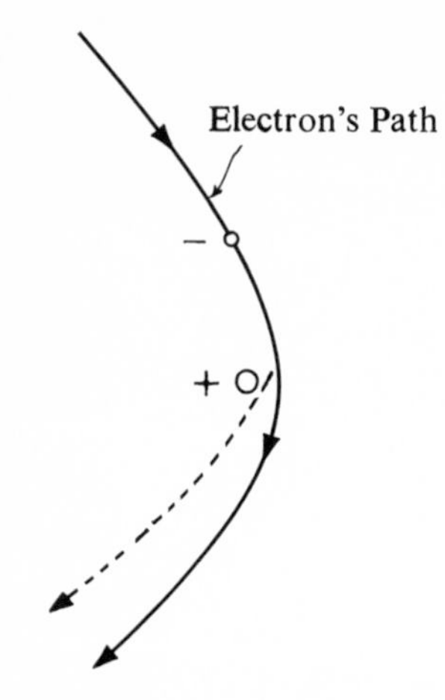

Figure 15.4. Free-free transitions. A free electron passes close to a proton without being captured. Energy lost in this encounter is radiated in the radio wavelengths.

electron and be able to emit. Throughout the nebula there will nevertheless be enough recombinations at any time to make the cloud glow brightly.

Within the HII regions there will also be many situations in which a free electron passes near a proton but does not come close enough to be captured. The path of the electron as it approaches the proton and passes it ought to be a hyperbola, as shown in Figure 15.4. It may actually happen that the electron will approach the proton along one hyperbola and leave along a different hyperbola. This is indicated by the dashed path in Figure 15.4. The electron must lose a small amount of energy in order to do this, however, and is said to have undergone a **free-free transition.** The electron is free before the transition and afterward as well. Since the energy of the electron is changed only very slightly, the energy emitted will be in the form of a photon with a wavelength of several centimeters. There can be great variety in the manner in which an electron approaches a proton, and so the energy lost in free-free transitions may vary considerably. From an HII region there will thus be radiation at many wavelengths, producing what is essentially a continuous spectrum. This radiation of centimeter wavelengths lies in the radio part of the spectrum. The HII regions have therefore received considerable attention from radio astronomers.

The most conspicuous optical features in emission nebulae are two bright lines in the green region of the spectrum (4959 Å and 5007 Å). For a long time these lines were not identifiable with the laboratory spectrum of any known chemical element, and the name *nebulium* was invented on the chance that this was a new element. Eventually it was realized that the two lines came from doubly ionized oxygen atoms which were undergoing transitions which could not occur under the then available laboratory conditions. Certain energy levels for OIII and other atoms are described as **metastable levels** because downward transitions from these levels do not occur very readily. An atom excited to a metastable level might remain that way for several hours, whereas an atom excited to a normal energy level would make a downward transition in a few millionths of a second. In the laboratory it is difficult to make an atom remain in a metastable level long enough for a downward transition to occur. Collisions with electrons or other atoms will raise the atom from the metastable level to a higher one from which downward transitions will occur more readily. For this reason, spectrum lines resulting from transitions starting in metastable levels are referred to as **forbidden lines.** The astrophysicist I. S. Bowen first recognized the "nebulium" lines as forbidden lines of doubly ionized oxygen. In the low-density condition of space and without any container walls with which to collide, atoms in metastable energy levels can remain undisturbed until they have had time to emit. In addition to the two prominent oxygen lines, there are other forbidden lines from oxygen and also from nitrogen, neon, and other elements which have been identified.

Reflection Nebulae

The basic difference between reflection and emission nebulae may be noted in their spectra. Reflection nebulae show a continuous spectrum and a similar, but slightly bluer, color than that of the illuminating star. These clouds must be composed of particles which reflect and scatter the light of the central star. Scattering by atoms may be ruled out because of the very high density needed for a gas cloud to scatter light as efficiently as the nebulae do. Scattering by molecules may also be eliminated on the grounds that molecular scattering is very selective of blue light. In the earth's atmosphere, molecular scattering causes the sky to appear blue. However, there is not enough difference between the color of a reflection nebula and its illuminating star for the scattering to be caused by molecules. We are left with the conclusion that the scattering is due to the presence of particles larger than molecules which may well be described as **dust.**

Some insight into the nature of these dust grains may be gained by comparing the total light scattered by the nebula to that radiated from the central star. This comparison indicates that the scattering particles reflect a very large percentage of the light incident upon them. This, coupled with the known high abundance of hydrogen in the universe, has led to the suggestion that the nebular particles may be frozen compounds of hydrogen. These frozen particles, only a few hundred thousandths of a centimeter in diameter, probably are much smaller than fine sand. It is also observed that the light from the reflection nebulae is polarized to a significant degree. If a polarizing filter were placed in a telescope and rotated as the nebula was being observed, it would be seen that the brightness of the nebula depends upon the orientation of the polarizer. In order that the nebula may polarize as well as scatter light, the dust grains must be elongated in shape and lie more or less parallel to each other. "Ice" crystals of some sort (possibly even frozen water) might well have the proper shape, and interstellar magnetic fields might bring about the necessary alignment provided that some iron compounds are present in the particles. Undoubtedly, scattering from some free electrons is also present.

Since we see reflection nebulae only because of their proximity to bright stars, there is little difficulty in determining the distances to the clouds. The distance to the illuminating star is essentially the distance to the bright nebula, whether it be composed of gas or dust. When the distance is known, the angular diameter of a nebula may be converted to an actual linear diameter in parsecs. In the next few paragraphs it will be shown that the presence of interstellar material makes stars appear to be fainter than they ought to be. As a result, one might overestimate the distances to the bright nebulae if the dimming by interstellar material is not considered.

15.2 Other Evidence of Dust and Gas

It takes a very bright star, such as one of spectral type B5, to illuminate a reflection nebula of such dimensions that it can easily be detected. Likewise, only the very bright, very hot stars can ionize the gas sufficiently to cause large HII regions. In some cases the limits of the HII region will simply be the actual edges of the gas cloud itself, but in other cases only that part of a very large cloud near the hot star will be seen. The existence of both the emission and the reflection nebulae in many parts of the Milky Way implies that there must be other huge quantities of both gas and dust which are not very obvious simply because they do not happen to have a nearby star which can illuminate them. Much of this nonluminous material is detectable through its effect on the light of background stars.

Not far from the great Orion Nebula of Figure 15.1 is another bright nebula pictured in Figure 15.5. Silhouetted against this glowing cloud is an irregular dark mass referred to as the Horsehead Nebula. This is one of the more obvious examples of what may be called **dark nebulae.** Other dark nebulae, the Gulf of Mexico and the Atlantic Ocean in the North America Nebula, appear in Figure 15.6. Less obvious dark nebulae appear as starless areas once called "holes in the sky." A good example is the Coal Sack, a dark blot on the Milky Way near the Southern Cross. Also, where the Milky Way runs through the constellation Cygnus in the summer skies, a large dust cloud blocks the center of the Milky

Figure 15.5. The Horsehead Nebula in Orion. It is a good example of dark clouds obscuring the light of more distant objects. (Hale Observatories photograph.)

Figure 15.6. The North America Nebula in Cygnus. The dark areas which outline the "continent" are due to clouds of opaque material which block the light of more distant stars. (Case Western Reserve University photograph.)

Way, causing it to appear to divide into parallel bands. A region like the Horsehead Nebula clearly indicates the presence of a nearly opaque mass. But how can one be sure that a starless area on the Milky Way is not in fact a tunnel through which we can see to the great emptiness beyond the limits of our entire galaxy? When we study the details of our galaxy in Chapter 17, it will be clear that such holes are not very likely to exist. There is, however, solid evidence to indicate that most of the dark areas in the Milky Way are dust clouds situated fairly close to the sun.

In many of the dark nebulae a few stars can be seen and classified as to spectral type. When studied photometrically, however, most of them will have color indices which are larger than they ought to be. We have seen already that spectral type and color index are both indicative of temperature. Therefore, when a star's spectral type is known, we know what its color index ought to be. That is, we can assign a **normal color** to stars of each spectral type. The difference between the color index actually observed and the normal color index assigned to a star on the basis of its spectral type is referred to as a **color excess.** When a star has a significant color excess, the star is too red, and we want to try to find out what might have caused this effect.

Most of the stars with large color excesses lie in or near the Milky Way, and we have just mentioned that the presence of bright nebulae implies the presence of nonluminous material. There is little doubt that the reddening is caused by dust grains of the same type found in reflection nebulae. Such clouds of grains will scatter the light of stars shining through them. The scattering will be greatest for the shorter wavelengths, with the result that the photons of blue light will be sent off in random directions. Most of the photons of red light will come through affected very little. This is the same effect that makes our setting sun appear redder and dimmer than the noonday sun. The path of sunlight in the atmosphere is longer at sunset than at noon. The scattered photons of blue light come to us from many directions and cause our daytime sky to appear blue. Similarly, it is scattered blue light that causes the reflection nebulae to seem slightly bluer than the stars which illuminate them. The blue color of the nebulosity surrounding the Pleiades is conspicuous in Color Plate 3.

By comparing reddened and unreddened stars of various spectral types, astronomers have acquired a good understanding of interstellar reddening as evidence of interstellar dust. It is clear that if the blue light has been removed from the measurable light of a star, that star is going to appear dimmer as well as redder than it otherwise ought to be. The relation between reddening and dimming is well established, so that when a star's color excess is measured, the effect of absorption or dimming may be computed. In visual magnitudes the absorption is approximately three times the color excess. We could write $m_v = m_{obs} - 3(\text{color excess})$, where m_v is the true visual magnitude of a star and m_{obs} is the observed visual magnitude. This is extremely important since any star which is even slightly reddened will appear to be farther away than it really is. The distances to the galactic and globular clusters discussed in the previous chapter must be corrected for the effects of absorption. Shapley's original estimates of the distances to globular clusters did not take this effect into consideration, since it was unknown at the time. It is for this reason that his original model of the size of our Milky Way system was too large. It is now realized that absorbing dust is so prevalent in the Milky Way that its effects must always be taken into account.

An excellent example of the dimming effect of interstellar material is demonstrated when one counts the numbers of stars of successively fainter magnitudes. In actual practice this method is seldom used today but may be considered one of the classical techniques in the study of interstellar absorption. In Figure 15.7 we imagine an observer looking at a square area on the sky of some angular size x. The observer will be able to see stars within a pyramid-shaped volume defined by the angular size of the area studied and the distance to the farthest star in the area. Let us assume for a moment that the stars considered are all of the same absolute magnitude and color, and let us assume further that the stars are uniformly distributed in space. On a photograph, then, the bright stars must be the closest ones. Again from Figure 15.7, the volume of space in which the bright stars fall must be relatively small, and only a few bright stars will be found. Thus, as one proceeds to count the numbers of stars in certain magnitude ranges, one should expect to count progressively larger numbers of fainter stars. The stars in each magnitude interval are farther away than the stars in the previous interval and lie, therefore, within a larger volume of space. The numbers counted in each magnitude interval should therefore progress in a predictable manner.

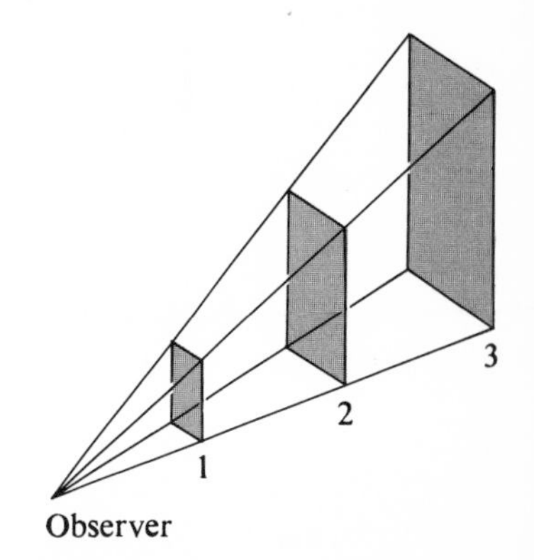

Figure 15.7. The volume of space increases with distance.

We know, of course, that the stars are not all of the same brightness and that some of the faint ones must really be close to us. But we also know how many stars of each magnitude to expect in a typical region of space. This may in turn be related to the distance-volume concept to enable us to predict the number of stars of any magnitude that would be expected in a given area of the sky. On a photograph of the sky one may then make actual counts and compare these with the predicted numbers. A graph on which the numbers of stars of successively fainter magnitudes are plotted is called a **Wolf diagram,** and a typical sample is shown in Figure 15.8. Here the solid line shows how the numbers of stars of fainter and fainter magnitudes should increase. The dashed line represents the actual numbers counted in some area. For the brighter stars there is good agreement between the two curves. The sudden divergence of the dashed line from the solid line shows that there are too few stars fainter than some magnitude. If there should be a dust cloud in this area, all the stars beyond the cloud would appear too dim, and the counted numbers would behave in this manner. From the actual star counts, astronomers developed analytical methods which revealed quite well the distances to the absorbing clouds and the actual thicknesses of the clouds. Star counts are some of the basic data in the area known as **statistical astronomy.** These methods reached their peak development just before World War II.

An extremely useful modern technique makes use of the two color indexes which may be derived from observations made at the three wavelengths of the UBV system. These two color indexes are B − V and U − B, and they may be determined with relative ease as

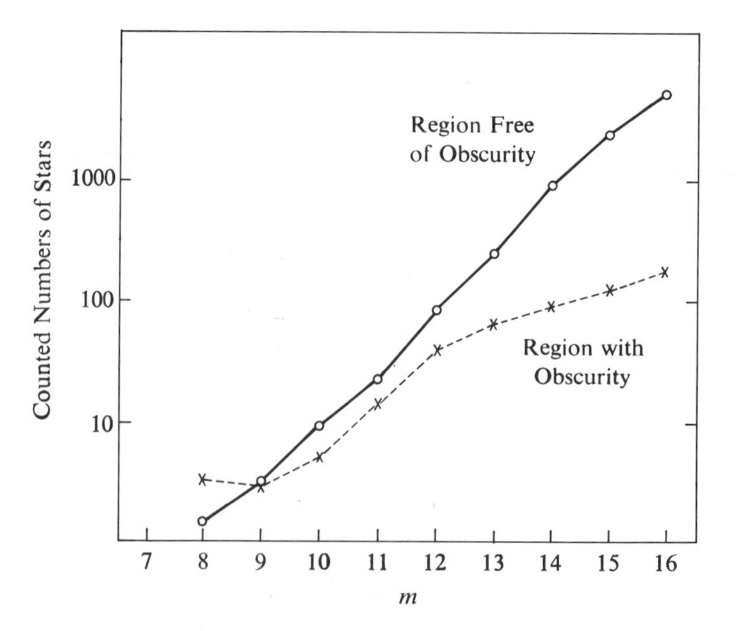

Figure 15.8. A Wolf diagram showing the effect of discrete clouds of obscuring material which affect the counted numbers of stars of successively fainter magnitudes.

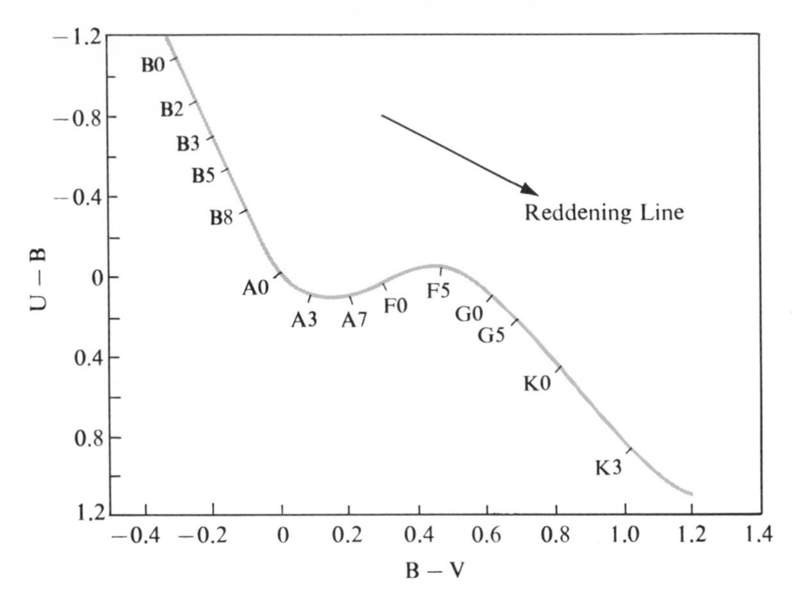

Figure 15.9. The color-color diagram in the UBV system.

described in Chapter 10. For a group of stars the values of B − V and U − B may be plotted against each other to form a color-color diagram as indicated in Figure 15.9. A normal, unreddened main-sequence star should fall somewhere on the solid line drawn in the figure, and the specific location of a star depends upon the star's spectral type. Since reddening affects both color indexes, a reddened star will be shifted away from this line. By means of both theory and observation, it has been established that the effect of reddening is to move a star in a direction parallel to the arrow in the diagram. Thus, the color-color diagram can be quite useful in the recognition of stars affected by interstellar dust. It should be pointed out that this diagram may be applied to stars regardless of their distances, whereas the color-magnitude diagram described earlier is valid only if the stars are all at nearly the same distance from the sun. Some ambiguities can arise in the color-color diagram because of the hump in the line for main-sequence stars and because giants and white dwarfs lie above the region of the main-sequence stars. Nevertheless, the color-color diagram has become an important tool in many studies of highly reddened areas of the sky.

In Chapter 12 we discussed the spectroscopic binary β Lyrae and pointed out in Figure 12.9 a narrow, well-defined absorption line present in all of the spectrograms but showing no periodic Doppler shifts. It was mentioned on page 253 that this line was due to interstellar gas and was used as a reference for the changes in position of the other lines due to orbital motion. The star's characteristic pattern of absorption lines arises in its own atmosphere. The light then must traverse great distances in space to reach us. Within this space there are extensive, low-density clouds which impress their own characteristic lines on the spectrum. The stellar lines tend to be considerably broadened by the pressure in the star's atmosphere, whereas the interstellar lines at very low temperature and pressure are quite narrow and sharp. As may be noted in Figures 12.9 and 15.10, the stellar and interstellar lines are quite distinct from each other in appearance.

The first interstellar lines detected were those of ionized calcium (CaII), and for many years there was little agreement as to the correct understanding of their presence. One of the problems was that these interstellar calcium lines were found only in the spectra of very distant hot stars. Astronomers tried, therefore, to relate the extra calcium lines to small regions in which the hot star could ionize the gas. Eventually it was shown that atmospheric CaII lines in cooler distant stars were so strong that the interstellar lines were simply lost in the broad stellar lines. Larger radial velocity differences between star and cloud could have helped to separate and distinguish the stellar and interstellar lines, but such velocity differences simply had not been detected in the early years of this century.

Today with very large telescopes and high-dispersion spectrographs, interstellar lines are detectable in the spectra of a great many stars. In addition, there are cases in which the interstellar lines are multiple, indicating that several distinct clouds lie along the line of sight to the star and that the clouds are moving at different radial velocities. Thus, the Doppler shift for each cloud is different. It has also been discovered that in the Milky Way interstellar lines are progressively stronger in the spectra of more and more distant stars, indicating a very general distribution of interstellar gas within the plane of the Milky Way.

Determination of distances to the absorbing dust clouds and to the gas clouds which reveal themselves as extra absorption lines is

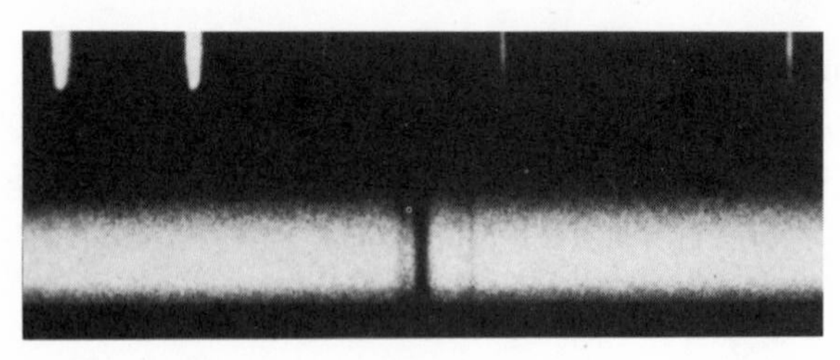

Figure 15.10. Absorption lines due to interstellar gas are often seen in the spectra of distant stars. Multiple lines of this sort indicate the presence of several clouds of gas moving with different radial velocities along the line of sight. This is the region of the K line in the spectrum of HD 47240. (Hale Observatories photograph.)

much less positive than for the bright nebulae. The best approach is to establish limits within which the interstellar material must lie in a given area. The distances to stars shining through the clouds may be determined by the usual methods, and we may know that the cloud lies in front of them. If a nearer star shows no evidence of interstellar obscuration, then obviously the clouds lie beyond that star. Thus, we may know with some degree of confidence that the cloud lies beyond certain stars and nearer than certain others.

In addition to those of ionized calcium, interstellar lines of sodium, potassium, and neutral calcium have been found. Absorption bands of interstellar CN and CH molecules have also been detected. As with the emission nebulae, the most abundant interstellar gas is hydrogen. The Balmer lines in the hydrogen spectrum would not be expected under the interstellar excitation conditions, so there is no optical evidence for interstellar hydrogen. Our knowledge of the vast quantities of interstellar hydrogen comes from an interesting and important emission line in the radio region of the spectrum.

15.3 Radio Emission from Neutral Hydrogen

In interstellar space atoms exist under conditions which are far different from those found in laboratory situations. In vast clouds of hydrogen the gas density is less than one atom per cubic centimeter, and collisions between atoms must be very rare. Under such conditions, hydrogen atoms may emit very low-energy photons which would be detectable in the radio region of the spectrum. It was shown theoretically by H. C. van de Hulst of the Netherlands that it should be possible for the hydrogen atom to exist in a metastable level only slightly above the ground state. This excited state represents the case in which the direction of spin of the orbital electron is the same as the direction of spin of the proton or nucleus. In the lowest possible energy state the spins are opposite to each other. Van de Hulst showed that the transition from the state of like spin to the state of opposite spin should result in the emission of a photon with a wavelength of 21 cm. He cautioned that collisional excitation to this level would be very rare and that the downward transition would also be very rare. For a single atom a 21-cm photon would be emitted only once in 10 or 11 million years. However, if there were large enough numbers of hydrogen atoms in space, a signal at 21 cm might be detectable. By 1951, radio detectors had been developed that could pick up the 21-cm hydrogen emission. Since the emission mechanism is a downward transition resulting in radiation at a discrete wavelength, it has become customary to borrow from the terminology of optical spectroscopy and refer to this radio emission as the **21-cm line of hydrogen.** Since the 21-cm signals come from neutral hydrogen atoms, the regions from which these signals come are often referred to as HI regions.

The actual detection technique involves measurement of signal intensity throughout a range of wavelengths centered on 21 cm. A plot of such measures is indicated in Figure 15.11a and is referred to as a **line profile.** Almost all of the radiation should be centered exactly at 21 cm, but because of random motions in each gas cloud, single emission features show somewhat broadened profiles as do most optical spectrum lines.

The effect of radial velocity on the 21-cm line is exactly comparable to the effect on optical lines, and displacements of the line maximum away from 21 cm can be correlated with the average radial velocity of the cloud. Figure 15.11b indicates a line profile shifted toward shorter wavelengths by a negative radial velocity of the emitting cloud.

The line profile in Figure 15.11c shows three peaks shifted away from 21 cm by varying amounts. Such a profile indicates that in the direction in which the radio antenna was pointed there were three hydrogen clouds moving at different velocities. There is no way to determine the distances to the emitting clouds from the line profiles themselves. In Chapter 17, however, it will be shown that line profiles from many directions in the Milky Way can be related to each other to plot the distance and distribution of the neutral hydrogen. This has been one of the most significant contributions of 21-cm radio astronomy.

15.4 Interstellar Molecules

The truly astounding success of 21-cm radio astronomy naturally led astronomers to ask whether or not there might be other emission "lines" which could be studied at radio wavelengths. A number of other atoms and some molecules were investigated theoretically to see if some mechanism, not just spin reversal, might not also produce radio emission. It was found that certain vibrations of the free radical OH^- would cause emission at 18 cm. Radio astronomers then looked for these lines in the direction of the Milky Way. In 1966 astronomers at Massachusetts Institute of Technol-

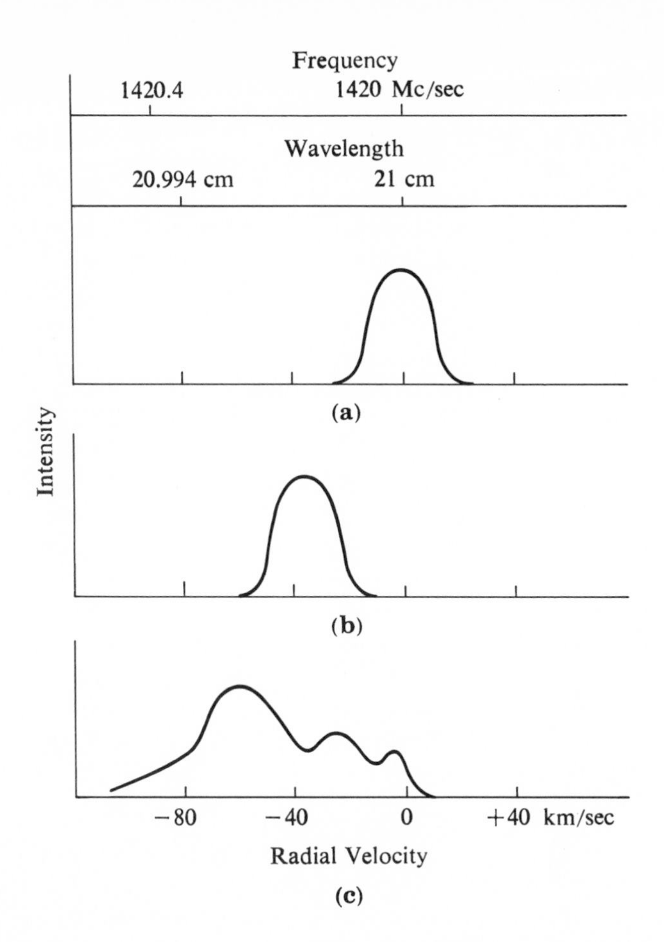

Figure 15.11. Hypothetical 21-cm hydrogen line profiles: (a) the profile of a single cloud of neutral hydrogen atoms with zero radial velocity; (b) the profile of a cloud moving toward the observer; (c) the line profile resulting when three clouds with different radial velocities lie along the line of sight.

ogy and at Harvard found the OH^- lines both in emission and in absorption. In the years since then the search for other interstellar molecules has become one of the most challenging and exciting areas of radio astronomy. The search usually begins with the assumption that a particular type of molecule might exist in space. The next step is the calculation, from complex theoretical considerations, of the wavelengths at which the molecule in question might be expected to emit. The radio telescope may then be tuned to the appropriate wavelength or frequency and pointed toward regions of the sky where gas and dust are likely to be present. Part of the excitement and surprise has come from the large number of different molecules which have been found from the radio signals which they emit. Also quite surprising is the fact that some of these molecules are complex and involve atoms of up to three different elements. Theorists had formerly thought that it would be very unlikely for large, complex molecules to form under the conditions

Table 15.1. Some interstellar molecules.

Symbol	Name	Wavelength
OH	hydroxyl radical	18 cm
H_2O	water vapor	1.35 cm
H_2	molecular hydrogen	ultraviolet
NH_3	ammonia	1.25 cm
CO	carbon monoxide	2.5 mm
CN	cyanogen radical	2.6 mm
CS	carbon monosulfide	2.0 mm
OCS	carbonyl sulfide	2.9 mm
HCN	hydrogen cyanide	3.5 mm
HNC	hydrogen isocyanide	3.4 mm
HCO^+	ionized HCO	3.5 mm
HNCO	isocyanic acid	3.4 mm
HC_3N	cyanoacetylene	3.2 cm
HCOH	formaldehyde	6 cm
HCOOH	formic acid	18 cm
CH_3OH	methyl alcohol	36 cm
CH_3CN	methyl cyanide	2.9 mm
NH_2COH	formamide	5.8 cm
CH_3C_2H	methylacetylene	3.2 mm

found in the vast clouds of interstellar gas. The formation of the molecules remains a puzzling and challenging question. A list of some of the interstellar molecules is included in Table 15.1.

15.5 Total Amounts of Interstellar Materials

The evidence for interstellar matter is certainly varied. In most respects the evidence is quite direct, as in the bright nebulae, the radio emissions, and the interstellar absorption lines. In the case of interstellar reddening the evidence is less direct, but quite convincing nevertheless. In all parts of the Milky Way the evidence is ample, and we must infer the presence of very large quantities of gas and dust. The effects of the gas and dust decrease rapidly with increasing angular distance from the center line of the Milky Way. Ninety degrees from the Milky Way we may feel confident that our view is completely unobstructed, and that we see to great depths in space. Toward the center of the Milky Way we may be sure that our optical view is badly obstructed within about 1000 pc. The view expressed in Chapter 14 that we are in a thin disc-shaped system seems to be confirmed. We may therefore conclude that when the angle between a cloud and the Milky Way is more than about 15°, the cloud must be fairly close to us. From diverse lines of evidence it has been possible to estimate the total amount of interstellar matter in the sun's neighborhood in space. In the most dense parts of our disclike system the dust and gas may amount to as much as 10 or 15% of all the mass present, with the remainder composed of stars. Throughout the entire disc of the galaxy the gas and dust probably represent 5 to 10% of the total mass.

QUESTIONS

1. How does the spectrograph reveal that the bright nebulae are really of two distinct types?
2. Describe the process by which the emission nebulae are made to glow.
3. Why are the green lines at 4959 Å and 5007 Å referred to as forbidden lines?
4. Why do the photons emitted from free-free transitions have longer wavelengths than those emitted from the types of transitions described in Chapter 3?
5. What is meant by the term *interstellar dust?*
6. How can the distance to a bright nebula be estimated?
7. How does the mere existence of bright nebulae imply that there must be vast quantities of nonluminous matter between the stars?
8. Define normal color and color excess. Why does a large color excess indicate the presence of interstellar dust?
9. Why is it that stars which have been reddened by interstellar dust are likely to be nearer to us than they at first seem to be?
10. What is a Wolf diagram and how does it indicate the presence of interstellar dust?
11. Account for the difference in appearance of the stellar and interstellar absorption lines.
12. Describe the transition which results in the emission of 21-cm wavelength radiation from HI regions.
13. How can the radial velocities of HI regions be determined from the observed line profiles?
14. Why is it safe to assume that if a cloud is 15° or 20° from the Milky Way, the cloud must be fairly close to the sun?

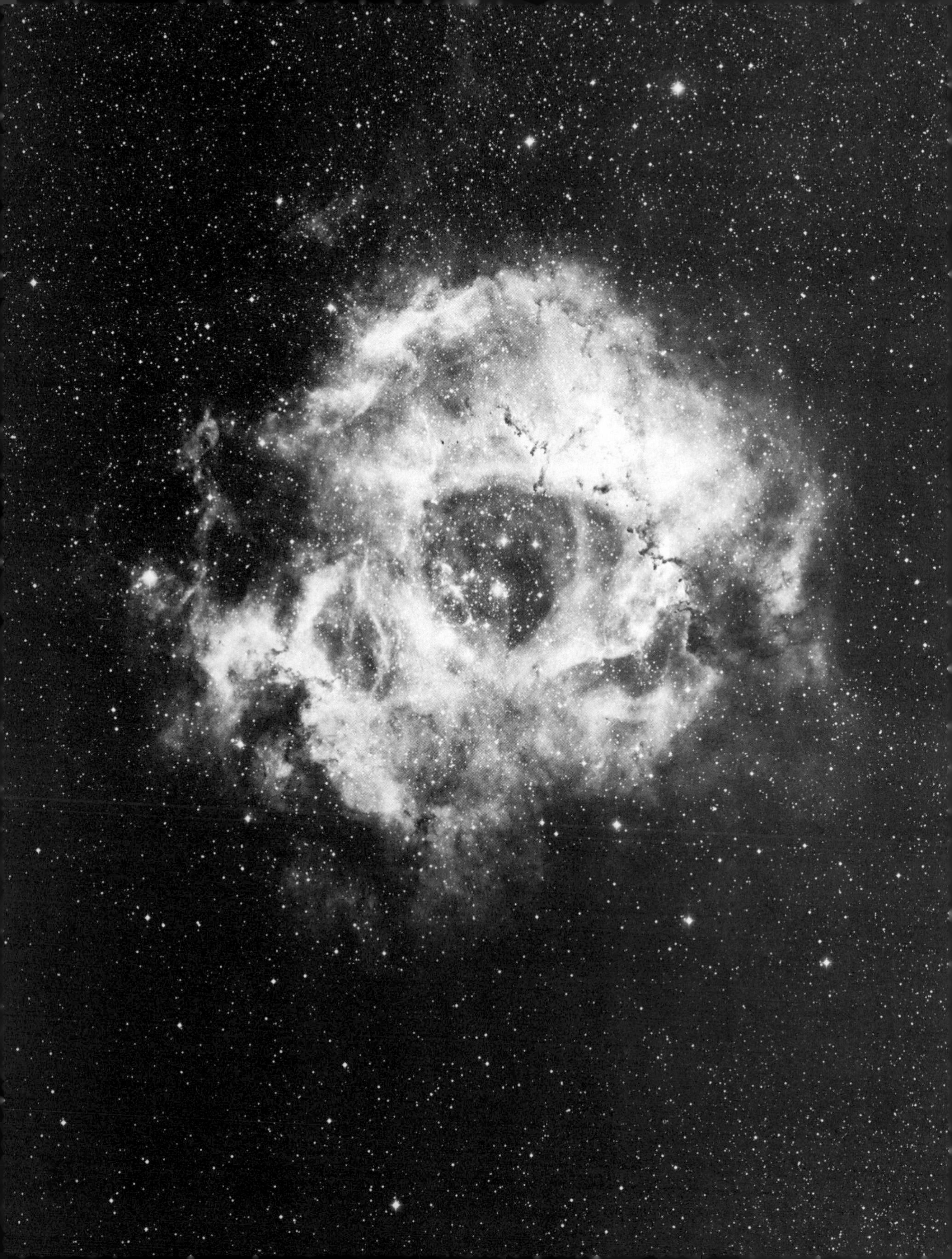

16

Stellar Evolution

In the last five chapters we have seen something of the great variety to be found among stars—some are stable; some are variable; some occur with companions apparently related to them. Many of the stars form clusters in which the stars may have originated at about the same time. Throughout much of the space occupied by the stars there are clouds of dust and gas.

From all of these observational data, astronomers have attempted to reconstruct the evolutionary cycles of the stars. It is not surprising that there have been false starts and radical changes in the theory over the years. For the sake of simplicity we shall consider only the current state of stellar evolution theory and the observations from which it has been derived. The present theory answers most of the questions pretty well, but there are still important problems which will require the cooperative efforts of both theoreticians and observational astronomers.

Having discussed the fusion of hydrogen into helium as the chief source of the energy of the sun and stars, we must recognize that there must come a time when the hydrogen available for the fusion reactions has been depleted. When hydrogen exhaustion occurs, the star must adjust itself in some manner. If there are fairly specific stages in the evolution of all stars, and if stars are being

◄**Figure 16.1.** The Rosette Nebula in Monoceros. The dark globules in this nebula may be areas in which stars are likely to form. (Hale Observatories photograph.)

formed at a constant rate, the length of time spent in each stage should be indicated by the number of stars which exist in each stage at a given time. If stars spend a large part of their lives in a particular stage, then we should expect to find many stars with the characteristics of that stage.

16.1 The Birth of Stars

Since a star must have a beginning, we might expect stars to evolve where the basic raw materials are found. Most of the material in any star is hydrogen and, as we know, there is plenty of hydrogen in the interstellar clouds. We might look for the birth of stars, therefore, in the interstellar medium. If the density of the interstellar material should become great enough within some volume of space, the gravitational attraction of the whole mass of gas for all of its parts might become sufficient to overcome the dispersive effects caused by the random motions of all the particles. At this point the mass of gas would begin to contract as a discrete unit within the nebula. Such a condensation would not be a star, but it could be called a **proto-star,** i.e., something that is to become a star. The critical density of such a mass is 10^{-18} grams per cubic centimeter, which means that a proto-star which would become a star like the sun would occupy an original volume with a radius of 0.1 light-year in the nebula.

Under the mutual gravitational attraction of all of its parts, the proto-star will contract. The actual attractive force on any particle acts as if all the rest of the material were at the center of the whole body, and the size of the attractive force depends upon the total mass of the body. The larger the initial mass, the greater the gravitational attraction and the more rapid the rate of contraction of the mass will be. A proto-star of approximately the sun's mass will contract to the sun's diameter in roughly 60,000,000 years, but a proto-star of 10 solar masses will contract to its stable diameter in only about 500,000 years.

As the proto-star contracts, the density and pressure in its interior increase. At the same time the potential energy of the particles, with respect to the center of the proto-star, is converted to kinetic energy, manifested by an increase in the temperature of the internal material. If the proto-star contains both gas and dust, the dust grains would soon be vaporized in the increasingly hot interior. The interior temperature will also soon be sufficient to dissociate any molecules so that the proto-star is mainly composed only of atomic gases. There will be a progressive increase in density, pressure, and temperature with depth from minimum values at the surface to maximum at the center. Eventually even the surface will become hot enough to be luminous at visible wavelengths, and for the first time the proto-star might become detectable as a very red object.

It was mentioned in the discussion of the sun in Chapter 8 that contraction could play an important role as an energy source in stars. This initial contraction of the proto-star provides the first source of the star's light and heat. The rate of radiation for most of the stars during the contraction stages is not sufficient to maintain the star in an equilibrium state, and the star will experience a continuing increase in temperature throughout. On the H-R diagram a star of about five solar masses ($5M_\odot$) will enter the diagram on the right as a cool object, but one that is luminous due to its large diameter. The large star then moves essentially from right to left across the diagram as its temperature increases. Eventually it will reach the main sequence. The Japanese astronomer C. Hayashi has pointed out that most stars will decrease somewhat in luminosity during the contraction stage and that the track on the H-R diagram depends very strongly upon the mass of the proto-star. In Figure 16.2, Hayashi tracks for stars of a number of different masses may be seen. For the low mass proto-star the track during contraction is essentially downward, indicating no real increase in surface temperature but a decreasing surface area.

16.2 The Main-Sequence Stage

Theoreticians such as Hayashi and L. G. Henyey have stated that during the initial contraction period there is considerable mixing of the proto-star's material due to convection. As the contraction proceeds, however, the star develops an interior region in which thermal energy is transported outward by radiation rather than convection. (Compare page 179.) Atoms are excited by collisions and by absorption of radiation, and they reradiate this energy. Slowly this energy finds its way outward to the bottom of a convective envelope, where it heats the gas. In this outer convective region the hot gas rises to the surface and radiates energy into space as light and heat. The size of the internal radiative region will depend upon the mass of the original proto-star. In large stars this

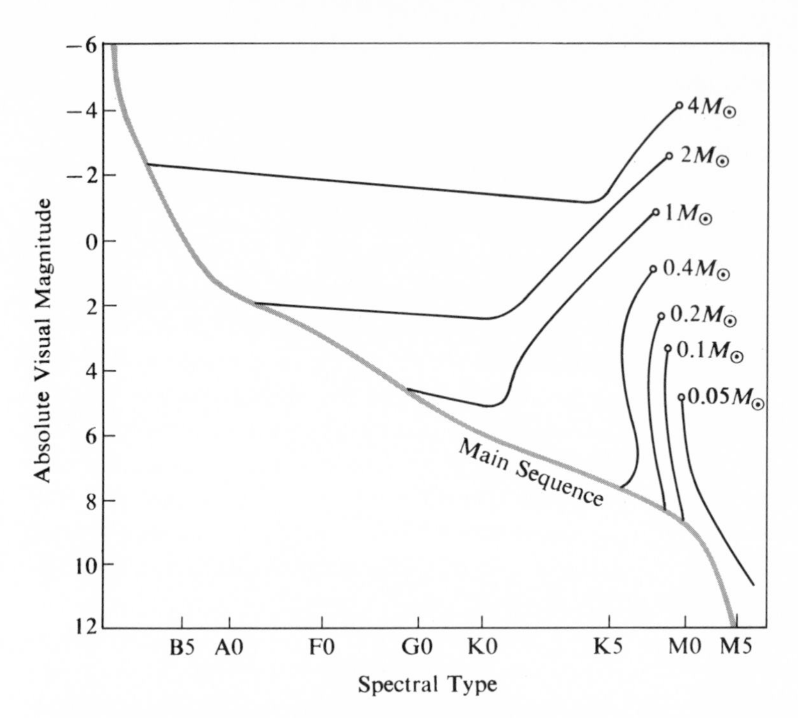

Figure 16.2. Hayashi tracks for contracting stars on the H-R diagram. Note the difference in the tracks for stars of different masses.

region may be a very large fraction of the total mass. Contraction will continue with a consequent increase in temperature, however, until some other factor can increase the internal gas pressure to balance the weight of the overlying gas.

It is the onset of nuclear reactions, chiefly the proton-proton reaction, which provides the necessary increases in temperature and pressure to balance the gravitational force. These reactions may begin slowly if the star contains light elements such as deuterium, lithium, beryllium, and boron. Reactions involving these elements may begin at temperatures as low as one million degrees Kelvin, and will serve to increase the temperature even more until hydrogen burning (carbon-nitrogen cycle for massive stars, proton-proton for less massive stars) can begin. As the temperature, pressure, and density continue to rise, the rate of hydrogen burning will begin to rise. Furthermore, since temperature, pressure, and density are greatest in the very center of a star, the rate of hydrogen burning will be greatest near the center and will fall off rapidly with increasing radius. In the sun, for example, 90% of the luminosity originates within $0.2 \times$ solar radius from the sun's center. With the fusion of hydrogen to helium proceeding actively in the central region, the gas pressure at every point in the star becomes sufficient to support the weight of the gas above that point. Thus, the contraction stops.

The surface temperature and luminosity of the star have changed during the contraction phase, and so the star's position on the H-R diagram will have changed. The succession of such changes results in the tracks shown in Figure 16.2, and each of these tracks ends on the main sequence. Thus, as contraction ends, the conditions of equilibrium depending upon mass dictate that stars have the properties of surface temperature and luminosity which define the main sequence. This equilibrium condition means that energy is being radiated from a star's surface at just the same rate at which it is being converted from matter in the star's interior. If the rate of hydrogen burning should increase or decrease, the star would have to readjust itself to reach a new equilibrium. As long as the supply of hydrogen remains sufficient, however, the hydrogen burning should proceed at a nearly constant rate, and no changes should occur.

Eventually, in the central region responsible for the production of most of a star's energy, the amount of hydrogen available for fusion will have been depleted to the point where a change must occur. For all stars it takes a long time to reach this point, but the most massive stars come to it most quickly. It is estimated that when the sun reached the main sequence, there was enough hydrogen in its most reactive regions to maintain it as a stable star for 12 billion years. As the age of the solar system is estimated at about 4.5 billion years, the sun is not likely to run out of fuel very soon. In a very bright star of spectral type B0 with a mass of about $20M_{\odot}$, energy is being produced at such a prodigious rate that hydrogen readily available for reactions will be depleted seriously in a time period of 10 million years. At the other end of the scale a small red star of spectral type M5 might have a mass of only $0.2M_{\odot}$. Such a star could maintain its main-sequence equilibrium for the incredibly long period of 10^{12} years. These smaller stars have an extra advantage in that there is convective mixing of all of their material, making it possible for all of the hydrogen to come into the central region in which reactions can take place. For stars between 5 and $0.3M_{\odot}$, the convective region comprises only the outer layers, and there is a radiative region in the interior in which there is little mixing. For the most massive stars, the region in and near the core is probably completely mixed. And, although the envelope is radiative, the high rates of stellar rotation for a number of these stars may produce *some* mixing even in the radiative zone.

16.3 The Giant Stage

With the significant depletion of hydrogen in the central region, the rate of hydrogen burning must necessarily decrease, and a star must become unstable. The gas pressure decreases and gravitational forces again become dominant. Just as the star should become dimmer it again begins to contract. Surprisingly perhaps, this contraction actually causes the star to become brighter rather than dimmer. The central regions, already at a very high temperature and density, become further compressed by this new contraction, and hydrogen burning spreads to a region outside of the original core. The star must adjust itself to a new equilibrium appropriate to its new rate of energy production. If the mass is large enough, the star may actually produce energy faster than before and become brighter than before. More important, however, the gas in the outer regions becomes hotter and expands. The star as a whole acquires a diameter very much larger than that which it had on the main sequence. With the larger diameter the star can radiate at a lower surface temperature. On the H-R diagram the star must leave the main sequence and become a **red giant.**

The central region of the star, now rich in helium, eventually reaches a temperature of 100 million degrees Kelvin, and a new series of nuclear reactions involving helium can begin. These new reactions (Figure 16.3) essentially involve the fusion of two, three, four, or five helium nuclei to form nuclei of beryllium, carbon, oxygen, and neon. From the theoretical point of view, the most difficult part of this process is that the Be^8 nuclei are very unstable and quickly split again into two helium nuclei. This is indicated by the double arrows in the diagram. Some of the Be^8 nuclei may capture another He^4 nucleus before having time to break up. This capture produces a carbon nucleus (C^{12}) and some gamma radiation. Capture of another He^4 produces an oxygen nucleus (O^{16}), and capture of a fifth He^4 produces a neon nucleus (Ne^{20}). In this manner the synthesis of heavy elements begins in the interiors of the red giant stars.

One can imagine a repetitive situation in which a source of nuclear energy comes into operation at some central temperature. When the "fuel" for this reaction has been used up, contraction

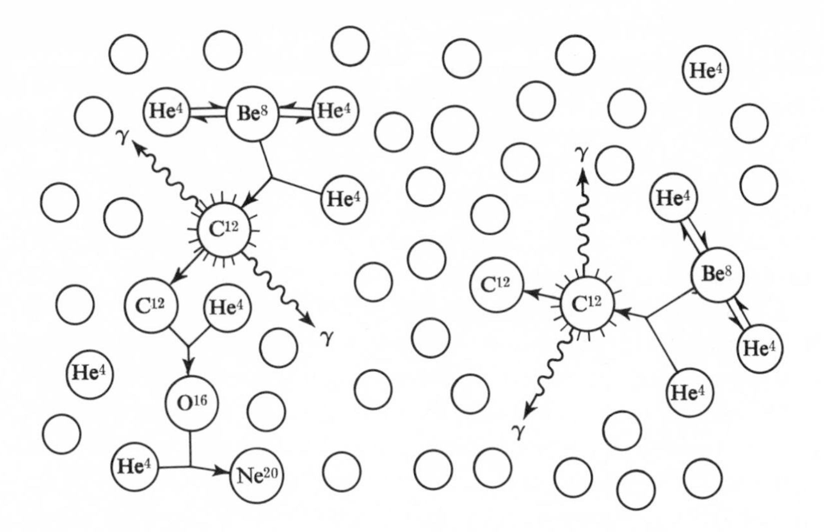

Figure 16.3. In the interiors of the red giant stars, helium nuclei combine into nuclei of heavier atoms.

raises the central temperature. At some higher temperature the fusion products of the first reaction become the fuel for another series of reactions. In this way a number of contractions and stable phases could occur. Nuclear physicists have predicted that, by means such as this, all of the elements up to about iron in the periodic table could be created in the interiors of aging stars.

Some doubt remains as to just what would happen to a star after it has become a red giant. We shall see later in this chapter that there is some observational evidence for the further evolution of red giants; however, the observations cannot be well understood until they are supported by complete theoretical work. The most widely accepted ideas today suggest a general rise in surface temperature with a slight initial decrease in luminosity. On the H-R diagram the star would move back toward the left, approaching the main sequence somewhat higher than its original position. On the horizontal portion of this track the star would be exhausting its last sources of nuclear fuel in the final creation of iron, vanadium, nickel, and so forth in an extremely hot, dense core.

16.4 White Dwarfs

There must come a time when the nuclear reactions within the star's core have gone as far as they can go. Then, contraction must again provide the only source of energy remaining for the star. No longer can the release of nuclear energy put a temporary stop to contraction, and nothing can prevent the collapse of the star into a white dwarf. Unlike proto-stars, energy transfer in white-dwarf interiors is principally carried on by conduction rather than convection. As mentioned in Chapter 12, white dwarfs are known to have densities greater than that of any known terrestrial material. Again there are gaps in the theory, so that astronomers cannot be sure of the exact track of a star on the H-R diagram as it becomes a white dwarf. In spite of this, however, two important matters related to white dwarfs are well understood on theoretical grounds. First, the behavior of such highly compressed gases has been carefully investigated, and second, the properties of white dwarfs of various masses have been studied.

The central temperatures in white dwarfs are millions of degrees Kelvin and the densities are 10^6 g/cm^3. Under such conditions all of the atoms have been completely ionized, and the nuclei and electrons have been pressed very close together. The spaces occupied by the orbiting electrons in normal gases are lost. In spite of the high density, the material in a white dwarf must still be considered as a gas, and the properties of this gas will be controlled principally by the motions of the electrons rather than the nuclei. For this reason we refer to this material as a **degenerate electron gas,** or simply as degenerate matter.

In 1939, S. Chandrasekhar of the University of Chicago made important studies of the properties of white dwarfs and showed that there was an interesting relationship between mass and diameter of these superdense stars. The greater the mass, the smaller would be the diameter. A white dwarf of one solar mass should be about the size of the earth, while a white dwarf with a mass greater than $1.2M_{\odot}$ cannot exist. About 100 white dwarfs have been discovered, but masses are known for only three. These three are all less massive than Chandrasekhar's upper limit of $1.2M_{\odot}$. Once a star has reached the white-dwarf stage, it radiates because of its high surface temperature. The low luminosities result from the small diameters. When it has contracted to its smallest possible diameter, it can only become cooler, dimmer, and less luminous. On the H-R diagram there should, therefore, be a **white-dwarf sequence** more or less parallel to the main sequence, but some 10 magnitudes fainter. The final product must be a **black dwarf,** a cold, nonluminous body of degenerate gas.

In the interior of a white dwarf the transfer of thermal energy toward the surface will be chiefly by conduction. In comparison to radiation and convection as heat-transfer mechanisms, conduction is a relatively slow means of heat transfer. As a result the temperature is fairly uniform throughout the interior of a white dwarf, and the star cools very slowly. The time scale of this cooling off is such that it does not seem likely that in our galaxy any white dwarfs have yet had time to become black dwarfs.

16.5 Neutron Stars

The stars of largest mass evolve through their main-sequence and giant stages most rapidly and so they should be the first ones to reach advanced evolutionary stages. Such stars are likely to be more massive than the limit of $1.2M_{\odot}$ for white dwarfs, and it is interesting to consider what would happen to such stars when they finished burning all of their nuclear fuels. They might eject mass into space through nova or planetary nebula mechanisms until they were sufficiently small to become white dwarfs. From what is known of such processes, however, it does not seem likely that mass loss goes on at a sufficient rate in most stars. In recent years much theoretical work has been done on the possibility that stars too large to become white dwarfs but smaller than $3.0M_{\odot}$ might become **neutron stars** instead.

Gravitational contraction of such a star would produce central densities of the order of 10^9 g/cm^3. Such densities would be so great that all atomic nuclei would be destroyed, and only neutrons could exist. A star of $2.0M_{\odot}$ would shrink to a radius of the order of 10 km. The surface temperature of a neutron star could be as high as 2,000,000°K, but the diameter would be so small that the luminosity would be very low. Detection of such a star by normal optical means would be extremely difficult. A further hindrance to optical detection would be that at such a high temperature most of the radiation of the neutron star would be at wavelengths too short for optical detection. The neutron star may, however, be a strong emitter of x rays through several mechanisms.

As mentioned in Chapter 13 (page 289), the pulsar in the Crab Nebula is thought to be a neutron star. Data from rockets and satellites have shown that this object emits x-ray pulses as well and that the period is the same as that of the radio and optical pulses. This is probably the most direct evidence that we now have of the existence of a neutron star. Like the white dwarf, the neutron star will be quite stable. It can contract no further, and will cool off only very slowly.

16.6 Black Holes

It is currently recognized that an original star larger than $3.0M_{\odot}$ is too large to become a neutron star. It would have to contract into something even smaller and more dense, and it is interesting to speculate on what type of object this might be. When a very large mass collapses into a very small object such as a white dwarf or a neutron star, the surface gravity of the final object becomes enormous. Going beyond the conditions of a neutron star to stars of even greater mass and smaller size, one can predict an object with such great surface gravity that nothing could escape from it—not even the photons emitted from its hot surface. Light would not leave such an object, so it would not be visible to us. Anything that came too close to it would fall into it and disappear. These hypothetical bodies have been given the rather appropriate name of **black holes.** We may someday find a star which seems to have an invisible companion of large mass. The companion might then fit our definition of a black hole. Such a system might be a strong emitter of x rays; several variable x-ray sources have been suggested as possible black holes. Or we may learn to detect and locate strong sources of gravity waves which could emanate from a black hole. For the present, however, the black hole remains a fascinating possibility at the end of the existence of stars of very large mass.

16.7 Degenerate Stars of Very Small Mass

In the discussion of proto-stars and contraction of stars to the main sequence, considerable emphasis was placed on the role of mass in determining the characteristics of the stars. Over quite a wide range, mass determines the location of a star on the main sequence. There is, however, a lower limit to the masses of stars with the characteristics which have been described as normal. If the mass is too low, the evolutionary cycle is entirely different and is governed *not* by hydrogen burning and other fusion reactions but by the properties of degenerate matter. In the discussion of white dwarfs it was implied that the gas in a stellar interior would become degenerate when the density became greater than some specific value. Actually, the criterion of degeneracy depends upon both temperature and density, and a gas can become degenerate at a relatively low density if the temperature is also sufficiently low.

In the interior of a contracting star of very low mass, i.e., less than about $0.07M_{\odot}$, the density increases at such a rate that the gas begins to become degenerate at a relatively low temperature. In becoming degenerate the gas absorbs energy at such a rate that the temperature stops rising before it has become high enough for hydrogen burning to begin. Gradually more and more of the gas becomes degenerate, until it is completely degenerate and contraction ends. A star of about $0.07M_{\odot}$ would then be about the size of Jupiter with an absolute magnitude of about +25 and would be very red. Such a star would be impossible to detect unless it were very close to the sun. With stable diameters these stars would become dimmer as they cooled off. The total lifetime for stars of mass less than $0.07M_{\odot}$ would be less than one billion years from the beginning of contraction to black dwarf.

The best possibility for discovering stars of small mass probably lies in observations of these stars while they are still contracting and are comparable in brightness to the sun. With the increase in infrared astronomy, contracting low-mass stars will probably be found in or near clouds of interstellar material.

16.8 Observations Supporting the Theory

The times involved in each of the stages outlined above are such that we cannot hope to see a single star evolve through its whole cycle from proto-star to black dwarf. In the cosmic time scale our view is "instantaneous." We can see a great variety in stars and we can try to fit all of the stars into one scheme. We already know that the course of stellar evolution depends greatly upon mass. Although not discussed here, chemical composition of the stars plays a role. The evolutionary theory at its present state is based upon crucial parts of our "instantaneous" observational data analyzed according to the most modern physical theories. Some of these observations have been mentioned in earlier chapters, but the manner in which they support the evolutionary theory could not be discussed until we had considered all the data.

Figure 16.1 is a photograph of the Rosette Nebula in the constellation Monoceros. (See also Color Plate 5.) Quite distinct against this emission nebula are to be seen a number of relatively small dark areas. These condensations of dust indicate that areas of high density can occur in association with nebulae. It is generally believed that proto-stars may form within areas such as these since a dark globule of this sort contains considerably more than enough material to form a star.

In its earliest stages, a proto-star would be undetectable by optical means since it is nonluminous, but there is some evidence of very young stars which may still be contracting. These are the stars in associations mentioned in Chapter 14. The color-magnitude diagram for an association is shown in Figure 14.9. We learned that the *brighter* stars in associations may actually be on the main sequence while the fainter ones are still contracting and becoming hotter. This behavior is exactly what should be expected. Our knowledge suggests that the more massive a star, the farther up the main sequence it must lie, and, according to theory, the faster it should contract to the main sequence. The youthfulness of the association members is further supported by the observed fact that associations are found in regions where there is clear evidence of considerable interstellar matter for star formation. The Orion Association is the most notable example.

Evolution away from the main sequence is best supported by the color-magnitude diagrams of the galactic clusters. In Figure 14.7, the diagrams show clearly that the brightest stars evolve from the main sequence sooner than the fainter stars. If all of the stars in a cluster were formed at about the same time, the more massive ones will evolve the most rapidly and will be the first to leave the main sequence. As the cluster ages, stars of lesser mass will leave the main sequence. If we compare the color-magnitude diagrams for a number of clusters, then, we should see the relative cluster ages indicated by the shape of the main sequence. In a young cluster, stars should be found along most of the zero-age main sequence. In progressively older clusters, the upper end of the main sequence should show a bend. The lower the position of this bend away from the zero-age main sequence, the older will be the cluster. Judged in this manner, the cluster M67 in the constellation Cancer would be described as very old. (See Figure 16.4 for M67's color-magnitude diagram.) The lower end of the cluster's main sequence is well populated since the stars of small mass lie here and evolve very slowly. Of particular importance is the pattern traced by the stars which have already left the main sequence. Their pattern suggests the kinds of giants into which main-sequence stars of various masses will evolve.

On the basis of this interpretation of the color-magnitude diagrams, the stars in the globular clusters must be very old. The

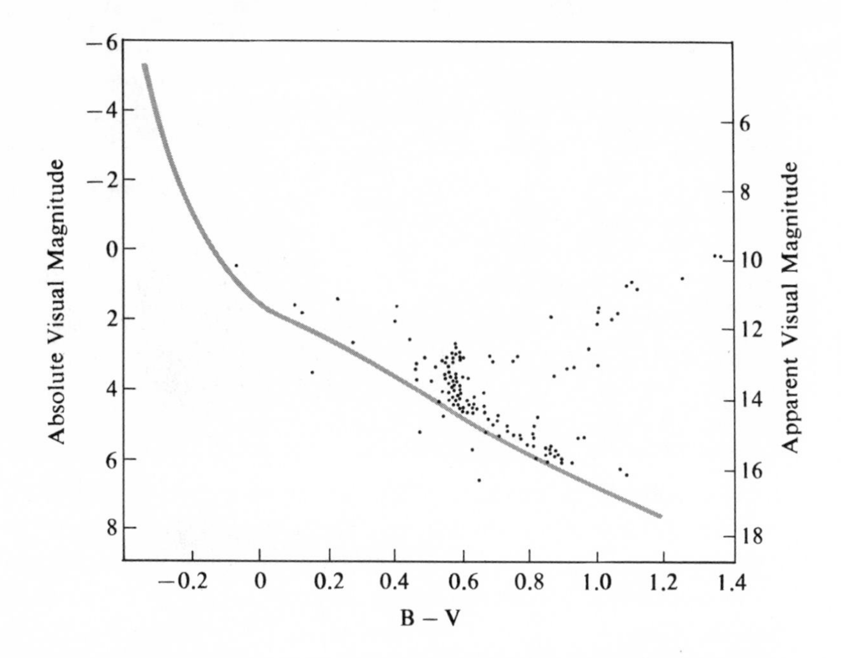

Figure 16.4. The color-magnitude diagram for the very old galactic cluster M67. Available data show that the main sequence actually extends at least to V=20. (Plotted from results published by R. Racine.)

color-magnitude diagram for the globular cluster M13 was shown in Figure 14.11. The main sequence is highly evolved, and the giant branch well populated. The presence of stars along the horizontal branch has helped support the idea that stars become hotter again after the giant stage and move to the left on the diagram. A gap appears in the horizontal branch of the diagram if the RR Lyrae stars are omitted. This has suggested that as stars evolve to the left along the horizontal branch, they become unstable and thus variable as they cross the "RR Lyrae gap." RR Lyrae variables, then, are very old stars.

The Cepheid variables, on the other hand, are more likely to be fairly young. In general, Cepheids tend to be found close to the Milky Way, where the raw materials are found. It has been shown theoretically that a period of instability should occur for stars after they have left the main sequence. Cepheids have been found in a few galactic clusters, which further supports the idea that these stars are considerably younger than the RR Lyrae variables.

There are now a number of cases in which a star that has become a nova has been identified on photographs made before the nova outburst. From such data the characteristics of pre-nova stars can be described. Usually the pre-nova star has a high surface temperature and lies slightly below the main sequence. If a post-giant

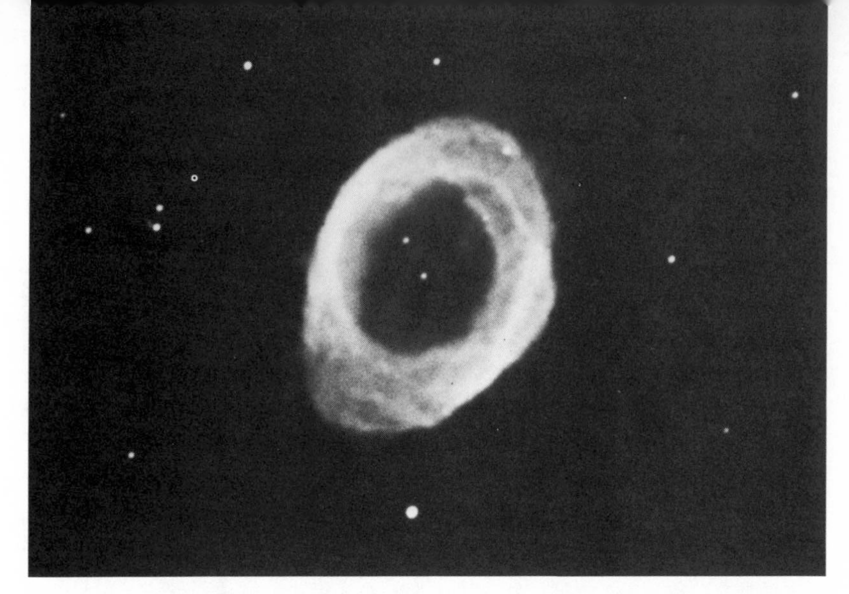

Figure 16.5. The Ring Nebula, a planetary nebula in the constellation Lyra. The gas in this expanding shell is excited by radiation from the star in the center of the ring. A background star also appears to be within the ring. (Hale Observatories photograph.)

star has evolved to the left on the horizontal branch, a slight decrease in luminosity could give a star the outward characteristics of the pre-nova stars. Then, in a series of nova outbursts, the star could lose enough mass to become smaller than the white dwarf limit of $1.2M_{\odot}$.

Further important evidence related to the origin of white dwarfs and to mass loss is suggested in recent studies of **planetary nebulae.** These objects are typified by the Ring Nebula in the constellation Lyra (Figure 16.5 and Color Plates 6 and 7). There is some variety in the appearance of the planetary nebulae as a group, but all of them have somewhat the appearance of the one pictured here. The ringlike shape is derived from the fact that we view a spherical shell of gas expanding from a central star. Radiation from this star (usually a spectral type O star), excites the gas, causing it to emit a bright-line spectrum. As we look toward the shell, the path length of the line of sight within the gas is longest at the edges of the sphere. Thus the shell appears brightest at its rim and less bright at its center. Approximately 1000 of these nebulae are known, and they lie chiefly in the direction of the Milky Way in the sky.

In recent years, detailed studies of the diameters of the planetary nebulae have been correlated with studies of their central stars. It now appears that in the smaller gas shells the central stars are brighter than $M_V = 0$. In the progressively larger planetary nebulae the central stars are progressively less luminous. For those cases in which the shell has a radius of about 0.7 pc, the central star has the temperature and luminosity of the hotter white dwarfs. From the emission lines which are found in the spectra of the gaseous shells, the expansion velocities are found to be on the order of 20 km/sec. This velocity means that the shell could reach its radius of 0.7 pc in a period of some 35,000 years. Since the central stars show very little range in temperature, the differences in luminosity must be indicative of a considerable range in diameter for the central stars. The evidence certainly seems to suggest a very rapid decrease in diameter during the period of the expansion of the gaseous shell. The fact that so many planetary nebulae are known when the phenomenon is so rapid means that this process must be quite common. It may well be that not all stars eject a planetary nebula shell, but it now seems reasonably clear that this is an important mass-loss mechanism in which the end product is a white dwarf.

An evolutionary track of a star of 5 solar masses on the H-R diagram is indicated in Figure 16.6. The changing characteristics of the star are manifested in changes in its surface temperature and luminosity. These quantities, of course, determine the star's location on the diagram as it changes from a main-sequence star to a giant and eventually to a white dwarf. Estimation of the time involved in the whole cycle is difficult and requires careful and extensive theoretical studies. The contraction phase might last 2 million years for the $5.0M_{\odot}$ star, and it might remain on the main sequence for 500 million years. In terms of the time spent on the main sequence, a star may be a giant for a relatively short time,

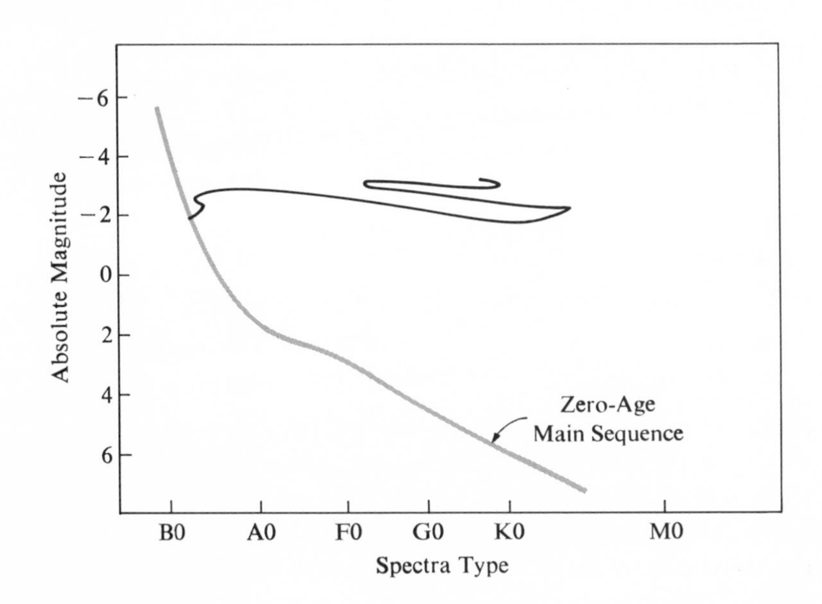

Figure 16.6. Track on the H-R diagram for a star of five solar masses.

and the loss in luminosity to the white-dwarf stage may occur rather rapidly. The final cooling off to a nonluminous body may take an extremely long time. It should be emphasized again that the time involved, especially in the earlier stages, is considerably longer for the stars of smaller mass.

Another observational aspect related to the stellar evolution theory is chemical composition. As heavier elements are created in stellar interiors and then returned to the interstellar medium through nova, supernova, and planetary nebula mechanisms, a gradual enrichment of the material in terms of heavier elements must occur. If all the original matter were only hydrogen, after one generation of massive stars had gone through their lives there should be a percentage of heavy elements, as well as hydrogen, in the interstellar medium. Second-generation stars forming from this material should have a noticeably higher metal content than the first-generation stars. Spectroscopic techniques permit the determination of relative abundances of elements, and it is in fact confirmed that younger stars are richer in metals than the older stars. The sun, for example, has a higher percentage of metals and heavy elements than a star of comparable mass in a globular cluster. The stars to come after the sun will contain even higher percentages of heavy elements.

The principal phases in the life of a star may be summarized as follows:

1. Contraction of a proto-star from the interstellar gas and dust.
2. Stable life as a main-sequence star.
3. Expansion to a red giant followed by new reactions which replace hydrogen burning in the star's interior.
4. Period of heavy-element synthesis with increase in surface temperature.
5. Loss of mass as an explosive variable star. (Uncertain.)
6. Stable life as a white dwarf, neutron star, or black hole which cannot contract further and must slowly cool off.

The concept of "age" in stars is so related to the mass of a star that it is perhaps more meaningful to speak of evolutionary age than it is to speak of age in years. Thus, a star of large mass becomes "old" in a relatively small number of years. Stars of spectral type O and B, of large mass and high luminosity, are using their nuclear fuel at very rapid rates. They cannot keep this up for very long. If there is no mixing at any time, the O and B type stars, therefore, can only be young. For a star of moderate mass, on the other hand, the picture is not so clear. There is no way to estimate the age of a single star of spectral type G, K, or M. Such a star could have just reached the main sequence or it could already have been there for billions of years. In the clusters we believe that we are seeing the effects of stellar evolution clearly displayed in groups of stars which formed within a relatively short span of time.

QUESTIONS

1. Explain why proto-stars of large mass will contract to a stable diameter more quickly than proto-stars of smaller mass.
2. At what stage may a contracting mass of gas meaningfully be called a star?
3. What is the reason for the belief that stars like the sun have a distinct region in which nuclear reactions are taking place?
4. Why does the contraction phase in the life of a star end? Where will the star be at this point on the H-R diagram?
5. What will cause a star to leave the main sequence?
6. What are the sources of energy during the period in which a star is a red giant?
7. Why does a star become a white dwarf? What is the nature of the material in the interiors of white dwarfs?
8. What is the relation between mass and diameter for white dwarfs? Explain.
9. Describe the evolution of a white dwarf star to later stages.
10. What is a neutron star and how can neutron stars be detected?
11. Why do stars of small mass become degenerate without having ever initiated nuclear reactions in their interiors?
12. How does the shape of the main sequence in the color-magnitude diagram of a galactic cluster indicate the age of the cluster and support the theory of stellar evolution outlined in this chapter?
13. At what stage in the evolutionary cycle is a star likely to become a Cepheid variable?
14. Why should we expect the older stars to show lower metal content than the younger stars?

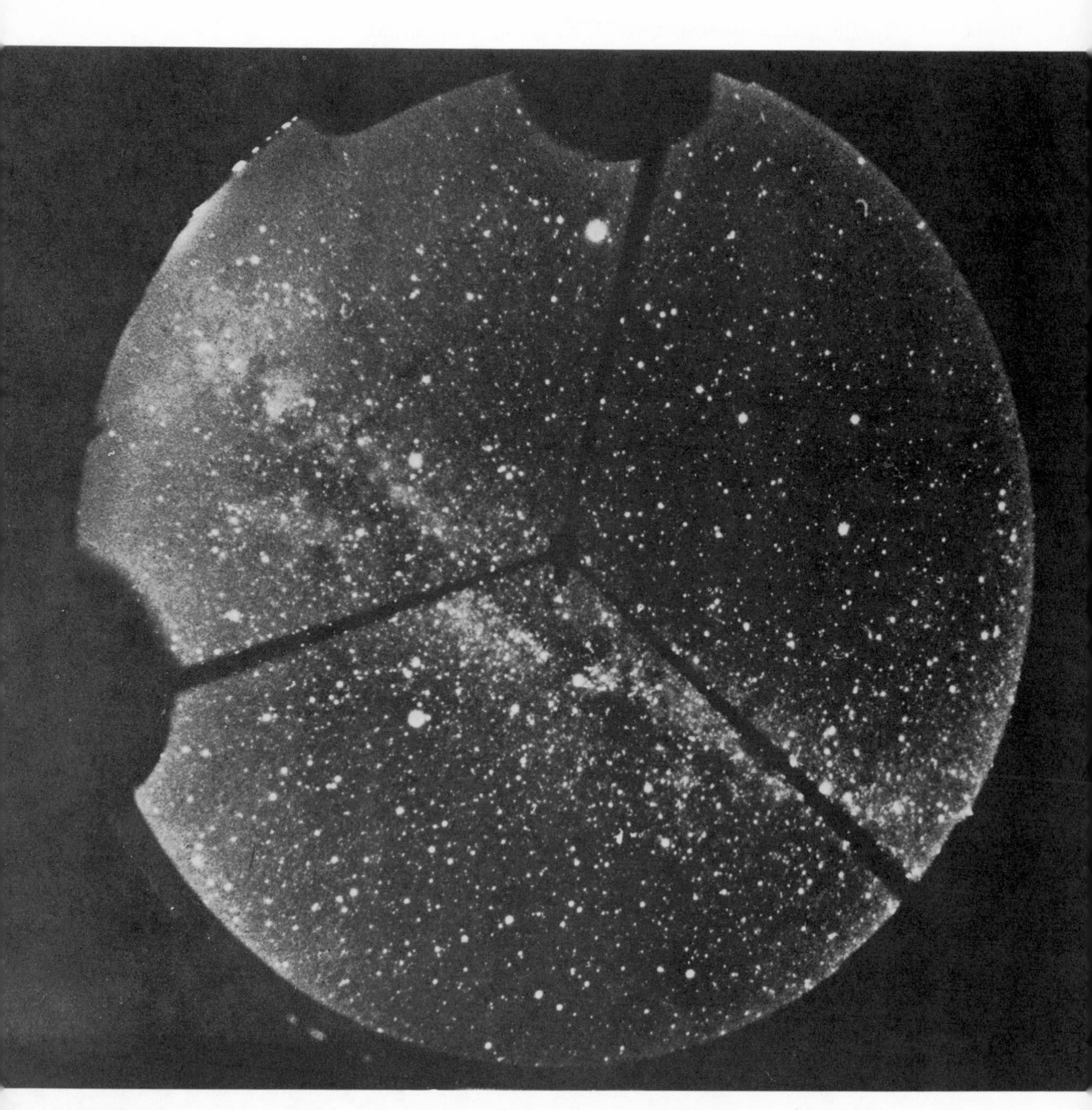

Figure 17.1. A wide-angle view of the southern Milky Way. The center of our galaxy is believed to lie in this direction. (Yerkes Observatory photograph.)

17

The Milky Way

The broad belt of the Milky Way is a spectacular sight on a dark summer night in the northern hemisphere. At such times, the Milky Way stretches through the constellations Cassiopeia, low in the north, and Cygnus, high overhead, to Sagittarius in the south. In Cygnus, the Milky Way divides into two streams for a time, while in Sagittarius it has its greatest width and brilliance. In the skies of the southern hemisphere (Figure 17.1), the Milky Way continues past the Southern Cross northward through Orion. In the northern sky once more the Milky Way runs through Taurus and Perseus back again to Cassiopeia.

When plotted on a celestial globe, the Milky Way divides the sky into two nearly equal portions. Galileo was the first astronomer to note that this belt of light owes its appearance to the presence of large numbers of faint stars all along its course. Since the faint stars are likely to be at greater distances from the sun than the bright stars, the realm of stars surrounding the sun must somehow be more extensive in the directions toward the Milky Way than in other directions on the celestial sphere. Careful and complete analysis of the distribution of stars in the direction along the Milky Way has led to the present concept that our sun is part of a great spiral galaxy or system of stars comparable to many of those which can be seen at great distances in space. This vast system is referred to as the Milky Way Galaxy or simply the Galaxy.

Originally the term **galaxy** had the same meaning as Milky Way. With the recognition that we are part of a distinct system of stars, comparable to large numbers of similar systems far away in space, the word "galaxy" has come to mean any such independent stellar system.

17.1 Early Models of the Stellar System

The first person to attempt a systematic analysis of the Milky Way was William Herschel. In 1785, Herschel completed a series of counts of the numbers of stars in various parts of the Milky Way. Reasoning that the largest numbers of faint stars within limited areas should be found where the system was most extensive along the line of sight, Herschel produced a three-dimensional model of the system of stars around the sun. This model suggested an ellipsoidal distribution of stars extending about five times farther in the direction of the Milky Way than in the directions 90° from the Milky Way. Unfortunately, parallaxes had not been measured at this time and Herschel was unable to calibrate his model in any absolute units.

Much later, in the early years of this century, the Dutch astronomer Kapteyn produced a model quite similar to Herschel's but with a diameter of some 7000 pc and a thickness of about 1320 pc. Kapteyn's model was also based on star counts, but involved a much more complex statistical analysis than did Herschel's.

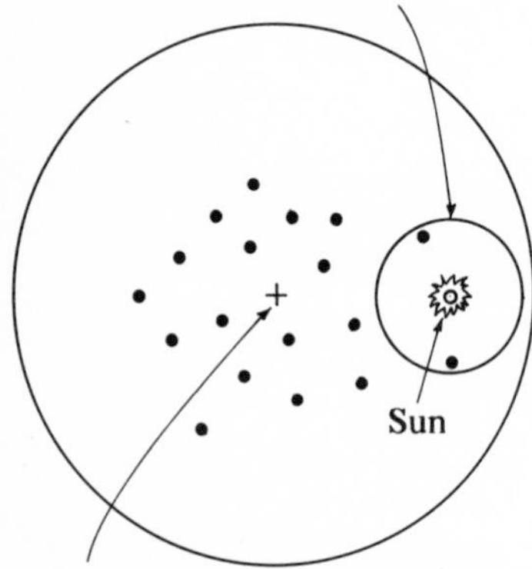

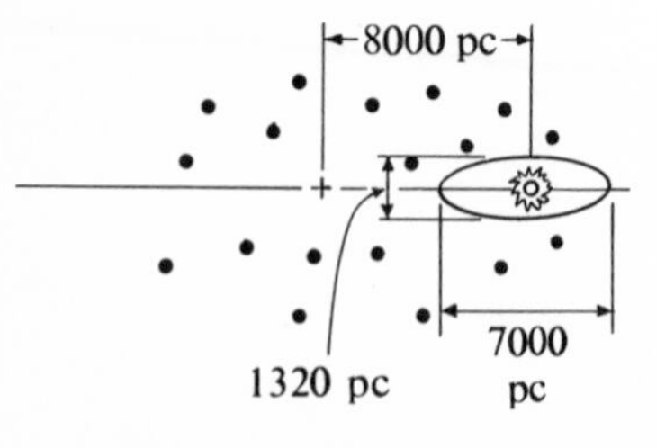

Figure 17.2. The Kapteyn universe as part of the larger system outlined by Shapley. The ellipsoid indicates the distance at which Kapteyn's star density has decreased to 1/16 of the star density near the sun.

It was just shortly after Kapteyn's work that Shapley revolutionized our whole concept of the size and extent of the Milky Way system. Shapley's studies of globular clusters (see Chapter 14) showed that the sun was not really the center of the stellar system at all. At present we recognize that if the position of the galactic center is indicated by the distribution of the globular clusters, the sun must be some 10,000 pc from that center. The models of Herschel and Kapteyn represented only small parts of the actual picture. Figure 17.2 indicates Kapteyn's system in relation to the new concept of a larger overall system.

Later, in 1930, the presence of interstellar absorbing material was also convincingly demonstrated. It was clear then that our view into the Milky Way was limited, and therefore our optical view of the whole system of stars must remain limited. If the system had the dimensions suggested by Shapley, there would be vast areas which could never be seen. The interstellar absorption will forever restrict our view in the direction of the Milky Way. In other directions, however, we are able to see great distances beyond the limit of our system to numerous individual galaxies of a variety of sizes and shapes. From our position inside of the Milky Way Galaxy we have tried to look for evidence in our vicinity which will permit us to compare our system with the distant ones.

17.2 Galactic Coordinates

It is possible to pass a plane through the earth and the celestial sphere in such a way that it cuts the Milky Way all the way around the sky. The intersection of this plane with the sphere is a circle called the **galactic equator** (Figure 17.3). Positions of stars on the celestial sphere may be defined with respect to the galactic equator just as they were previously referred to the celestial equator. In Figure 17.3 the basic reference points of the galactic coordinate system are indicated. The zero point for the measurement of **galactic longitude** is the direction toward the center of the galaxy, while **galactic latitude** is measured northward or southward along a vertical circle passing through the object. Such a vertical circle must also pass through the north and south **galactic poles,** which are 90° from the galactic equator. Thus the position of any object on the celestial sphere may be specified in terms of its galactic latitude and longitude instead of its declination and right ascension. The main difference is simply in the choice of the galactic equator rather than the celestial equator as the fundamental reference circle on the celestial sphere.

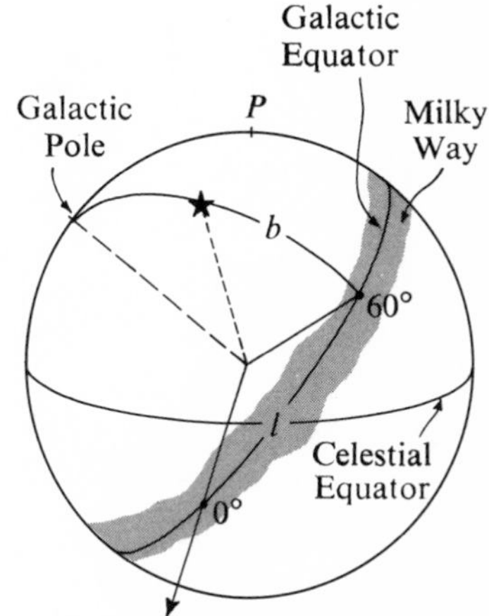

Figure 17.3. The galactic equator runs through the central part of the Milky Way and is the fundamental circle of the galactic coordinate system.

Galactic coordinates are never measured directly as are equatorial coordinates, but are always computed from the measured right ascension and declination of a star and the equatorial coordinates of the galactic pole. The labor involved in such calculations is well spent, however, because a star's galactic coordinates convey a good deal of information by themselves. Thus if a star's galactic longitude is near 0° or 360° and its latitude is low, we know that the star must lie in the general direction of the galactic center. Whenever a star's galactic latitude is low, the star must lie in or near the Milky Way. If the galactic longitude and distance of some Milky Way star are known, the position of that star in the plane represented by the Milky Way will be known.

The letters l and b are conventionally used by astronomers to designate galactic longitude and latitude, respectively. The symbols l^{II} and b^{II} have been used in astronomical literature since 1958 to indicate that the present reference frame was adopted by international agreement at that time. In this book, however, l and b mean the same thing as l^{II} and b^{II}.

17.3 Galactic Rotation

The spiral pattern seen in many distant galaxies suggests that they are rotating. This possibility of rotation around a center in other galaxies led astronomers to wonder if our own galaxy might not be rotating around its center. Such rotation might take either of two forms. First, the galaxy might be rotating as a solid wheel, and second, the rotation might actually consist of the separate revolutions of all of the individual stars around the galactic center. If the latter should be the case, and if the stars all moved around this center according to Kepler's laws, there should be systematic motions of the stars with respect to each other. The inner stars would be overtaking and passing the outer stars.

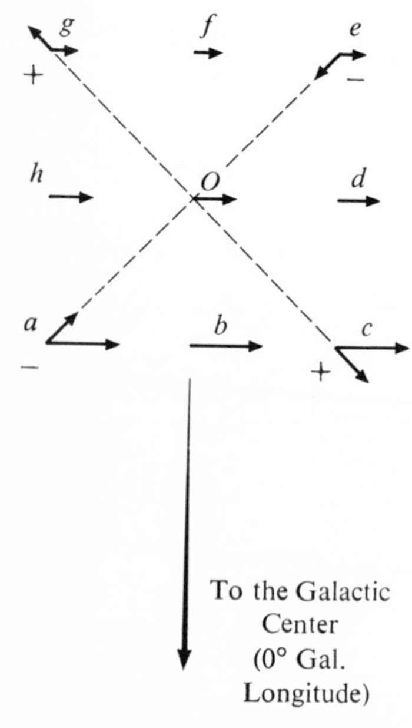

Figure 17.4. Differential galactic rotation results in systematic effects in the radial velocities of stars at various galactic longitudes.

In 1927, J. H. Oort of the Netherlands demonstrated that the inner stars did in fact revolve more rapidly than the outer ones, and thus he showed that the Milky Way Galaxy must actually be in rotation. Oort's proof lay in his discovery of systematic effects in the radial velocities of large numbers of stars in the Milky Way. In Figure 17.4 three possible orbits around the galactic center are indicated. One of these is the sun's orbit, one is closer to the center, and the third is farther away. Stars are indicated in the diagram at three points along each orbit. The position of the sun is at O, and the plane of the diagram represents the plane of the Milky Way. In other words, the observer looking at the stars successively from a to h would be looking along the entire Milky Way. A star at point a, being closer to the center than the sun, will be moving faster than the sun and will be overtaking the sun. The distance between the sun and this star must necessarily become shorter until the star passes the sun, and the relative position of sun and star become O and b. Thereafter, the distance increases. In the general direction of point a ($l = 315°$) the stars should be approaching the sun, and their radial velocities should be negative (spectral shifts toward the blue). In the general direction of point c the reverse should be true. Stars there should be moving away from the sun, and the radial velocities ought to be positive (spectral shift toward the red). In the direction of point *b* the stars are moving essentially parallel to the direction of the sun's motion, and there should be no radial velocity.

Stars at *d* and *h* would be essentially the same distance from the center as the sun, so their galactic revolution periods ought to be the same as that of the sun. The positions of these stars relative to the sun will be constant, so their radial velocities should be zero. The sun will be catching up with the stars in the direction of point *e*, and as a result those stars will have negative radial velocities. Again at *f* the situation is the same as it was at *b*. The sun is passing the stars in this direction, but there is no radial velocity. Finally, at *g* the stars are being left behind by the sun and must, therefore, show positive radial velocities.

Thus, if the stars revolve around the galactic center, there should be a systematic pattern in their measured radial velocities. When radial velocities of a large number of stars are plotted as a function of galactic longitude, a two-peaked curve such as that of Figure 17.5 should result. This was in fact exactly the curve found by Oort. Astronomers have since added data from fainter stars at greater distances and confirmed Oort's original finding. Thus, astronomers today are confident that the stars move essentially according to Kepler's third law in accounting for the overall rotation of the galaxy. The galaxy rotates not as a wheel, but "differentially." The revolution periods of the stars are progressively larger for stars at greater distances from the center.

It was mentioned in Chapter 15 that clouds of interstellar gas are detectable from their absorption lines, which are radial-velocity shifted to separate them from the lines originating in the atmospheres of the stars. In many cases the presence of more than one interstellar line indicates more than one cloud occurring along a particular line of sight. The interstellar lines in the quadrant

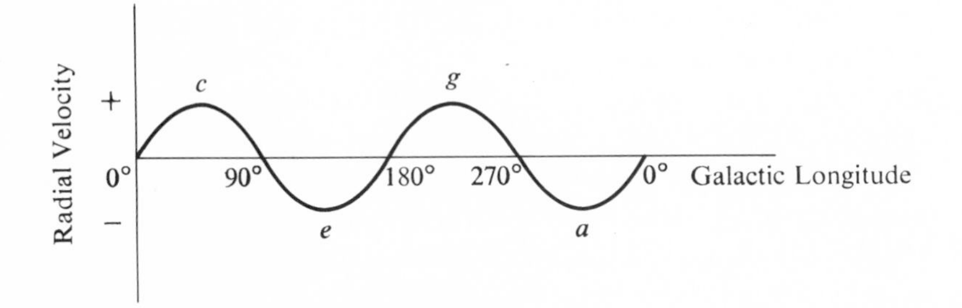

Figure 17.5. The observed effect of galactic rotation on radial velocities. This is the evidence that all stars move approximately according to Kepler's laws in orbits around the center of the galaxy.

from galactic longitude 90° to 180° have been studied in detail by G. Münch of the Hale Observatories. In this quadrant the radial velocities of stars are generally negative as a result of galactic rotation, and the effect is greatest at $l = 135°$. Münch has shown that the same effect occurs in the radial velocities of the interstellar gas. Radial velocities increase with increasing galactic longitude from 90° to 135° and then decrease to 180°. In most of the spectra the interstellar lines are double, and the same longitude effect is noticeable in the radial velocities of both components. Since the effect of differential galactic rotation on radial velocity becomes greater as the distance becomes greater, the evidence from the interstellar gas suggests that there are two relatively discrete clouds in this Milky Way quadrant. One cloud is relatively near us, and the other is quite far away. This finding is substantiated by evidence that the stars as well as the gas fall into two distinct bands in this quadrant.

It should be realized also that a rotation effect should be noticeable in proper motions of stars. At the points b and f in Figure 17.4 the radial velocity is zero but the proper motion should be largest. The effects of differential galactic rotation are greater with increasing distance from the sun, but proper motions become less with increasing distance. Therefore, proper motions cannot provide the vivid demonstration of galactic rotation that is found in radial velocities.

The rotation of the galaxy means that the sun itself must also be in an immense orbit about the galactic center. This motion of the sun must be determined relative to some reference frame outside of our galaxy. The motion referred to in Chapter 11 was determined with respect to the stars in the sun's immediate neighborhood. On the scale of galactic rotation the motions of the sun and the nearby stars with respect to each other are relatively minor. They are all moving in much the same manner around the galactic center. This new motion of the sun is determined from the radial velocities of galaxies and globular clusters in the direction of the sun's motion and in the opposite region of the sky. In its galactic orbit the sun seems to be traveling at about 250 km/sec towards a point at 90° galactic longitude.

This galactocentric velocity of the sun may be used with the currently accepted value for the distance from the sun to the center (10,000 pc) to compute the period of the sun's revolution around the center. Computations show that this period is 200,000,000 years, a sizeable fraction of the lifetime of a star such as the sun. This value for the period also permits astronomers to make a further application of Kepler's third law by which the total mass of the galaxy may be computed. In the third law, written $P^2(M+M_\odot)=D^3$, P will now be the revolution period of the sun around the galactic center or 200,000,000 years, and D is the distance of the sun from the galactic center expressed in astronomical units. M represents the mass of the entire galaxy under the assumption that this mass acts gravitationally as if it were concentrated at the center of the galaxy, while $M_\odot$ is the mass of the sun. Making the proper substitutions, $M+M_\odot=2\times10^{11}$ or $M=2\times10^{11}$. This means that the total mass of our entire galaxy is of the order of 200 billion solar masses. A significant percentage of this total mass is in the clouds of dust and gas. This is necessarily a rather crude method for estimating the mass of our galaxy, but it does give reasonable results.

17.4 Rotation and the 21-cm Data

The measurement of radial velocities of hydrogen clouds from the 21-cm line profiles was discussed in Chapter 15. It was mentioned there that in many directions in the Milky Way the line profiles showed several peaks, indicating the presence in a particular direction of several emitting clouds moving with different radial velocities. With the basic pattern of differential galactic rotation established from optical studies, radio astronomers were able to analyze their own radial velocity data in a most significant manner. The important result has been that radio astronomers are now able to relate radial velocity to distance and thus locate the hydrogen clouds in space.

In Figure 17.6 two concentric circles represent possible orbits around the galactic center. A line has been drawn from the sun tangent to the inner orbit. This line could represent the direction in which a radio telescope might be pointed when a line profile was recorded. If the gas clouds revolve in the same pattern as the stars do, material moving along the inner orbit should be moving faster than material moving along the outer orbit. This difference in orbital velocities is indicated by the lengths of the two arrows a and b, representing the velocities of gas at points 1 and 2. Since the line of sight from the sun is tangent to the inner orbit at point 1, all of the motion represented by a is directed along the line of sight and is therefore detected as radial velocity. The motion represented

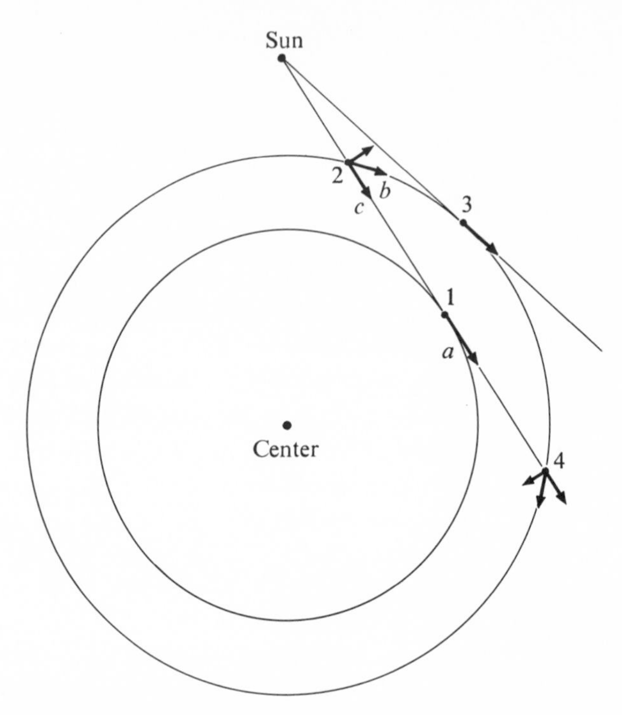

Figure 17.6. The relation of radial velocity of the hydrogen clouds to their distances along the line of sight.

by b may be divided into a component along the line of sight and one at right angles to this direction, so only a fraction of b can be detected as radial velocity. This radial component is indicated by arrow c. For obvious reasons, a must be larger than c. Therefore, upon examing a line profile, radio astronomers attribute the peak with the largest radial velocity to the presence of an emitting cloud at point 1. Any other peaks on the same line profile represent other emitting clouds along the same line of sight, but at some other distance. The distances to such clouds can be determined from the geometry of the situation, provided that the value of b has been determined from an observation made with the radio telescope directed toward point 3.

Another peak with a radial velocity between c and a would represent a gas cloud situated between point 1 and point 2. A smaller radial velocity would have to represent a cloud closer to the sun than point 2. Thus, once the line profiles have been observed in a great many directions, the tangent-point velocities will be known from the maximum radial velocities indicated on each profile. The distance, related to any other radial velocity, may then be computed. In this manner there is a relation between radial velocity and distance, and this relation differs with each direction in which the radio telescope is pointed.

Finally, it should be noted that this velocity-distance relation is double-valued. The radial velocity of a cloud at point 4 will be the same as that for the cloud at point 2. The radio astronomer must find some means by which he can decide which of the two possible locations to choose. He might assume that the intensity of the signal was an indication of the distance. In Figure 17.6 the signal from the cloud with velocity c is stronger than that from the cloud with velocity a. This suggests that the first cloud is closer than the second, but requires the assumption that the emitting clouds are all somewhat similar. A safer procedure is to make additional line profiles slightly out of the plane of the Milky Way. Since the thickness of the layer of hydrogen may be fairly uniform, the angular size of a gas cloud will depend upon its distance. A distant gas cloud will not be in the beam of the radio telescope pointed a few degrees out of the plane, while the closer cloud will. With this extra step, the ambiguity of the distance estimates can be eliminated.

Complete analysis of the 21-cm line profiles gives the distances to hydrogen clouds in many directions in the plane of the Milky Way. A map was drawn showing the distribution of hydrogen in the galactic plane. One of the most significant contributions of radio astronomy, the map (Figure 17.7) was compiled from the observations of pioneer Dutch radio astronomers in the northern hemisphere and Australian workers in the southern hemisphere. The map clearly reveals a spiral distribution of the hydrogen clouds throughout most of our galaxy. Since the radio waves are not absorbed by interstellar material, the map shows the positions of the clouds at the far side of the system and confirms our concept that the sun is situated quite far from the center. In the directions toward the center of the galaxy and away from the center there is no radial velocity resulting from the rotation of the galaxy, as shown by the two blank areas in galactic longitudes 0° and 180° in Figure 17.7.

The method of analysis described above assumes that the hydrogen clouds move essentially in circular orbits around the galactic center. If the gas has some radial motion with respect to the galactic center or if there is streaming motion along the spiral arms, then the interpretation of the 21-cm data becomes very difficult. Recent 21-cm surveys of the Milky Way have been made with more

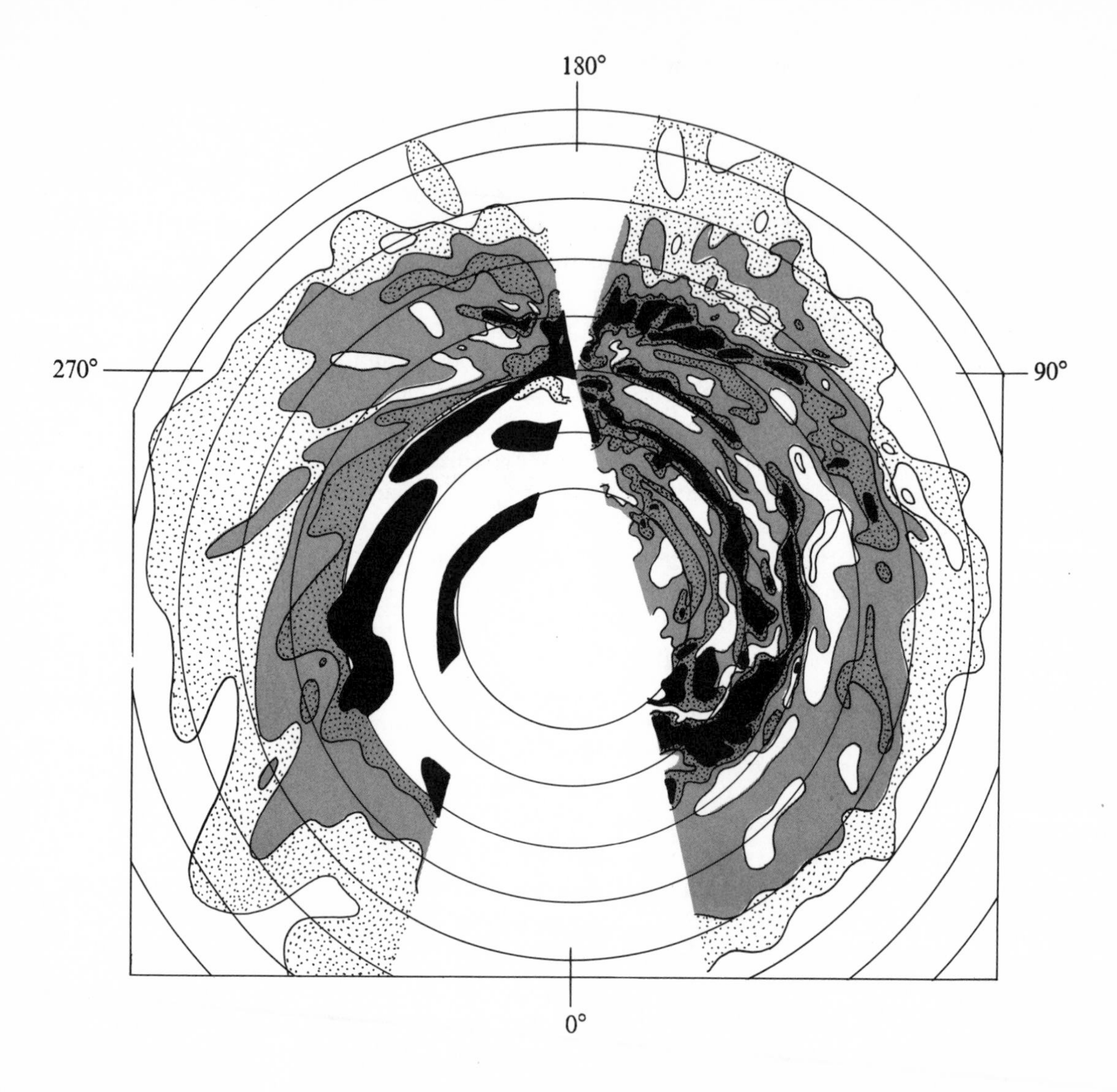

Figure 17.7. The distribution of hydrogen in the plane of the galaxy as determined from 21-cm observations. The darker areas represent more dense concentrations of gas. The concentric circles are 2 kpc apart and the sun is 8 kpc from the center. (From a diagram by J. Oort, F. Kerr, and G. Westerhout.)

refined equipment than that used in the early studies. The newer results show that the structure in the gas is more complex than originally thought and introduce the possibility of distant spiral arms which lie above the central plane of the galaxy. The initial analysis, which seemed to be quite straightforward, is now clearly inadequate. For the radio astronomers, however, the problem remains much the same. The observed radial velocities of hydrogen clouds must still be interpreted in terms of the distances to the clouds.

17.5 Optical Evidence of Spiral Structure

Because of interstellar absorption, we can never hope to obtain from optical evidence a picture of the galaxy which can compare in extent with the radio map just discussed. Nevertheless, within a limited area around the sun, patterns showing spiral structure have recently been discovered. The means by which the spiral structure of our galaxy was discovered indicates generally the kinds of objects to be expected in spiral arms and is further related to the theory of stellar evolution.

The first study revealing evidence of spiral structure in our galaxy was based on distance estimates to HII regions. When these were plotted on a map representing the plane of the galaxy, the pattern seen in Figure 17.8 was revealed. The distances were estimated from the magnitudes of the stars which are ionizing the gas. This map tells us that the HII regions seem to fall into three fairly distinct groups in the area within roughly 3000 pc of the sun. If these groups indicate the presence of spiral arms, then the sun seems to be on the inner edge of one arm. Another arm seems to lie between the sun and the galactic center, and a third arm is suggested beyond the sun's arm. These are meager data, but the implications are confirmed from some other similar data.

In Figure 17.9 the positions of a large number of galactic clusters are shown on the plane of the galaxy. The distances of the clusters were obtained from their color-magnitude diagrams. There is no very obvious pattern in these data as they appear here; however, from the color-magnitude diagrams one may also note the position at which the cluster main sequence turns off from the zero-age main sequence. In the previous chapter the position of this turn-off point was cited as an indication of the age of the cluster. From their color-magnitude diagrams, then, clusters which must be relatively young may be selected. Such clusters would be ones in which the earliest spectral types were O to B2. In these young clusters the turn-off point would be at a color index of about -0.3 on the color-magnitude diagram.

When the sample of galactic clusters is limited in this manner to young clusters, and these are plotted on the galactic plane, the map indicated in Figure 17.10 results. Here, in a striking manner, three arms are again indicated, matching those indicated by the HII regions. In addition, some other important ideas are suggested. The fact that only the young clusters lie in the arms means that the random motions of the clusters cause them to move out of the spiral arms if a large enough time elapses. Furthermore, if the young clusters lie in the arms, then there is the implication that stars form in the spiral arms of galaxies. We might therefore expect this pattern of arms to be indicated by other objects known to be young.

The same sort of analysis as that described for the clusters may be applied to the Cepheid variables with very similar results. The distances of the Cepheids are determined from their observed periods through the use of the period-luminosity relation. When the distances and galactic longitudes of large numbers of Cepheids are used to plot the Cepheids on the galactic plane, no particular pattern is apparent. When, however, the sample is limited to Cepheids with periods on the order of 100 days, the spiral arms are again apparent. From the period-luminosity law these are brightest of the Cepheids. They are also probably the Cepheids which have evolved from stars of very large mass, and for this reason they will have evolved fairly rapidly. If their parent stars were main-sequence stars of spectral type B0 which were born in spiral arms, the bright Cepheids have not yet had time to move out of the spiral

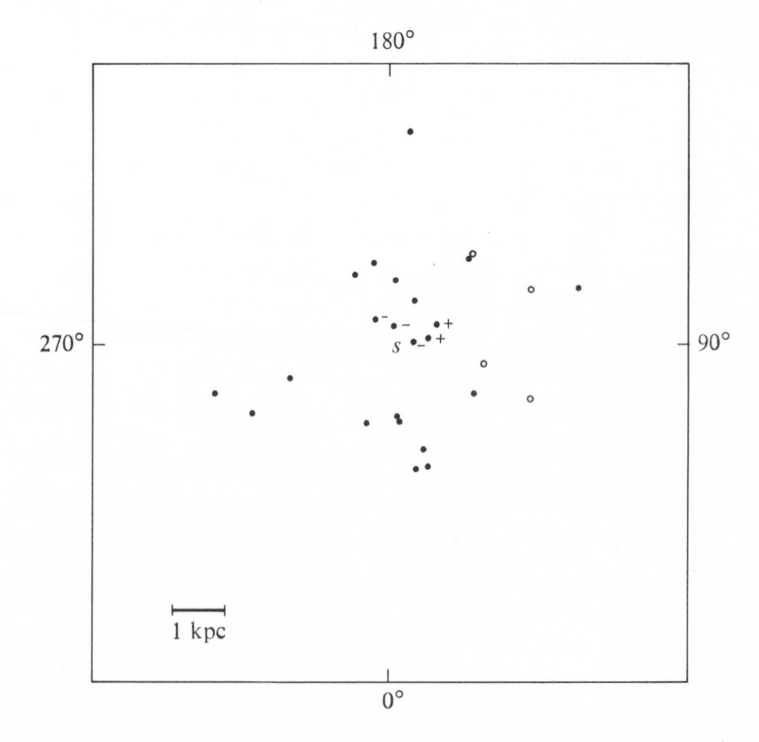

Figure 17.8 Suggestions of parts of three spiral arms as revealed from the distribution of HII regions in the plane of the galaxy. (After a diagram by S. Sharpless.)

arms. The shorter-period Cepheids can be somewhat older, and many of these have had time to move far enough from their places of origin to smear out the spiral pattern.

Since these independent optical studies all indicate the same general pattern, astronomers today are convinced of the reality of spiral arms in the sun's vicinity. Names for these arms have come into use from the constellations which lie in their general direction. Thus, the arm nearest to us which lies between the sun and the center is referred to as the **Sagittarius Arm.** Similarly, the well-defined arm beyond the sun's position is called the **Perseus Arm,** since it lies beyond the foreground stars which form the constellation Perseus. The sun definitely seems to lie on the inner side of a third arm whose outline and extent are less definite. Some astronomers choose to call the third arm the **Carina-Cygnus Arm** because it extends away from us in the direction of Carina on one side and in the direction of Cygnus on the other side of the sky. In the direction of Cygnus ($l = 80°$) the arm seems to be better defined than in the opposite direction. There seems to be a spur extending off the Carina-Cygnus Arm in the direction of Orion.

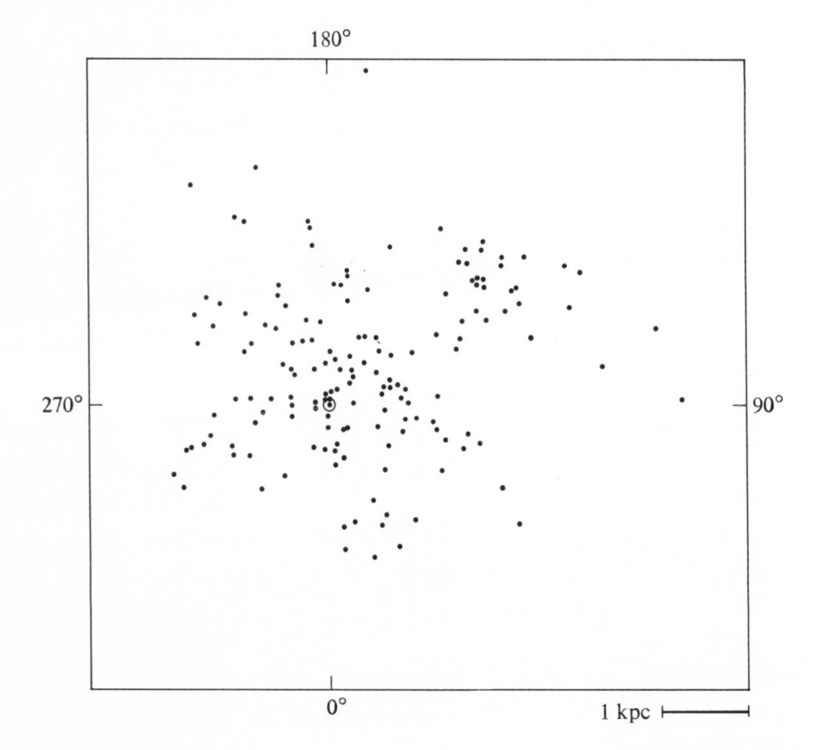

Figure 17.9. The distribution of galactic clusters on the plane of the Milky Way. The circled dot represents the location of the sun, and galactic longitudes are marked around the edge. (Diagram by Prof. Wilhelm Becker, University of Basel, Switzerland.)

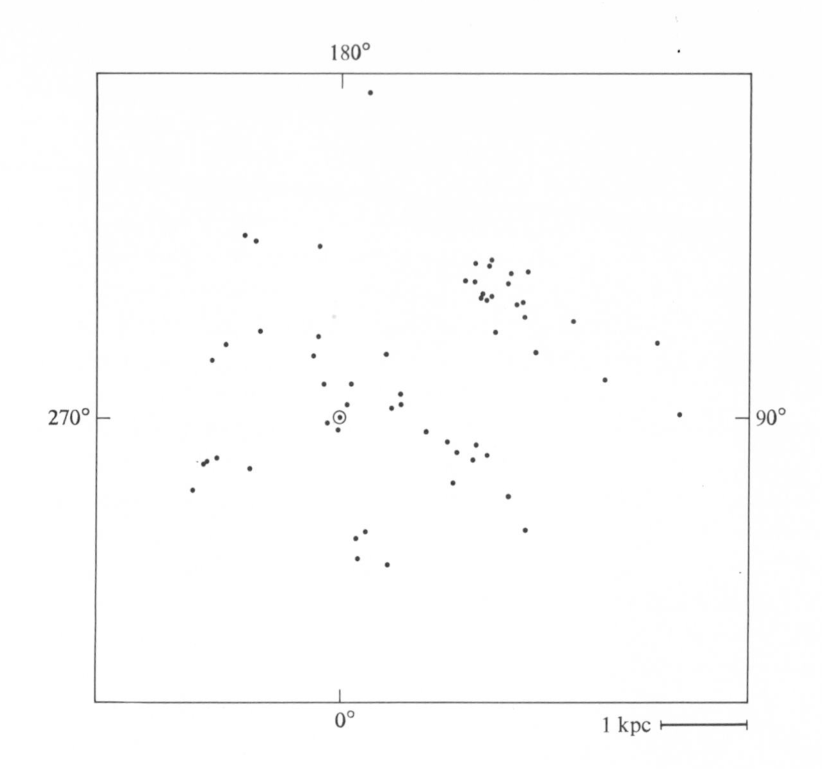

Figure 17.10. The distribution of galactic clusters which contain O and B type stars and are probably young. Portions of three "arms" are suggested. (Diagram by Prof. Wilhelm Becker, University of Basel, Switzerland.)

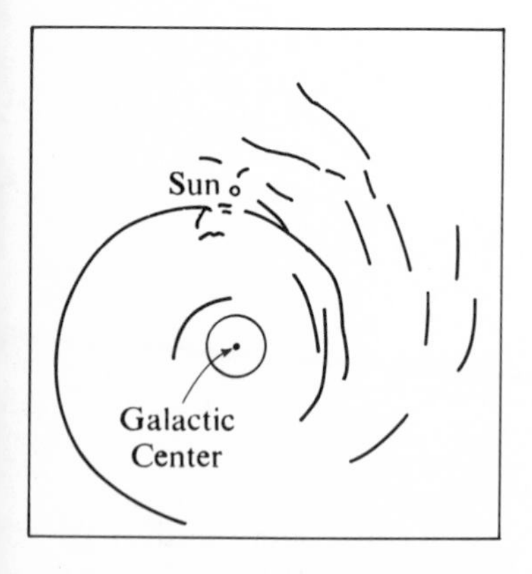

Figure 17.11. The distribution of neutral hydrogen from 21-cm observations as interpreted by Harold Weaver in 1969. The Sagittarius and Perseus Arms are clearly indicated, but the Orion Arm in which the sun is located is less clearly defined. (After a diagram by H. Weaver.)

It is reasonable to ask how this optically determined pattern matches that derived by the radio astronomers. Comparison of Figures 17.8, 17.9, and 17.10 with Figure 17.11 shows that the agreement is not as striking as one might expect. There is no simple explanation of this. The error is not likely to be in the radial velocities of the hydrogen clouds, since these are known with greater accuracy than most spectroscopic radial velocities. The difficulty lies in interpreting the 21-cm data in terms of an assumed model of the rotation of our galaxy.

Further support for this pattern of arms could be obtained if similar patterns could be seen in some other galaxies. Here the possibilities are somewhat more encouraging. Figure 17.12 is a photograph of the spiral galaxy NGC 1232 in the constellation Eridanus. Superimposed on this photograph are a number of white dots which represent essentially the young objects of Figures 17.8 and 17.10. The data have been increased to include the most luminous Cepheids, which also help to delineate the spiral arms. Clearly, the degree of curvature and the number of arms in our part of the

Figure 17.12. The spiral pattern defined by young objects in the neighborhood of the sun superimposed on a photograph of the spiral galaxy NGC 1232. (Photograph by Prof. Wilhelm Becker, University of Basel, Switzerland.)

Milky Way galaxy are similar to what can actually be found in at least one other galaxy.

With some imagination, one may explain the appearance of the Milky Way from within our arms of this spiral galaxy. In two directions, i.e., toward Cygnus and toward Carina, the observer is looking lengthwise into the arm which contains the sun. In both of these directions the Milky Way appears to be very bright. At slightly different directions the observer is looking between two arms and sees, at a great distance, part of another arm. Because of the great distance, the Milky Way here looks dimmer and narrower. Analyzed in this manner, the variations seen along the Milky Way are completely consistent with the spiral pattern described above.

17.6 The Galactic Nucleus

On most photographs of the Andromeda Galaxy, such as the one in Figure 18.1, the central region is considerably overexposed so that the detailed structure in the outer portions can be recorded. Short exposures in which the central region is properly exposed, however, reveal a bright central nucleus of small angular diameter. The exact nature of this nucleus is unknown, but astronomers have speculated that a similar nucleus might exist in our own galaxy. Unfortunately, such a nucleus would be invisible to us behind the clouds of interstellar dust. Some evidence for the existence of a relatively small nucleus within our galaxy has nevertheless been gleaned from radio and from infrared observations.

On some of the early radio maps of the sky, intensity contours indicated that one section of the Milky Way was a particularly strong source of radio emission. In the nomenclature used by radio astronomers to specify discrete radio sources, this section came to be known as **Sagittarius A.** As the radio astronomers increased their ability to resolve small details, it became apparent that Sagittarius A was a rather complex radio source consisting of five or more discrete parts. This complicated source stretches for slightly more than one degree along the galactic plane. The interpretation of these and other data is that in the center of our galaxy there is a small spiral arm of hydrogen expanding outward toward the sun. Knowing the distance to the center and the angle at which the line of sight is tangent to this nearly circular arm, astronomers have calculated the radius of the arm as three kiloparsecs (3 kpc). Within this tightly curving arm there is a smaller source which may well be the small-diameter central nucleus suggested by our view of other galaxies. 21-cm radiation from this central source is actually absorbed as it passes through the 3-kpc arm.

To a significant degree, this picture of the nucleus of our galaxy is confirmed on recent infrared photographs which show that in the direction of the galactic center there is a bright region of small angular size. With the slow accumulation of more observational data of this sort, astronomers hope to understand the exact nature of the central nucleus and its role in the evolution of the galaxy.

17.7 Stellar Populations

Walter Baade was one of the first to notice a distinct difference between the stars in the central regions of the Andromeda Galaxy and those in the outlying spiral arms. He noted that in the arms the bright stars were blue, while in the nucleus the bright stars were red. In Color Plate 8 the blue color of the arms is quite apparent. By contrast, the central nucleus is yellow. To describe this situation Baade introduced the idea of two **populations** of stars. Population I consisted of the stars in the spiral arms and Population II consisted of the stars in the nucleus and in the halo. This rather clear distinction is very noticeable in color photographs of the Andromeda Galaxy and a number of other spiral galaxies.

In our own neighborhood in the Milky Way Galaxy the separation of the two populations is also well defined. The globular clusters and the halo contain stars which are distinctly Population II, and the spiral arms contain stars which are distinctly Population I. The concept of populations has proved to be very useful to astronomers and unquestionably helpful in the formulation of our stellar evolution theory. The population idea has developed to the point where five populations are recognized.

1. *Extreme Population I:* Gas, O and B stars, Cepheids, T Tauri stars, and galactic clusters, all of which define the spiral structure.
2. *Older Population I:* Type A stars and dwarf M stars which show emission lines among other types.
3. *Disc Population:* Stars of the galactic nucleus, planetary nebulae, novae, and RR Lyrae variables with periods less than 0.4 day.
4. *Intermediate Population II:* Long-period variables and "high-velocity" stars which have elongated orbits around the galactic center.
5. *Halo Population II:* Subdwarfs, globular clusters, and RR Lyrae variables with periods longer than 0.4 day.

From the previous discussion of stellar evolution, it should be clear that population type is closely related to evolutionary age. The Extreme Population I stars concentrated in the galactic plane are the very young ones, and the Halo Population II stars are the very old ones. That there must be some important relationships between the distribution of stars and their ages is well supported by the evidence that only young stars lie in spiral arms.

The gradual development of our knowledge of the size and shape of our galaxy has revealed a vast and complex system containing billions of stars and enough material to make billions more. The present picture relates discoveries in many areas of astronomy to each other. Advances in each area have been dependent upon advances in other areas, and so in a real sense the modern concept of our galaxy represents great achievement in all fields of observational and theoretical astronomy. The years to come will see the filling in of detail, but the overall structure of our galaxy is now well established. With the understanding of the type of stellar system in which we live has come an understanding of the position of our galaxy in the universe of other galaxies.

QUESTIONS

1. Why does the presence of large numbers of faint stars in the direction of the Milky Way indicate that our entire system of stars must be most extensive in the direction of the Milky Way?
2. How did Shapley's analysis of the distribution of globular clusters reveal the direction and distance to the center of our galaxy?
3. Define the galactic equator and show how star positions are defined in the galactic coordinate system.
4. How is the rotation of our galaxy proven through the observed systematic variation of radial velocity with galactic longitude?
5. What galactic rotation effect is to be seen in the radial velocities of the absorption lines due to interstellar gas?
6. How is Kepler's third law used to estimate the total mass of our galaxy?
7. Describe the reasoning by which radio astronomers are able to relate radial velocity, as measured from 21-cm line profiles, to distance along the line of sight.
8. Why is the above relationship between radial velocity and distance different for each different galactic longitude?
9. Describe the main optical spiral arms in the sun's vicinity.
10. What evidence here lends support to the concept that young stars form in spiral arms?

11. What are the basic differences between the two stellar populations?

12. How do the differences in the populations lend support to the ideas that stars form in spiral arms and the theory of stellar evolution set forth in Chapter 16?

Figure 18.1. The spiral galaxy M31 in Andromeda. The central region in this photograph is somewhat overexposed. The two large objects seen near the spiral are smaller elliptical galaxies which accompany M31. (Hale Observatories photograph.)

18

Galaxies

Astronomers find some of their greatest challenges in the study of galaxies beyond the limits of our own Milky Way. Theories of stellar evolution are called upon in attempts to understand galactic evolution. The observed geometrical forms, radiations, distributions, and motions of galaxies are analyzed concurrently by astronomers in order to reconstruct, in the most general way possible, the origin and evolution of the universe as a whole. The research which supports such investigations is very expensive and must necessarily be carried on by the few astronomers who have access to the world's largest optical and radio telescopes. Even for these men, the knowledge of galaxies has been gained only with great difficulty. At the telescope, spectra of galaxies require long exposure times, while in the laboratory the analysis of the data requires patience and an imaginative approach. All areas in astronomy are interdependent, and advances in some fields are possible only after advances in related fields. This is particularly true in the study of galaxies.

18.1 Classification

There are literally millions of galaxies within reach of our large telescopes (see Figure 18.2). From these vast numbers three basic

Figure 18.2. Part of a large cluster of galaxies in the constellation Hercules. Many of the principal types of galaxies are represented here. (Hale Observatories photograph.)

types of galaxies appear—**elliptical galaxies, spiral galaxies,** and **barred spirals.** A fourth type, of which there are relatively few examples, is described simply as **irregular.** We can see galaxies in "random" orientations, from a polar view to a side view. What is believed to be a spiral galaxy seen edge on is pictured in Figure 18.3.

It is natural that as astronomers examined galaxies in greater and greater detail they should have recognized similarities and differences which would lead to a fairly detailed classification scheme. Today's rather complex classification of galaxies is basically an elaboration of the scheme first used by Edwin Hubble in 1926 and modified slightly some years later. The newer classification schemes simply recognize the progression of details and substructure among the basic types. More detailed recent studies have also shown that small numbers of galaxies really represent additional distinct classes. Although the numbers of galaxies in these new classes are small compared to the numbers of "normal" galaxies, some of our most perplexing problems are presented by these objects. In particular, these new classes are the Seyfert galaxies, the qua-

Figure 18.3 The spiral galaxy NGC 4565. This galaxy is so oriented in space that we look almost exactly into the edge. The negative inset in the corner of the photograph shows this galaxy on the same scale as the galaxy M31 in Figure 18.1. (Hale Observatories photographs, © copyright by National Geographic Society.)

sars, and the radio galaxies. We shall discuss the normal galaxies in detail and then proceed to the more recently discovered types.

Elliptical Galaxies

The elliptical galaxies vary in appearance from those which appear circular to those which are quite elliptical. Hubble designated such galaxies with the letter E followed by a number specifying the degree of ellipticity. This number ranged from 0 for galaxies which appeared circular to 7 for the most elliptical galaxies. The number was computed from the measured dimensions of the elliptical image of a galaxy on a photograph, and so required little subjective judgment by the classifier. Hubble noted that elliptical galaxies were never flatter than E7, but he found also that the E7 galaxies showed outlines which were pointed on the two ends. For this reason Hubble later separated the E7 galaxies from the ellipticals and classed them as S0, extreme cases of spiral galaxies.

In their general overall appearance the elliptical galaxies are quite uniform, as may be readily seen in Figure 18.4. The outer parts of some of the closer ones, such as the companions to the Andromeda Galaxy, may be resolved into stars.

Figure 18.4. The elliptical galaxy NGC 147 in Cassiopeia. (Hale Observatories photograph.)

Spiral Galaxies

In Hubble's scheme the normal spiral galaxies were designated Sa, Sb, or Sc on the basis of their general appearance. (See Figures 18.1 and 18.5.) The Sa galaxies have a large nucleus and well-defined arms which are tightly wound around the nucleus. Absorbing clouds are obvious in the arms and may be said to define the arms. The Sb galaxies show a slightly smaller central nucleus and arms more loosely wound than the Sa galaxies. In addition, there is often evidence of "clumpiness" in outer portions of the arms. That is, the arms are not quite as smooth as the arms in the Sa galaxies. In the Sc galaxies the nucleus is often quite ill defined and the arms show a very definite patchy appearance. The curvature of the arms is such that they have become tangent to the galaxy after about one-half turn.

In all of the spiral galaxies there are usually only a few major arms in spite of the confusion caused by silhouetted dust lanes and the branching of the arms. The identification of major arms is even more difficult because most galaxies are viewed more or less obliquely.

Barred Spiral Galaxies

Barred spirals are characterized by a bar-shaped nuclear region with spiral arms extending from the ends of the bar. Hubble designated these galaxies as SB and further classified them on the basis of arms and general appearance in the same manner as used for the normal spirals. There are, therefore, SBa, SBb, and SBc types among the barred spirals, depending upon the tightness of the arms and the degree of patchiness in the arms. In general, it may be stated that there are fewer barred spirals than normal spirals.

Irregular Galaxies

Designated Irr by Hubble, the irregular galaxies have no characteristic pattern as a class. Also, the shape offers no suggestion of symmetry due to rotation. A good example of an irregular galaxy is

Figure 18.5. The spiral galaxy NGC 3031 (M81). Again, the negative print in the inset shows this galaxy on the same scale as that of M31 in Figure 18.1. (Official U.S. Navy photograph; inset, Hale Observatories photograph, © copyright by National Geographic Society.)

Figure 18.6. The Large Magellanic Cloud, an example of an irregular galaxy. (University of Michigan photograph courtesy of Karl Henize.)

the Large Magellanic Cloud pictured in Figure 18.6. The same type of patchiness is observed in the irregulars as in the outer parts of the arms of the Sc spirals.

Eventually Hubble recognized two types of irregular galaxies. Type I irregulars were those like the Magellanic Clouds and Type II irregulars were those similar to M82 pictured in Figure 18.7. M82 shows a great deal of matter and dust and is now believed to have undergone a violent explosion of some sort. Unfortunately, little is presently known about Type II irregulars.

Ring Substructure

The spiral galaxies and barred spirals may be further separated by the presence or absence of a rather definite ring around the nucle-

Figure 18.7. The irregular galaxy M82. (Hale Observatories photograph.)

us. The ring may be seen in the photograph of NGC 1398 in Figure 18.8. According to the notation suggested by Alan Sandage of the Hale Observatories, this galaxy is classified as SBb(r), where (r) indicates the presence of a ring. Sandage has also called attention to the complexity of many of the former E7 galaxies which have been reclassified as S0. Enough variety, including the presence of ring structures, has been noted within the group of S0 galaxies to warrant the recognition of subclasses.

Figure 18.8. The ring substructure is easily seen in this barred spiral galaxy (NGC 1398) in the constellation Fornax. (Hale Observatories photograph.)

It is quite natural that astronomers should have looked for some means to arrange the galaxies of all types in a logical sequence. Hubble did this in his well-known "tuning-fork" diagram sketched in Figure 18.9. Hubble felt that this sequence of types of galaxies might well indicate something of the evolutionary development of galaxies. At that time there was not enough known about stellar evolution to do more than speculate on the evolution of galaxies. Today we have a more complete theory of stellar evolution, but we still have difficulty in relating it to galactic evolution.

It is possible that all of the elliptical galaxies are really highly flattened spheroids which look like E6 galaxies when seen from the side and E0 galaxies when seen from the top. If this should be the case, and if the galaxies are randomly oriented in space, the numbers to be expected in each possible orientation should be predictable by straightforward statistical techniques. Hubble was the first to show that the actual statistical counts of the E galaxies did not support the notion that the E galaxies were all the same shape. The counts indicate that there are probably similar numbers of galaxies of all the real degrees of flattening indicated from E0 to E6. There are as many galaxies which are spherical as there are galaxies which are highly flattened ellipsoids.

The same sort of statistics applied to the measured ellipticity of the photographic images of the spiral galaxies indicates, on the other hand, that these galaxies are in reality all very flat. This is certainly consistent with the intuitive impression gained from examination of a large number of spirals. If these galaxies are flat, the angle of tilt between the line of sight and the plane of the galaxy may be estimated from the measured shape. Because of many irregularities and actual lack of symmetry in individual galaxies,

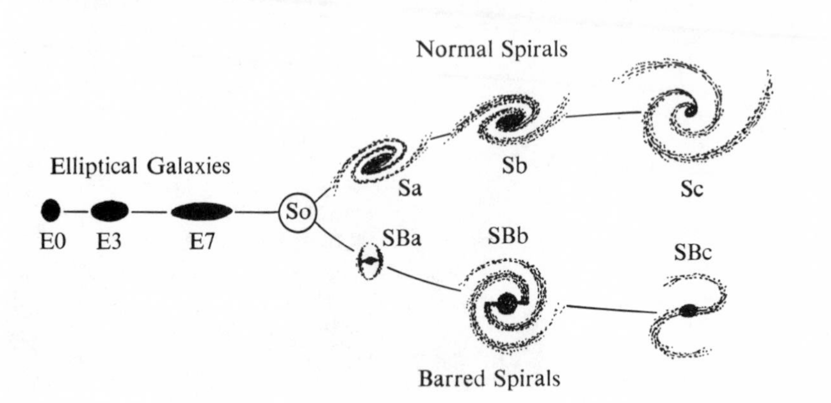

Figure 18.9. Hubble's scheme for the classification of galaxies.

the true angle of tilt is usually not accurately known. Equally difficult to determine in most cases is the direction of the tilt. That is, it is hard to say for sure which edge is closest to us. In a number of cases, however, the dark material of the arms is almost certainly seen silhouetted against the nucleus, and the direction of tilt may be derived. From the direction of tilt and from radial velocity data, it has been concluded that the spiral arms are trailing with respect to the direction of rotation.

18.2 Apparent Magnitudes

It is rather difficult to specify the apparent magnitude of a large extended object such as the Andromeda Galaxy. By integrating the entire light of the whole galaxy, however, a magnitude can be assigned which represents the equivalent apparent magnitude that a star would have if we received the same total amount of light from it as from the galaxy. The problem is a bit simpler for the more distant galaxies, whose photographic images are relatively small, since the image of the entire galaxy may be studied at one time by normal photometric methods. For the Andromeda Galaxy, the integrated apparent magnitude in visual light is 3.5. There are roughly 20 galaxies brighter than ninth magnitude. The numbers of faint galaxies seem to increase steadily as the sky is examined with successively larger telescopes. So far, there seems to be no decrease in number at the faint end of the scale. The larger the telescope, the fainter the galaxies which may be seen. Also, the numbers counted are proportionally greater for the fainter and more distant galaxies as we look to greater distances and therefore to greater volumes of space. The absolute magnitudes of the individual galaxies may be computed if the distance to each galaxy is known.

18.3 Distribution

The Andromeda Galaxy is, of course, situated in the constellation Andromeda. Its location in relation to the bright stars in Andromeda and Pegasus may be seen on the star maps of Chapter 1. Except for the two bright Magellanic Clouds of the southern hemisphere, the Andromeda Galaxy is the only one which can easily be detected with the naked eye. Other galaxies brighter than ninth magnitude are scattered over the sky somewhat randomly. As one looks to fainter magnitude limits, the number of galaxies increases rapidly and the apparent distribution of these on the sky begins to take on a significant pattern. In a zone centered on the galactic equator, large areas are completely devoid of the images of galaxies. It was for this reason that Hubble referred to the region as the "zone of avoidance." Today we know that near the plane of the Milky Way our view is blocked by the huge clouds of obscuring matter which lie in the central plane of our galaxy. One may count the number of faint galaxies per square degree of the sky and find that the largest numbers per square degree are found in the directions of the galactic poles. This is not surprising, since in these directions the path through the interstellar medium is shortest and the effects of absorption should be least.

One of the very important points related to the distribution of galaxies is their tendency to occur in clusters. Our own Milky Way Galaxy is a member of a cluster containing 19 or more individual members. This cluster, generally referred to as the "Local Group," includes samples of most important types of galaxies. Spirals are represented by the Milky Way and the Andromeda Galaxy. Ellipticals are represented by the two companions of the Andromeda Galaxy, while the Magellanic clouds, companions of our own galaxy, are irregular galaxies. A portion of a large cluster of galaxies in the constellation Hercules is shown in Figure 18.2. Not all clusters show such variety in their membership, however. Some particularly compact clusters contain exclusively elliptical galaxies ranging from E0 to S0.

The number of cluster members also shows great variety from one cluster to another. The smallest cluster contains only four members within a volume of space 50 kpc in diameter. The largest clusters, such as one in Coma Berenices, may contain thousands of members filling a volume millions of parsecs in diameter.

There are a number of clusters of galaxies which seem to form, with our Local Group, an even larger aggregation, which has been referred to as a supercluster or a cluster of clusters. This ensemble has been estimated to be some 50 million pc across. Beyond the vague limits of this huge system there is a definite decrease in the space density of galaxies. Furthermore, there is other evidence of clustering of clusters among the very distant galaxies. The concept of the superclusters seems to be well established, but the evidence is not as clear-cut as in the case of the more obvious clusters of galaxies. At the present time the largest related collections of matter in the universe are believed to be the clusters of clusters of galaxies.*

18.4 Colors and Spectra

By using the same techniques and instruments as those available for the stars, the "color" of a galaxy or parts of a galaxy may be determined. The color of an entire galaxy measured this way should indicate something about the average stellar composition of the galaxy. If the majority of the bright stars are red, the overall color of the galaxy will be red; if most of the bright stars are blue, the color of the galaxy as a whole will be blue. Thus the color index, B − V, of a galaxy is indicative of the types of stars which compose the galaxy. The mean colors of galaxies of various types are indicated in Table 18.1. For each type of galaxy there is a range of colors through various individuals. The values tabulated here are mean colors. It should also be pointed out that the colors of individual parts of the same galaxy can show differences.

Table 18.1. Mean colors of various types of galaxies.

Type	E	S0	Sa	Sb	Sc	Irr
B − V	0.93	0.93	0.82	0.81	0.53	0.50

Since the light of a galaxy represents a mixture of the light of many stars of a variety of spectral types and luminosity classes, it seems surprising that any features would show in the integrated spectrum of an entire galaxy. Nevertheless, some identifiable features do show up and indicate that the galaxy must contain many stars with these absorption features in their spectra. For example, the H and K lines of ionized calcium may be seen in the spectra of many galaxies, and we know that these lines are prominent in the spectra of many individual stars. In other galaxies the sodium D lines and the TiO bands may be identified. Enough features are detectable to permit the classification of spectra of galaxies in somewhat the same manner as used for stars. As one might expect, the range of spectral types is very limited and runs only from about A to G. As with the integrated colors, the integrated spectral

*The student is cautioned not to confuse the terms *galactic cluster* and *cluster of galaxies*, since a galactic cluster is composed of stars, as discussed in Chapter 14.

types must indicate the nature of the mean stellar content of the galaxies.

The spectra of galaxies can also confirm the rotation of the galaxies suggested by the flattened appearance of the ellipticals and the pattern of the arms in the spirals. The rotation of our own galaxy leads us to expect rotation in other galaxies as well. For galaxies seen obliquely, rotation is confirmed and measured by noting the radial velocity in many different areas of a galaxy. Since proximity determines angular size, rotation has been measured best for the closer galaxies. For galaxies seen pole-on, rotation causes no radial velocity differences. Some differences in radial velocity across galaxies can be determined from the absorption lines, but the best results come from emission objects, such as HII regions and emission nebulae in the vicinity of supergiant stars. Also of great value in rotation studies, especially in the nuclear regions, is the discovery of an oxygen emission line at $\lambda 3727$, which seems to be present in the spectra of most galaxies.

Analysis of radial velocity data to find the manner of rotation of galaxies is complex and requires accurate knowledge of the angle of tilt of the galaxy to the line of sight. Nevertheless, the complete analysis has been done for a number of galaxies with results which are consistent with the observed rotation of the Milky Way. Radial velocity measures made at many points across spiral galaxies show that the central nucleus rotates essentially as a solid disc. That is, the velocity increases as one moves from the center of the galaxy toward the edges of the nucleus. In the area of the spiral arms, however, the rotation is Keplerian: the velocity becomes less as one moves outward from the nucleus toward the edges of the spiral pattern. This is the manner of the rotation of the Milky Way Galaxy. The elliptical galaxies are observed to rotate in the same manner as the nuclear regions of the spirals, that is, as if they were solid disclike objects.

The observed radial velocities, corrected for the axial inclination, may be used to compute orbital periods at various distances from the center provided that the actual dimensions of the galaxy are known. This sort of analysis has been applied to a sufficient number of galaxies so that a general pattern may be seen. The elliptical galaxies seem to rotate most rapidly, completing a turn in about 5,000,000 years. The centers of the spirals rotate more slowly and may take as much as 80,000,000 years per rotation. The periods of objects in the outer parts of the spirals may take hundreds of millions of years for a single revolution around their galactic centers.

Certainly one of the most significant facts to come from the study of galactic spectra has been the discovery that the spectra of the most distant galaxies are all shifted systematically toward the red. Beyond the limits of the Local Group there is an inverse relationship between the apparent magnitude of a galaxy and the amount by which the spectrum has been shifted toward the red. This implies a direct relationship between distance and red shift, as indicated in Figures 18.10 and 19.2. The most obvious effect which can cause spectrum shifts of this sort is the positive radial velocity of the source. The data for galaxies imply, then, that all galaxies are moving away from us and that the most distant ones are moving away the most rapidly. This basic observational material has led to consideration of dynamic rather than static models of the universe. Figure 18.10 shows spectra of a number of galaxies. To the left of each is a photograph showing how the galaxy appears in the sky.

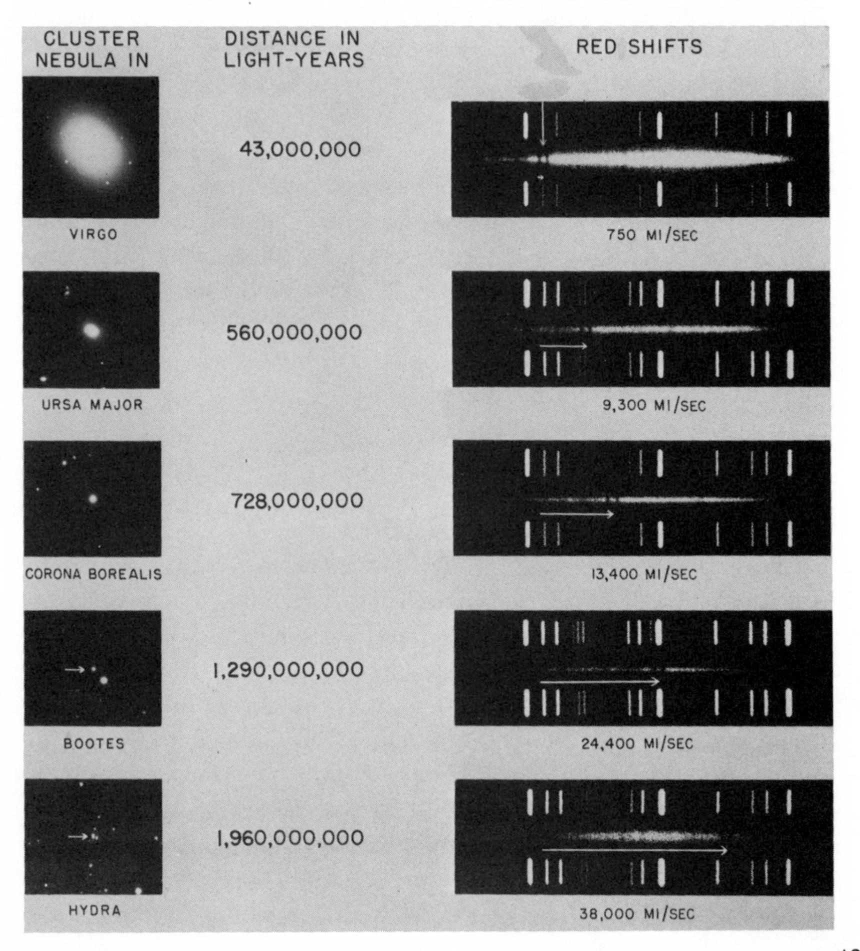

Figure 18.10. Spectra of distant galaxies showing the relation between the amount of red shift and the distance. The red shifts indicate radial velocities ranging from 1150 km/sec to 60,000 km/sec. Arrows indicate shift for calcium lines H and K. (Hale Observatories photographs.)

These few spectra represented a considerable effort since each one required many hours of careful work with the 100-inch telescope. Today, spectra of other very distant galaxies have been obtained with the 200-inch telescope. These later spectra confirm the original basic relationship between distance and amount of red shift, within the observational uncertainties.

18.5 Distances to Galaxies

One of the most significant pieces of work done with the 100-inch telescope when it was first put into operation at Mt. Wilson was the photographing of the Andromeda Galaxy. For the first time its outer portions were resolved into individual stars. This work solved the problem of the "spiral nebulae" and proved that these objects were situated at vast distances from us. The recognition of individual stars also made possible the determination of a distance to the Andromeda Galaxy, using methods similar to those employed for stars within the Milky Way Galaxy. Cepheid variables, for example, were soon identified in the Andromeda Galaxy because they are the most luminous variable stars showing well-defined periodicities. From a series of the early photographs, light curves of individual Cepheids were constructed, and from the light curves the periods were determined. Using the period-luminosity relation for Cepheids in the Milky Way Galaxy, the periods could then be used to find the distance to the Andromeda system (see Chapter 13). Comparing the absolute and apparent magnitudes of an individual Cepheid in the expression

$$M = m + 5 - 5 \log r$$

the distance r to the Cepheid in question could be calculated. This distance to one Cepheid is for all practical purposes the distance to the galaxy itself, since the distance to the galaxy is very large compared with the dimensions of the system as a whole. In this manner the distance to M31 was first found to be some 750,000 light-years. Later when the zero point of the period-luminosity relation was revised, the distance had to be doubled to 1,500,000 light-years. It is now generally accepted that the distance is actually slightly more than 1,800,000 light-years (570 kpc).

In general, then, the most luminous individual stars can be resolved in the nearer galaxies, and we make the basic assumption that the stars in all galaxies are essentially the same. Thus, we assume that the Cepheids in the Andromeda Galaxy obey the same period-luminosity law as those in our galaxy. The bright stars in the Andromeda Galaxy have spectra very similar to those in our own galaxy. They are assumed, therefore, to have the same mean absolute magnitude. These assumptions are reinforced by the general agreement obtained when distances to individual galaxies are calculated on the basis of more than one type of star identified in a galaxy. Thus the calibration of distances to galaxies depends upon the calibration of absolute magnitude for special types of stars within our own system. Ultimately, the trigonometric parallaxes of the nearby stars form the basis of the entire distance scheme.

In addition to stars, several types of familiar objects in our galaxy may be recognized in others as well. Such objects would include planetary nebulae, globular clusters, galactic clusters, novae, and supernovae. The usefulness of each type depends upon its known range of luminosity. Galactic clusters, for example, show much more range in absolute magnitude than do globular clusters or planetary nebulae. Supernovae may have peak absolute magnitudes in the range −14 to −18 and are obviously useful to the greatest distances. Supernovae may be used to gauge distances as great as 16,300,000 light-years (5000 kpc).

Relatively few galaxies are so close to us that individual objects, except supernovae, are identifiable in them. For distances greater than about 5000 kpc, distance estimates are based on the mean photometric and geometric properties of each type of galaxy. Thus it becomes necessary to assign a mean absolute magnitude to a particular class on the basis of the previously determined absolute magnitudes of the closer ones. In practice, this is difficult to do since there are considerable deviations from the mean absolute magnitude for galaxies of a particular type. Overlap in absolute magnitude between types is not unusual either. Generally the Sb spirals are the brightest, with some as bright as $M_V = -21$. Some of the faintest dwarf elliptical galaxies with $M_V = -12$ have luminosities only hundreds of times brighter than *single* supergiant stars in our own galaxy. The irregulars can vary in visual absolute magnitude from −13 to −16. Most ellipticals and spirals lie somewhere in the range from $M_V = -14$ to -19. Choice of the proper value from this range may be slightly more reliable if the galaxy in question is a member of a cluster of galaxies. In such a case the whole range of absolute magnitudes may then be seen within the cluster. Distances as great as 326,000,000 light-years (100,000 kpc) have been estimated by such means.

Finally, one may try to measure, from the spectrum of a distant galaxy, the amount of that galaxy's red shift. An estimated distance to the galaxy may then be established from the velocity-distance relationship as discussed earlier in this chapter. Even this technique has a limited application, since at very great distances it is likely that the straight-line velocity-distance relationship no longer holds.

Because of the many uncertain quantities which enter the problem, it is not surprising that distance estimates become less reliable for progressively greater distances. In a sense it is surprising that distances so vast are known with any degree of reliability at all.

18.6 Neutral Hydrogen in Galaxies

The presence of vast quantities of neutral hydrogen in our own Milky Way Galaxy implies that free hydrogen might be present in other galaxies as well. As a consequence, radio astronomers were eager to look for 21-cm signals coming from outside our own galactic system. The observations were extremely difficult because of the great distances involved and the small angular diameters of most galaxies, but success was at last achieved. It has been found that most spiral and irregular galaxies do indeed emit the 21-cm signals and must, therefore, have large quantities of neutral atomic hydrogen. Surprisingly (and significantly), neutral hydrogen has been detected in only one elliptical galaxy. When a reliable distance is known for a galaxy, the strength of the 21-cm signal can be related to the number of hydrogen atoms in the interstellar clouds of that galaxy. This can then be expressed as a fraction of the total mass of the galaxy. Results of this work done, among others, by Morton Roberts of the National Radio Astronomy Observatory are summarized in Table 18.2. The percentages are roughly the same for the barred spirals as for the normal spirals. Clearly, the galaxies with the highest proportion of young stars are the ones with the most hydrogen available—presumably for the formation of more young stars.

The distribution of neutral hydrogen within galaxies has also been studied for a few cases. Because the angular sizes of galaxies

Table 18.2. Neutral hydrogen in galaxies.

Galaxy Type	Ellip.	Sa	Sb	Sc	Irr
% of Total Mass	< 1	5	5	10	20

are small, it takes very high resolving power to locate the specific areas from which the 21-cm signal is being emitted. This situation will improve as the techniques of long-base-line interferometry are brought to this area, and the angular resolution will become far better than that of optical telescopes. Present data indicate, however, that in irregular galaxies the hydrogen is concentrated in the center, while in a large spiral such as M31 the hydrogen is in a broad ring around the central region. It will be important to see whether or not future results confirm the preliminary concept of the relationship of hydrogen distribution related to galaxy type.

The 21-cm observations have also provided a good means by which radial velocities of galaxies may be determined for comparison with optical radial velocities. The Doppler effect causes the received wavelength to be slightly greater than 21 cm in the case of receding sources (compare with page 379); as the optical data show, most galaxies are receding from ours. The results of the two methods are in excellent agreement over the velocity range from −400 to +5000 km/sec. Over this broad range of wavelengths the Doppler principle seems to give perfectly consistent results.

18.7 Classification of the Milky Way Galaxy

Having established classification criteria for other galaxies, we logically re-examine our own galaxy to see how it should be classified with the others. The appearance of the Milky Way implies a flattened system, and the evidence of spiral structure in the sun's neighborhood is described in Chapter 17. We may confidently say that the Milky Way Galaxy is a spiral. So far, neither optical nor radio astronomers have found any evidence of a central bar. We may conclude that it is very likely that the Milky Way is a normal spiral. We may then try to choose among Sa, Sb, and Sc. Figure 17.12 shows the spiral pattern in the sun's vicinity superimposed on a photograph of NGC 1232, a large galaxy seen in plan view. NGC 1232 is classified as an Sb galaxy, and so we may conclude from the comparison that the Milky Way is probably an Sb galaxy also. The established small size of the nucleus and the fairly open pattern of the arms are certainly in agreement with this classification.

18.8 Population Types in Other Galaxies

Stellar populations were first apparent in the studies of the Andromeda Galaxy, and the subtypes were developed from studies of the Milky Way. It is particularly interesting to trace the populations through galaxies of all classes.

Elliptical galaxies are essentially composed of Population II stars. Therefore, the bright stars are, for the most part, red giants. The presence of red giants accounts for the larger color indexes $(B-V)$ for these galaxies and helps to explain their late composite spectral types. The ellipticals are also notably free of obscuring dust clouds, and the distribution of stars within them seems to decrease quite uniformly from the center outward.

It is somewhat surprising to find positive evidence of gas in elliptical galaxies, since we have come to associate gas with Population I. Nevertheless, it has been found that about 15% of all elliptical galaxies show the emission line of ionized oxygen at $\lambda 3727$. The gas seems to be concentrated principally in a small spherical volume at the center of the galaxy. A second surprising feature in the central regions of a few elliptical galaxies is the presence of bright blue individual stars, which again seem to be inconsistent with the general concept that elliptical galaxies are composed of Population II type stars. A possible explanation is that the bright blue stars are "horizontal branch" stars. That is, they are stars which have evolved through the giant stage and are proceeding toward the left on the H-R diagram of the galaxy.

In the spiral galaxies the population types are generally as they were described for the Milky Way and Andromeda galaxies—the youngest bright stars in the arms and the older types of objects in the halo and nucleus. It is hard to trace significant differences in star types from Sa to Sb to Sc and decide that one has arms which are more Population I than another. In the gas and dust, however, there are interesting progressive differences. The amount of gas which is revealed by emission lines becomes greater as one examines galaxies from Sa, Sb, Sc, to Irr in that order. The gas is also seen to be more and more widely distributed in the galaxies in the same order. In the irregulars the gas is spread fairly evenly throughout. The dust is most conspicuous near the central regions of the more tightly wound spirals and is more difficult to detect in the open types and the irregulars. Certainly this is partly due to the fact that, when silhouetted against the large bright nucleus of an Sa galaxy, the dust clouds are most conspicuous. For the irregular galaxies and the more open spirals, the distribution of the dust must be studied from polarization measurements and from estimates of the reddening of objects which shine through the galaxy.

A final significant point regarding the types of objects found in various galaxies is the presence in all galaxies of a large number of very old stars. Even in the irregular galaxies which exhibit all of the attributes that we have come to associate with youth and Population I, there are large numbers of very old stars.

18.9 Evolution of Galaxies

It was mentioned earlier in this chapter that Hubble's "tuning fork" diagram had been suggested as an indicator of the progress of galactic evolution. Hubble felt that the more flattened types could develop from the spherical E0 galaxies as a result of rotation. The spiral arms would form later as material moved outward into the arms. Eventually the stars would be dispersed into a formless irregular galaxy. From our present knowledge of stellar evolution and from the actual observed situation, we can see that Hubble's theory cannot be correct. The elliptical galaxies are definitely not younger than the other types, as Hubble's scheme would require. If anything, the elliptical galaxies may actually be the oldest types. However, the presence of very old stars in all galaxies may really mean that all galaxies are the same age. The present formation of stars in the spirals and the irregulars could result from a situation in which the gas and dust were not all used up in the original period of star formation. An alternative might be that the elliptical galaxies were somehow able to rid themselves of their excess gas and dust. At any rate, some other process of change in galaxies must be sought.

In recent years much attention has been given to dynamical differences which might account for the variety of types of galaxies. If the galaxies formed from some vast primeval clouds of gas, there would have been presumably an initial period of contraction in the life of every galaxy. At some stage in this period the density would have reached the level at which star formation could have begun at localized areas. If there were some angular momentum in the contracting gas of the proto-galaxy, the size of the eventual rotating galaxy would be determined by the amount of this angular momentum. In a case in which there was a large initial angular momentum, contraction would end before the gas density became very high. In this low-density situation, star formation would occur at a slow rate. This idea implies, then, that if the angular momentum of the contracting mass were small, contraction would produce a high density from which an elliptical galaxy would form. In this case a very large fraction of the available gas would go into star formation in a short time. In the opposite case, where the initial angular momentum was large, contraction would have ended at a stage of low density in the rotating mass of gas. The low-density situation with star formation proceeding at a much slower pace would give the result that we see today in many galaxies—stars of all ages from young, blue giants to highly evolved, very old stars. The distribution of stars of various ages would indicate the areas which were most dense in the early stages. Galaxies formed from this second case would be the spirals, and their highly flattened shapes would be indicative of their high angular momenta.

In spite of the reasonableness of this theory in accounting for some aspects of the evolution of galaxies, there are other areas which remain completely puzzling to astronomers. Recent computer solutions of the dynamical problem of galaxies show that "barred" spiral patterns can form. It is difficult for astronomers to predict the continued existence of these bars without postulating forces other than gravity. The implication is that magnetic fields may play an important role in the formation and stability of the barred systems. Further, the persistence of the spiral arms in normal spirals through many rotations of the galaxy is a subject actively studied at the present time. Since the rotation period is short compared with the lifetime of a star, we would expect to see either a large number of turns or the spiral pattern completely smeared out. It may turn out that much more will be understood regarding galactic evolution when magnetic fields in our own and in other galaxies have been thoroughly investigated.

18.10 Quasars

With the advent of high-resolution radio telescopes both in the form of large-aperture paraboloids and as interferometers using more than one receiver, astronomers began to find powerful radio sources of extremely small angular size. Some were less than one second of arc across. Because of their small dimensions, these objects became known as "quasi-stellar radio sources." This term was later shortened to **quasar.** Quasars have proved to be extremely interesting objects, but they are at the same time very puzzling. Many of these radio sources have been identified with optical objects, and some general statements may be made regarding them.

Optical Appearance

Optical quasars are less than one-half second of arc in angular diameter. Therefore, on photographs they do not differ much in appearance from star images. The brightest, 3C 273, has an apparent visual magnitude of 12.8. (See Figure 18.11.) The faintest found so far are about twentieth magnitude. All of the quasars are extremely blue. 3C 273 is a double radio source and the optical image shows a faint "jet" extending from the starlike image. This

Figure 18.11. The quasar 3C 273 showing its "jet." (Hale Observatories photograph.)

jet is at the location of the second source. Several others show faint nebulosity within a few seconds of arc of the quasar. A few quasars have been observed photometrically over a long period of time and have been found to fluctuate in brightness over the years. In some cases these variations have been rather sudden. Long-exposure photographs have also shown that some quasars seem to be the bright nuclei of very faint but more extended galaxies.

Spectra

Because of their faintness, the optical spectra of quasars are difficult to study. Very few details have been found, and the spectra certainly do not look like those of stars or normal galaxies. A few emission lines were discovered, but these seemed difficult to identify. Then in 1963 it was pointed out by Maarten Schmidt that some of the unidentified emission lines in the spectrum of 3C 273 could be the Balmer lines of hydrogen which had been shifted toward the red by a very large amount. As a Doppler shift this displacement of the hydrogen lines would indicate radial velocity of 45,000 km/sec. This interpretation of the spectrum of 3C 273 was confirmed shortly afterward when infrared spectrograms revealed that the Hα line had been shifted far out of the visible region of the spectrum. Spectra of other quasars showed even larger red shifts. Spectra of about 30 quasars were photographed between 1962 and 1967, and all of these showed similar very large red shifts.

The quasars are all too far away in space for any distance estimates based on the methods used for single stars. The only method that has been feasible so far rests on the assumption that the red shifts of quasars fit Hubble's velocity-distance law for galaxies. This relationship is indicated in Figure 19.2, in which an approximately linear relation between radial velocity and distance is shown. The line defined here by the galaxies must be extended toward greater velocities and greater distances in order to accommodate the quasars with their very great red shifts. By means of this extension of Hubble's law it is possible to estimate the distances to the quasars. Two important facts become clear when this is done. First, on the basis of the measured red shifts the quasars must be extremely far away. The brightest one, 3C 273, must be more than one billion light-years away. The faintest ones must be the most distant objects known in the universe. Second, calculated from their apparent magnitudes and assumed distances, the absolute magnitudes of the quasars must be greater than about −25. This would make the quasars more than 100 times brighter than the brightest galaxies. The great distance also means that the quasars must radiate almost unbelievable amounts of energy at radio wavelengths. The total optical and radio energies of the quasars are so great that none of the conventional energy-release mechanisms seems adequate.

The optical and spectroscopic data taken together indicate a considerable variety in the intrinsic luminosities of the quasars. For a given value of red shift, the range in apparent magnitudes among the quasars can be more than two magnitudes.

The quasars have presented so many problems that many astronomers have sought a simpler picture of quasars as objects outside of the Milky Way Galaxy but still relatively close to us in space. This theory eliminates the requirement for great radiated energy in the quasars, but it makes difficult the interpretation of the red shifts in their spectra. The red shift could be due to a strong gravitational field, but this would require a mass too great to be likely. Astronomers are devoting great effort to the study of quasars in the hope that soon they will be able to say more about the true nature of these perplexing objects.

18.11 Seyfert Galaxies

In the 1940's the astronomer Carl Seyfert compiled a list of galaxies which had very small, bright nuclei. Most of these galaxies were spirals, and their spectra showed broad emission lines. For many years nothing much was done with the objects on Seyfert's list. With the discovery of quasars, however, interest in Seyfert galaxies was renewed, for it was seen that in a number of ways their nuclei were similar to quasars. Both are bright in the ultraviolet and in the infrared and both show emission lives in their spectra. In both groups most members emit at radio wavelengths. Some Seyfert nuclei and some quasars have been found to be variable both in their optical and in radio outputs.

Since the Seyfert galaxies seem to be basically spiral galaxies, it is possible to make reasonable estimates of their distances. With an estimate of the distance, the angular size of the nucleus can then be converted to a linear dimension. By such means the nuclei of the Seyfert galaxies are estimated to be 14 to 16 pc in diameter. This is comparable to the 10 pc estimated for the nucleus of our own galaxy. It is interesting to note also that very short-exposure photographs of the great Andromeda galaxy M31 (Figure 18.1) reveal a small faint nucleus which is probably 15 pc in diameter. It is the great luminosity of the Seyfert nuclei that sets them apart from the nuclei of normal galaxies.

The Seyfert galaxies still number only a few dozen, and astronomers are by no means certain of the physical processes which cause the emission from the central nucleus. It is tempting to imagine that the Seyfert galaxies represent some intermediate stage between spiral galaxies and quasars. As is often the case, however, such a step leads to another long series of unanswered questions.

18.12 Radio Galaxies

With large-aperture radio telescopes and interferometers, astronomers have been able to map the sky with high resolution. As a result it has been found that in many areas on the celestial sphere there are radio sources which are quite small in angular size. As the coordinates of these sources became known within narrow limits, astronomers began to make optical identifications of objects which could be reliably considered to be the source of the radiation. The Crab Nebula in Taurus is mentioned in Chapter 15 as an optical object which was later identified as a strong radio source. In this particular case the optical identification was not difficult, and the general outline of the radio source resembled that of the optical source. In a large number of cases, however, the identifications are not so simple. In many cases of positive optical identification the radio sources have been found to be galaxies, and these galaxies very often have some peculiar characteristics. For example, the galaxy NGC 5128, pictured in Figure 18.12, is a strong radio source and is definitely not normal compared with the types of galaxies discussed above. Silhouetted against what looks like an E0 galaxy is a very broad and irregular dark lane. Figure 18.13 shows another strong radio source (NGC 4038–39) which is quite clearly a very unusual galaxy also. This object may be either a pair of galaxies in collision or a single galaxy in which some cataclysmic event has taken place to drastically alter the original form. Other strong sources seem to be elliptical galaxies from which jetlike structures protrude. And, finally, some strong sources look like normal galaxies except for an intense central nucleus of very small dimensions.

While the strongest radio sources are associated with peculiar

Figure 18.12. The peculiar galaxy NGC 5128, a strong radio source in the constellation Centaurus. (Hale Observatories photograph.)

Figure 18.13. NGC 4038–39, another peculiar galaxy which is a strong radio emitter. (Hale Observatories photograph.)

galaxies, most normal galaxies are found to radiate (essentially a continuous spectrum) in the radio region also. This is indicated by the fact that for the nearby systems radio energy is detectable at wavelengths from shortwaves to microwaves. Among the normal galaxies, the irregulars and the Sc types are the strongest sources. The radio source surrounding the Andromeda Galaxy is very much larger than the galaxy itself.

Most of the strong radio sources are also larger in angular size than their optical images. Quite puzzling is the observed duplicity of many of the radio sources associated with the peculiar galaxies. A typical situation shown in Figure 18.14 is the irregular, but decidedly double, source associated with NGC 5128 (Figure 18.12). Enough similar double cases are known so that there is no danger that this is a mere coincidence.

When the distance to a galaxy is known, the absolute intensity of the radio signals may be calculated. The total radio energy is found to be extremely large, and astronomers have been unable to explain the physical mechanism by which the radio energy is emitted. The signals have some of the characteristics of synchrotron emission (see Chapter 15), but if this is true the means of accelerating electrons in a magnetic field and of maintaining the magnetic field must be discussed. Other suggested processes for the great release of energy have been gravitational collapse, collisions of two galaxies, supernovae, and the annihilation of matter. It is safe to say at this time that there are difficulties with all of the attempted

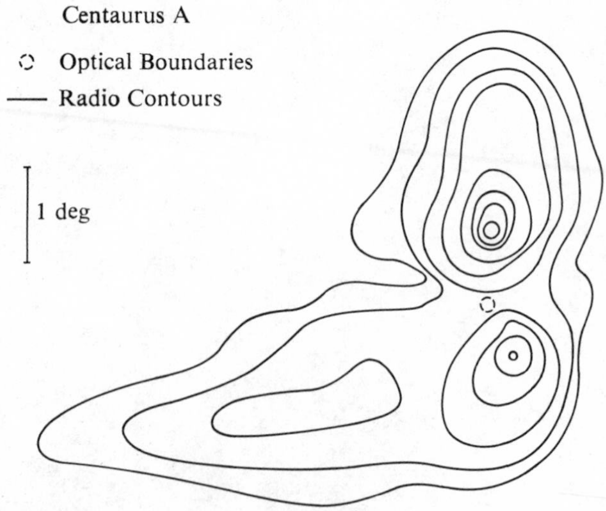

Figure 18.14. The radio emission from NGC 5128 reveals a double source extending some 10° across the sky. The dark band in the optical galaxy is at right angles to the line between the double sources. (After J.G. Bolton and B.C. Clark.)

Figure 18.15. The spiral galaxy M51 with an irregular galaxy as an appendage at the end of one of the spiral arms. (Hale Observatories photograph.)

explanations. It is likely that further advances in the physical theory of radio emissions will precede or accompany the eventual solution of this problem.

18.13 Peculiar Galaxies

In the preceding section we have discussed the fact that the strong radio sources outside our own Milky Way Galaxy are often associated with galaxies in which some unusual characteristic sets them apart from those galaxies which appear more normal. In others the spiral arms are distorted from the conventional pattern. Bright "knots" are occassionally found in the arms of galaxies. Since there are many examples of this kind, it becomes meaningful to speak of "peculiar" galaxies. These are not always to be considered as a separate class. Rather, they are often objects which could fall into one of the Hubble classes but which have some unusual feature which sets them apart. Photographs of a large number of such galaxies were assembled in the Atlas of Peculiar Galaxies by

H.C. Arp of the Hale Observatories. The approximately 350 galaxies in this collection show great variety, but there are certain characteristics which occur frequently enough to warrant mention:

1. Many peculiar galaxies are strong radio emitters over a broad range of wavelengths.
2. Many are spiral galaxies with a smaller galaxy on the end of one arm. M51 is an example of this (Figure 18.15). In some cases a condensation or smaller galaxy may be seen on the end of both spiral arms.
3. Peculiar galaxies often occur in chains or groups and often show the effects of tidal interaction with those galaxies close to them in space.
4. Galaxies which appear to be connected by a "bridge" of stars sometimes have very different red shifts in their spectra.

In many areas of science it is the peculiar objects which turn out to be the most interesting and which sometimes lead to more complete understanding of the phenomena being studied. It may turn out that this also will be the case with the peculiar galaxies. At present the peculiar galaxies leave us with more questions than answers.

QUESTIONS

1. Briefly describe the characteristics of each of the principal types of galaxies.
2. How can the angle of tilt to the line of sight be estimated for a spiral galaxy? What assumptions are required?
3. What is the "zone of avoidance," and how do astronomers account for it?
4. Describe two lines of evidence for the fact that galaxies occur in clusters.
5. Comment on the observed color indexes of galaxies and the significance of the color index of a galaxy.
6. How has the rotation of distant galaxies been confirmed?
7. Why does the number of galaxies per square degree on the celestial sphere increase with increasing galactic latitude?
8. What factors will determine the pattern of absorption lines which are seen in the spectra of galaxies?

9. How has the distance to the Andromeda Galaxy been estimated?
10. How can the distance to a cluster of faint galaxies be estimated?
11. How do the studies of population types in other galaxies compare with the populations in our galaxy?
12. Describe the distribution of gas and dust in spiral and elliptical galaxies.
13. What are some of the problems encountered in trying to develop a theory of galactic evolution which does not conflict with the theory of stellar evolution?
14. Describe the types of galaxies which are known to be strong radio sources. What sort of emissions are received from these galaxies?
15. What is a quasar and how are the distances to quasars estimated?

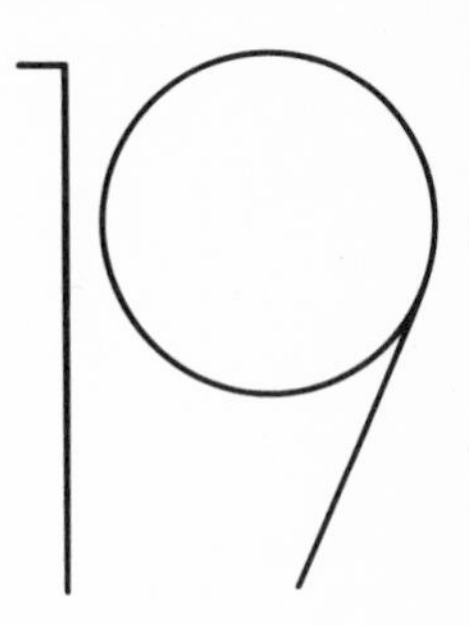

The Expanding Universe

In the preceding chapters the interdependence of observation and theory in studying planets, stars, and galaxies has been stressed. We know the kinds of observable material in the universe and how this material is organized and arranged. The universe, however, seems to be expanding, and in the context of our everyday lives an understanding of this expansion poses a difficult challenge. We are accustomed to the idea that a "common sense" approach can be used in most situations. In the study of the expanding universe, however, the old approach is no longer sufficient, and scientists must turn to the most advanced physical and mathematical concepts. The present theories of the expanding universe are somewhat general and many details are lacking. Nevertheless, these theories often point to crucial observations which in turn lead to the extension and refinement of the theories.

19.1 Observations

The basic observational data of the theory of the expanding universe are quite simple and are mentioned briefly in Chapter 18. The spectra of all distant galaxies are shifted toward the red end of the spectrum, and the amount of this red shift is proportional to the distance to each galaxy. The greater the distance to a galaxy,

◄**Figure 19.1.** Part of a cluster of galaxies in the constellation Corona Borealis. The distance to this cluster is believed to be about 120 million light-years. (Hale Observatories photograph.)

the more its spectrum is shifted toward longer wavelengths. This shift has been measured for lines at a number of wavelengths in some spectra, and the results are in good agreement in all parts of the spectrum. Attempts have been made to find radio emission lines which might prove that the amount of the shift was the same throughout the entire electromagnetic spectrum, and the 21-cm line has provided the best check so far. Radial velocities determined at optical wavelengths and at 21 cm are in close agreement for the galaxies in which this comparison has been possible.

Again, the only consistent way in which physicists and astronomers have been able to account for the red shifts of galaxies is to recognize them as the result of radial velocities of the sources. Thus, if all distant galaxies have red-shifted spectra, all distant galaxies must be moving away from us. Furthermore, the observations must mean that the acceleration of the galaxies is nearly constant in all directions. This is characteristic of an expansion under a constant repulsive force.

The initial work on radial velocities of galaxies was done by V.M. Slipher, using the 24-inch refractor of Lowell Observatory. By 1925 Slipher measured most of the 45 radial velocities available at that time. From these observations the existence of the velocity-distance relationship became clearly established. In order to extend the observations to fainter galaxies, a larger telescope was needed, and so the bulk of the subsequent work was accomplished by Milton Humason using the 100-inch telescope at Mt. Wilson. By 1935 Humason had added the radial velocities of about 150 additional galaxies, the faintest of which were of apparent magnitude 21 and could not easily be seen through the telescope. (In order that these faint galaxies might be observed, the telescope had to be moved very slightly in right ascension and declination from the position of a star close to the faint galaxy on the celestial sphere.) A total of nearly 200 radial velocities formed the basis of the interpretive work of Hubble and many others which was to follow. The galaxies studied ranged from those nearby ones in which individual stars could be detected to those which were members of very distant clusters of galaxies. Figure 19.2 shows Hubble's original diagram in which red shift has been converted to radial velocity and has been plotted against apparent magnitude. Only points for the most distant galaxies, which were in clusters of galaxies (Figure 19.1), are shown. Hubble had plotted similar diagrams for galaxies resolvable into stars and for other galaxies at intermediate distances. From all of these data, Hubble was able to derive a

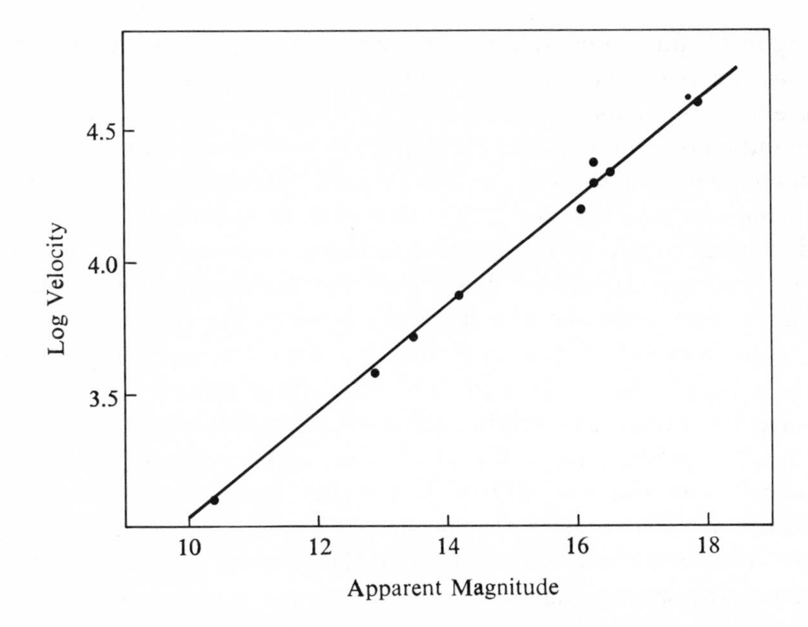

Figure 19.2. Hubble's original relationship between the logarithm of the radial velocity and the apparent magnitude. Apparent magnitude here is presumably an indication of distance. Each point represents a galaxy in some distant cluster of galaxies.

coefficient, since known as **Hubble's constant,** which expresses the relation between radial velocity and distance. Hubble's original value was 500 km/sec per million parsecs. Following the revision in the zero point of the period-luminosity law for Cepheids, it was recognized that the galaxies were twice as far away as previously imagined. A revision of Hubble's constant was clearly called for.

With the introduction of the 200-inch telescope the studies of spectra of galaxies have been extended to a total of about 800 galaxies. From this larger amount of data and the revised distance scale, a new value for the Hubble constant was obtained. Quite a range of values has actually been suggested, but there is today a fairly general acceptance of 50 km/sec per million parsecs as a usable value for the Hubble constant. In the course of the extension of these spectral studies, astronomers have also noted evidence that for very faint galaxies the line on the velocity-distance diagram is no longer straight but begins to curve upward. This curvature has important bearing on the choice of a possible model for the universe.

19.2 The Universe is Expanding

The data indicate, then, that all galaxies seem to be rushing away from us, with the most distant ones moving at tremendous speeds. The suggested interpretation has been that the universe is expanding in all directions. It seems at first glance that if all matter in the universe is moving away from us, we must be at the center of the universe. This is not necessarily so, however, for in an expanding universe, the evidence for expansion will be the same for any observer anywhere in the entire universe. This may be illustrated by considering a ball of dough with raisins in it. Before baking, the dough has a certain size and the distances between all of the raisins have certain values. When the dough is baked in the oven the entire mass expands. The distances between the raisins become greater, but, as indicated in Figure 19.3, the relative positions of all the raisins remain the same. The interesting point is that if we select one raisin as an origin and note the changes in position of all the other raisins, we find that all the raisins increase their distances from the one arbitrarily selected as origin. Two points should be noted. It does not matter which raisin is selected as origin. The recession of all others from the selected one remains the same. And second, the greater the distance between raisins in the dough, the more the distance has been increased in the bread. Therefore, during the expansion of the loaf, the raisin at the greatest distance from some origin will move away from that origin with the greatest velocity. On the average, the observed phenomenon will be the same if viewed from any origin within the loaf and in any direction relative to an arbitrary coordinate system.

In the actual universe, then, a general uniform expansion results in the motion of all parts of the universe away from any chosen origin. We must, of course, view the universe from our vantage point here in our own Milky Way Galaxy, where we see distant

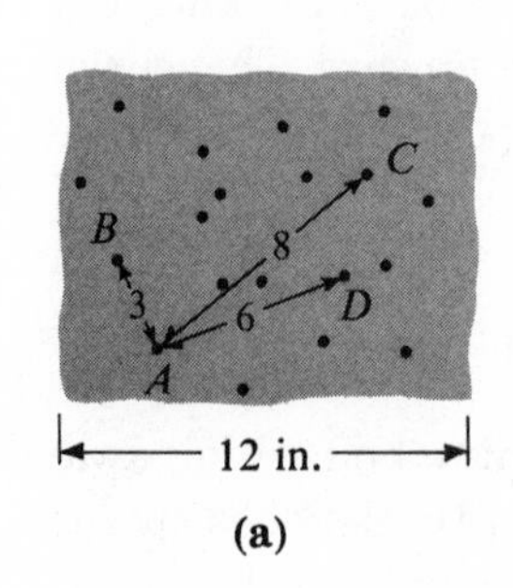

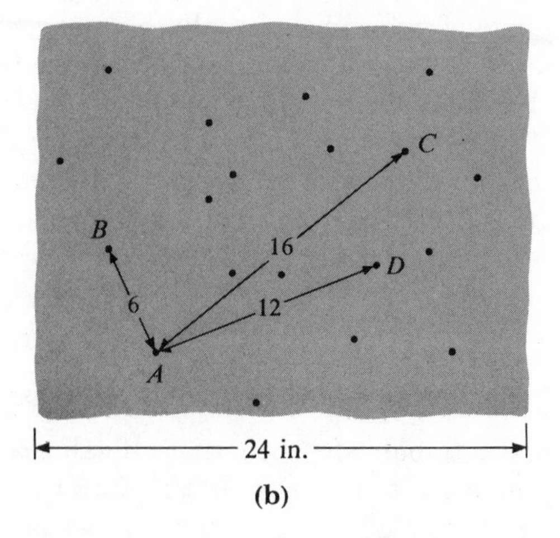

Figure 19.3. Expanding bread dough with raisins. As the dough expands during baking, the relative positions of the raisins remain the same, but the distances between them increase.

galaxies in all directions rushing away from us. From any other galaxy the average effect would be exactly the same. Otherwise, either the Hubble parameter would be a function of direction—which it is not—or we are really at the center of the universe. We are by no means obliged to feel that we are in some preferred place at the center of the expansion.

The implications of the expanding universe have been difficult for many people to accept. As a result, there have been efforts by many to find some substitute explanation for the red shifts of the galaxies. For instance, the general theory of relativity predicted that photons should be affected by locally strong gravitational fields. For objects of small radius and high mass (i.e., high density), one result is a **gravitational red shift** of light. Since the gravitational field near the photosphere of a white dwarf is very strong, a red shift which is interpreted as being gravitational has been observed in the spectra of some white dwarfs. Even though the reality of the gravitational red shift has been fairly well established by this and other experiments, there does not seem to be any way in which this mechanism can be invoked to account for the red shifts of galaxies.

Another explanation is that the light from distant galaxies has been traveling toward us for a long time. Some of the galaxies must easily be five billion light-years away. It has been suggested that photons may somehow lose energy in the course of such long journeys through space. If this were true, the galaxies would not really be moving away from us at all. Nothing in our observational or theoretical experience, however, suggests that this may be the case unless we are prepared to sacrifice some of our most basic physical principles. We still have no reason to interpret the red shifts as anything other than a result of the radial velocity of the source. On this rests the concept of the universe in a state of expansion.

19.3 Cosmology and Models of the Universe

Cosmology is the study of the universe: its past history, present structure, and probable future evolution. Theories or models which attempt to explain the major features of the universe are referred to as cosmological theories or, briefly, **cosmologies.** Adequate formulations of modern cosmological theories required the develop-

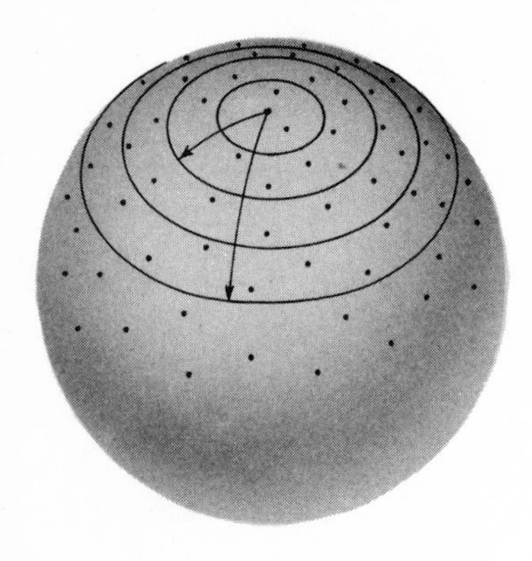

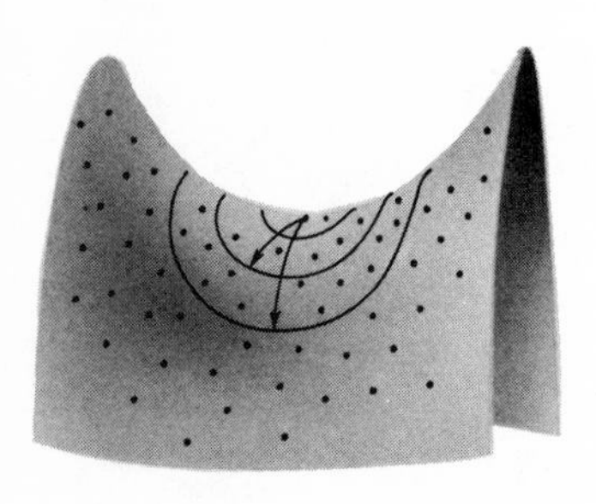

Figure 19.4. The concept of curved space illustrated by a spherical surface and a saddle-shaped surface.

ment of some very specialized mathematical concepts and generalized theories of the physics of space, time, and gravity.

The first of these developments was the recognition by mathematicians, such as Riemann and Lobachevsky in the latter part of the last century, that the geometry of Euclid was only a very restricted case of a much broader group of measurement systems. There are many possible geometries in which space and coordinates are really curved, but look straight over small distances. This is a very difficult notion to accept physically, but it has some interesting consequences. The curvature of space may be positive, in which case the universe has a finite extent but has no boundaries. Or the curvature may be negative, with the universe infinite and unbounded. Space with a positive curvature has been likened to the surface of a vast sphere. Lines which are parallel to each other in the realm of common experience can cross each other and return to their origin in the realm of positively curved space. Negatively curved space has been compared to a saddle-shaped surface on which parallel lines grow further apart when extended to great lengths. The concept of curved space is illustrated in Figure 19.4, which shows the spherical and hyperbolic surfaces. It must be remembered that these surfaces represent only two-dimensional analogues of three-dimensional curved geometries.

The second important advance was the development by Albert Einstein of the special and general theories of relativity, which brought about changes in the basic ideas concerning mechanics. Under most circumstances (slow velocities, small distances), moving particles would be governed by the laws of force, motion, and acceleration according to classical or Newtonian mechanics. In extreme cases of high velocity these laws no longer hold true, and the more complex relativistic ideas have to be invoked. For example, Einstein postulated the velocity of light as the limiting velocity of all motion. He suggested that accelerations experienced in motions are equivalent to being in a state of rest in a gravitational field. As a result, he predicted that the direction of light could be altered when the light passed near a body with a large gravitational field. In discussions of the universe it is necessary to consider large ensembles of self-gravitating bodies and the relation of electromagnetic signals emitted by them. It was natural for relativistic mechanics and cosmological theories to have parallel developments.

On the basis of the possible curvature of space and relativity, a number of cosmological theories were devised in the years before the red shifts of galaxies had been discovered. A cosmology credited to Einstein held that the universe was finite and positively curved. This theory mathematically eliminated the problems inherent to a finite system in Euclidean space, which would have to be surrounded by an infinite empty space. Einstein showed certain relationships between the total mass of the universe and the radius of curvature of the universe. If the mean density of matter in the universe could be known, from galaxy counts for example, the radius of the universe could be computed. In his equations Einstein had to make use of another factor which he called the **cosmological constant.** The effect of this constant was such that a repulsive force was introduced at very large distances. In order to make his model static (as the evidence then available seemed to indicate the universe to be) and avoid a gravitational collapse of all matter to a point, Einstein chose a value for the cosmological constant of such size that equilibrium would exist everywhere. Other mathematicians subsequently found solutions for other relativistic model universes which indicated that the universe could really be expanding, contracting, oscillating, or static. Much of this work was accomplished before the evidence for expansion was shown by Slipher, Humason, and Hubble. Since the time of the discovery of the red shifts of galaxies, all subsequent work has been an extension of the original mathematical possibility of an expanding universe.

19.4 Time-Varying Universe

In an expanding universe the average density of material must become progressively less, while the general large-scale characteristics of the universe must change with time. Two observers widely separated in the universe would see no important differences if their observations were simultaneous. But two observers widely separated in time would note significant differences, especially in density. Model universes based on this concept are described as **time-varying universes.** One of the most obvious implications of this idea is that at some time in the past all of the matter in the universe must have been concentrated at one point. The best known cosmology of this sort is the so-called "big bang" theory whose most important exponent was the physicist George Gamow.

Gamow suggested that expansion began from an original super atom which exploded at some very remote moment in time. In a very short time after this explosion, all of the chemical elements in their present abundances were formed from the original matter. Thereafter, condensations formed in this expanding matter, and from these condensations the galaxies and stars eventually formed. We believe now that most of the heavier elements are continuously being formed in stellar interiors and we are inclined to play down Gamow's assertions on the creation of the chemical elements.

This sort of an expansion, as we have said, implies a beginning, and various investigators have attempted to establish a time for this beginning. Originally an age of two billion years was derived from the Hubble constant on the basis of uniform expansion. On the basis of stellar evolution an age of about 10 billion years with nonuniform expansion has been suggested. This means that 10 billion years ago all matter was concentrated in the original location and that nothing present in the universe today can be older than 10 billion years. If we should ever find convincing evidence that some star or galaxy was substantially older than 10 billion years, there would have to be some significant changes in this theory. Recently, high-frequency radio radiation showing a uniform distribution over the sky was discovered and interpreted as a fossil remnant of the original fireball. This finding represents an important piece of evidence for the "big bang."

Another version of this theory, the oscillating model, implies that the gravitational attraction of the universe for all of its parts should cause the expansion to slow down and eventually stop. After that, the universe should begin to contract or fall back toward its original center. When all of the material again reaches the center there will presumably be another immense explosion which will mark the beginning of another expansion. In this manner the universe is imagined to pulsate endlessly over vast stretches of time. With the beginning of each new expansion, all traces of the previous cycle would be completely wiped out. An alternative theory suggests that the expansion does not slow down but continues at the presently observed rate indefinitely. In the latter case the density of matter in the universe must be continuously decreasing. There must come a time when each of the distant galaxies must disappear as the distance becomes too great for it to be detectable.

It is a well-known fact of mechanics that if the entire universe is even slightly rotating, total collapse will not occur. Thus it becomes possible for the pulsations of the universe to become less extreme. Contraction would then cease before all material had collapsed to some center. Explosions marking the beginning of each cycle would not occur, and there need be no age limit on bodies presently existing in the universe.

19.5 Steady-State Universe

The advocates of a steady-state universe add a further restriction to the notion that the universe should appear the same in its general broad aspects to observers everywhere at all times. In this theory the observations need not be simultaneous. An observer anywhere in the universe at any time will see essentially the same overall picture. The details of individual galaxies will, of course, vary with time. But the density of matter and other average properties within the universe as a whole remain constant. In order for the overall density to remain the same in an expanding universe, the authors of the steady-state theory postulate *a priori* the continuous creation of matter. In the emptiness between galaxies, hydrogen atoms simply materialize from nothing at a rate sufficient to offset the effects of the expansion. New galaxies and stars form from this newly created matter. Thus, in spite of progressive changes in certain respects, the overall universe remains the same.

19.6 Observational Tests

There are a number of interesting and important variations on the two general concepts of the expanding universe which are briefly described above. The proper choice of the best model from all of the possible ones is very difficult on the basis of the available observational material. Nevertheless, there is the hope that a choice can someday be made since the various models do, in fact, imply slight differences in the observable characteristics. Those data which can be found from observation are rather limited, of course. We observe apparent magnitudes and try to make accurate estimates of the distance. We measure the amount of the red shift from the spectra and compute radial velocities. We apply corrections to the observed magnitudes and distances because some of the light is shifted out of the normal wavelength region of our photometric systems, causing the galaxies to appear dimmer. We make counts of the numbers of galaxies of various magnitudes, and we look for variations in distribution and space density in these counts. We also examine the clusters of galaxies and relative numbers of galaxies of each type within the clusters. From radio observation we have a bit more information. We measure the radio energy at several wavelengths from galaxies known to be strong radio sources, and we try to map the distribution of radio flux for radio sources. Likewise, x-ray and cosmic-ray fluxes from space are important indicators of conditions. All of these observations must be incorporated in a self-consistent manner into any proposed model universe. When each cosmology has been sufficiently developed, one finds certain differences in the details required by each theoretical universe. It then becomes the task of the observer to try to look for details which will make possible the selection of one cosmology over all the others.

An important implication of the steady-state universe is that in any volume of space we should see galaxies of a variety of ages. Since galaxies become farther apart as they become older, there should be relatively few old galaxies in any volume of space. For the same reason there should be large numbers of young galaxies in any large volume. Thus, we might be able to confirm the steady-state theory by counting the numbers of galaxies of progressively greater ages. If there were only a single epoch of star formation, the oldest galaxies should be very red, since most of their stars would be highly evolved. However, a serious difficulty is that we still do not really have any means by which we can determine the relative ages of galaxies. All galaxies contain stars which are old, but the ellipticals contain the greatest percentage of old stars. The presence of even one galaxy of obviously great age would tend to support the steady-state theory on these grounds. The creation of matter as advocated for the steady-state also creates high-energy particles and radiations with uniform distributions over the sky. Predicted x-ray and cosmic-ray fluxes from space are much higher than those actually observed from spacecraft.

We know that the light from the most distant galaxies was radiated from its source a long time ago. Thus when we look toward the most distant galaxies, we are truly looking backward in time. We are seeing the galaxies as they were millions of years ago, and we have no way of knowing what the distant galaxies look like right now. If there were once a time when some aspects of the universe were different from what they are today, we should expect to detect such differences when we study the most remote reaches of the universe. In particular we might expect to see some evidence of the primordial "fireball," if it ever existed. Such differences ought to be helpful in the choice of one cosmological system over another.

Another such difference between the old and the current conditions of the universe must lie in the all important radial velocities themselves. In Figure 19.5 the logarithm of the red shift has been plotted against apparent magnitude corrected for the dimming by radial velocity and by intergalactic absorption. The solid lines drawn on the diagram indicate the expected values of a quantity q, known as the **deceleration parameter.** It is possible to relate q to the counted numbers of galaxies of successively fainter magnitudes. It is also possible to compute the values that q may have in each of several model universes. For example, in Euclidean space $q = 1/2$; in negatively curved space q is between $1/2$ and 0; in positively curved space q is greater than $1/2$; and in the steady-state universe $q = -1$. The steps required to draw these lines are rather complex and require a number of assumptions regarding the family of model universes to be considered for testing. Also plotted are points which represent the galaxies observed so far. These are the same points plotted in Figure 19.2. The scatter in these points indicates that no unique choice may be made from the data on hand. Over this range in magnitudes a considerable dispersion of the possible models is allowed by the scatter in the observed points. What is clearly needed, in the absence of more accurate data, is an observed point farther out along these lines.

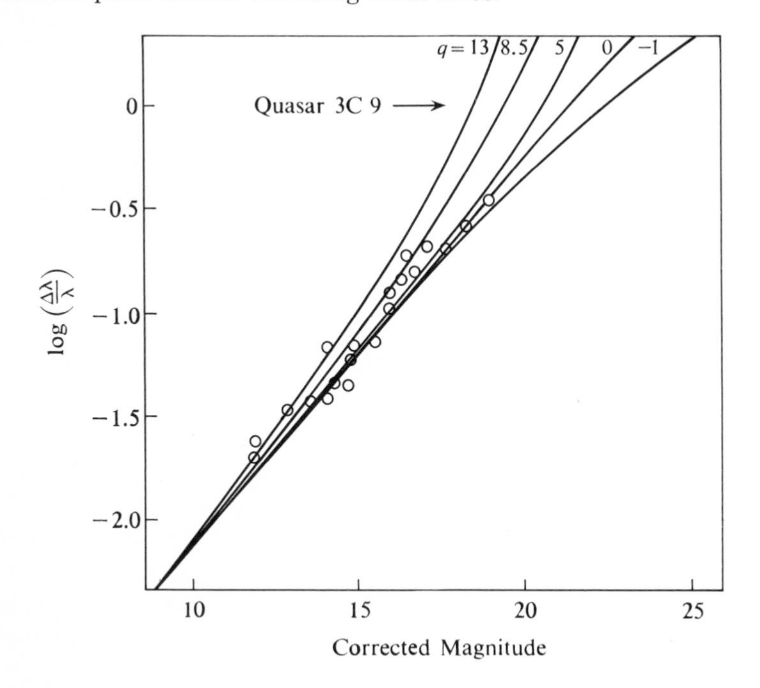

Figure 19.5. The red shifts of galaxies plotted against their apparent magnitudes. The solid lines represent various values of the deceleration parameter q. (After a diagram by A. Sandage.)

It is also possible to discuss in somewhat simpler terms the effect of a change in the rate of expansion on the observed radial velocities. If the universe is pulsating, the rate of expansion must be slowing down. Eventually the expansion will end and contraction will begin. Clearly, the rate of expansion must have been greater in the past than it is today; therefore, the red shifts of the most distant galaxies ought to indicate the rate of expansion at some much earlier time. Because of this the observed points on Figure 19.5 should curve upward for progressively fainter galaxies. Again, the observations on this point are inconclusive.

All one can hope to do at present is try to recognize the possible universes and continue to search for the crucial observational data. Cosmology is a challenging mathematical and physical study by itself, and its conclusions must always be in complete agreement with the entire body of astronomical knowledge. In the past, cosmologists have been forced to make gross simplifications as to the nature of the universe in order to deal with it at all. Advances in mathematics and in data analysis should permit progressively greater complexity in the models. Eventually the models may represent the true universe more closely.

QUESTIONS

1. Outline the basic observational evidence for the theory of the expanding universe.
2. Why is it that the positive radial velocities of faint galaxies in all directions do not necessarily imply that our galaxy is at the center of the universe?
3. What seems to be wrong with explanations other than radial velocity to account for the red shifts in the spectra of galaxies?
4. What are the basic concepts or "cosmological principles" on which the time-varying and steady-state theories of the universe are based?
5. What sort of possible observations might permit us to choose one model of the universe over the other?

Epilogue

Now that these things have been added, Syrus, and to my mind about all things which ought to be considered in such a treatise have been worked out as much as the time to the present affords for discovery and more accurate revision, and annals suggest as useful for the theory and not just as demonstration, it is therefore fitting and proper that this treatise end here.

Ptolemy, Book XIII

Appendix

A.1 Trigonometric Relations

Trigonometry is the study of the relationships between the sides and the angles in triangles. These basic relationships may be defined with respect to the right triangle in Figure A.1 as follows.

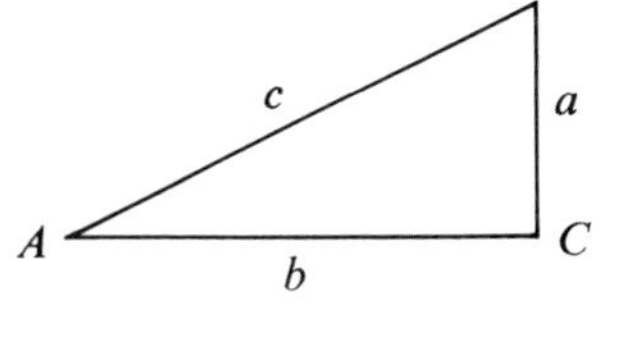

Figure A.1

sine	$\sin A = \frac{a}{c}$	cotangent	$\cot A = \frac{b}{a}$
cosine	$\cos A = \frac{b}{c}$	secant	$\sec A = \frac{1}{\cos A} = \frac{c}{b}$
tangent	$\tan A = \frac{a}{b}$	cosecant	$\operatorname{cosec} A = \frac{1}{\sin A} = \frac{c}{a}$

The values of these trigonometric functions are generally tabulated for angles from zero to 90°. An abbreviated table of this sort is included as Table A.1. It should be noted that the cosine and cotangent tables are to be read from the bottom upward, using the angles indicated to the right of the tabulated values.

As an example of the use of these formulae and tables, consider a greatly exaggerated case of the parallax of a star as in Figure A.2. In this figure the sun, earth, and star are at the three vertices of a triangle. From observations of the position of the star against the background stars, the parallax angle p is obtained. The earth-sun distance is known to be 1 A.U. Using the tangent formula then

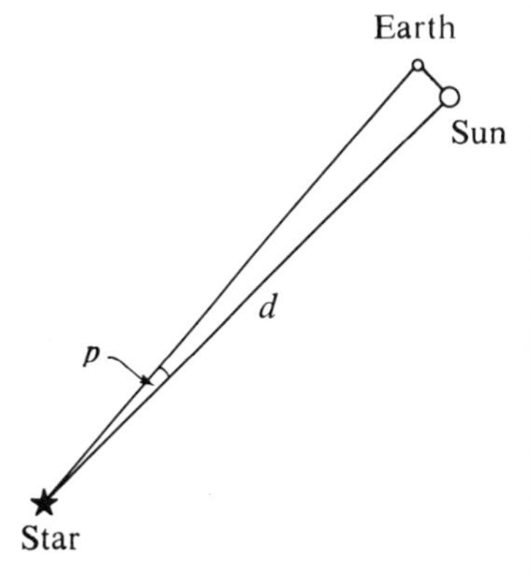

Figure A.2

$$\tan p = \frac{1 \text{ A.U.}}{d} \quad \text{or} \quad d = \frac{1 \text{ A.U.}}{\tan p}$$

Table A.1. Trigonometric functions.

Degrees	Radians	Sin	Tan	Cot	Cos		
0	0	0	0	—	1.0000	1.5708	90
1	.0175	.0175	.0175	57.290	.9998	1.5533	89
2	.0349	.0349	.0349	28.636	.9994	1.5359	88
3	.0524	.0523	.0524	19.081	.9986	1.5184	87
4	.0698	.0698	.0699	14.301	.9976	1.5010	86
5	.0873	.0872	.0875	11.430	.9962	1.4835	85
6	.1047	.1045	.1051	9.5144	.9945	1.4661	84
7	.1222	.1219	.1228	8.1443	.9925	1.4486	83
8	.1396	.1392	.1405	7.1154	.9903	1.4312	82
9	.1571	.1564	.1584	6.3138	.9877	1.4137	81
10	.1745	.1736	.1763	5.6713	.9848	1.3963	80
11	.1920	.1908	.1944	5.1446	.9816	1.3788	79
12	.2094	.2079	.2126	4.7046	.9781	1.3614	78
13	.2269	.2250	.2309	4.3315	.9744	1.3439	77
14	.2443	.2419	.2493	4.0108	.9703	1.3265	76
15	.2618	.2588	.2679	3.7321	.9659	1.3090	75
16	.2793	.2756	.2867	3.4874	.9613	1.2915	74
17	.2967	.2924	.3057	3.2709	.9563	1.2741	73
18	.3142	.3090	.3249	3.0777	.9511	1.2566	72
19	.3316	.3256	.3443	2.9042	.9455	1.2392	71
20	.3491	.3420	.3640	2.7475	.9397	1.2217	70
21	.3665	.3584	.3839	2.6051	.9336	1.2043	69
22	.3840	.3746	.4040	2.4751	.9272	1.1868	68
23	.4014	.3907	.4245	2.3559	.9205	1.1694	67
24	.4189	.4067	.4452	2.2460	.9135	1.1519	66
25	.4363	.4226	.4663	2.1445	.9063	1.1345	65
26	.4538	.4384	.4877	2.0503	.8988	1.1170	64
27	.4712	.4540	.5095	1.9626	.8910	1.0996	63
28	.4887	.4695	.5317	1.8807	.8829	1.0821	62
29	.5061	.4848	.5543	1.8040	.8746	1.0647	61
30	.5236	.5000	.5774	1.7321	.8660	1.0472	60
31	.5411	.5150	.6009	1.6643	.8572	1.0297	59
32	.5585	.5299	.6249	1.6003	.8480	1.0123	58
33	.5760	.5446	.6494	1.5399	.8387	.9948	57
34	.5934	.5592	.6745	1.4826	.8290	.9774	56
35	.6109	.5736	.7002	1.4281	.8192	.9599	55
36	.6283	.5878	.7265	1.3764	.8090	.9425	54
37	.6458	.6018	.7536	1.3270	.7986	.9250	53
38	.6632	.6157	.7813	1.2799	.7880	.9076	52
39	.6807	.6293	.8098	1.2349	.7771	.8901	51
40	.6981	.6428	.8391	1.1918	.7660	.8728	50
41	.7156	.6561	.8693	1.1504	.7547	.8552	49
42	.7330	.6691	.9004	1.1106	.7431	.8378	48
43	.7505	.6820	.9325	1.0724	.7314	.8203	47
44	.7679	.6947	.9657	1.0355	.7193	.8029	46
45	.7854	.7071	1.0000	1.0000	.7071	.7854	45
		Cos	**Cot**	**Tan**	**Sin**	**Radians**	**Degrees**

where d is the distance to the star in astronomical units. If p should be 1°, then from Table A.1 tan 1° = 0.0175. Then

$$d = \frac{1}{0.0175} \quad \text{or } 57.14 \text{ A.U.}$$

Of course, no star is this close to the sun and the parallax of a star is never as large as 1°.

At times it is also useful to show graphically the variation of the trigonometric functions for increasing angles. This has been done in Figure A.3, in which both the sine and cosine curves have been plotted.

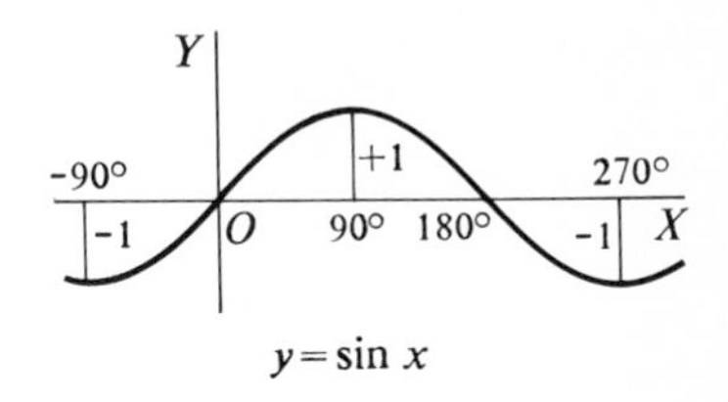

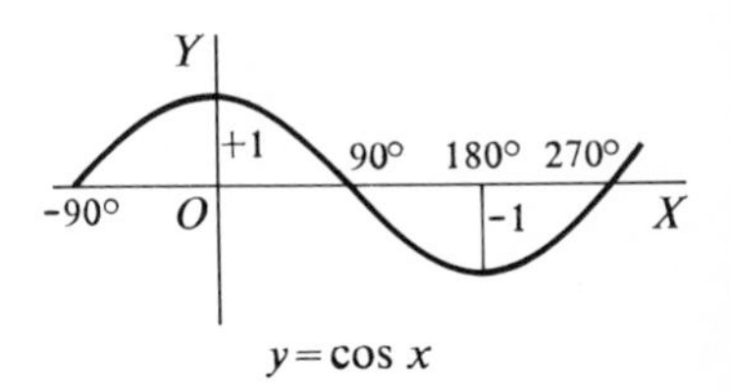

Figure A.3

A.2 Circular Measure

In many situations it is useful to measure angles in **radians** rather than in degrees. The radian is defined as the angle between two points on the circumference of a circle such that the distance between the two points is equal to the radius of the circle (see Figure A.4). Since the circumference is calculated from

$$C = 2\pi r$$

it is easily seen that 360° must be equivalent to 2π radians, or one radian is equal to $360°/2\pi$ or 57° 52′. If a length b along a circumference is known, then the central angle subtended by this length is found simply by dividing b by the radius of the circle.

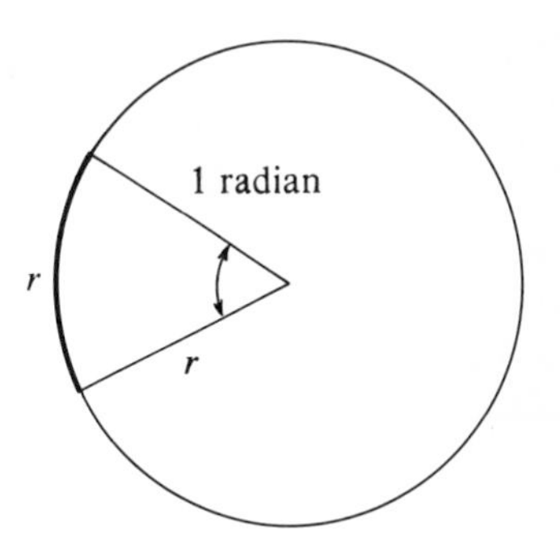

Figure A.4

The use of radians is particularly advantageous when very small angles are being measured. This is often the case in astronomy. Consider again, for example, the measurement of stellar parallax. The parallax angle at the star is so small that there is really no difference between the earth-star and the sun-star distances. We may imagine a circle with a radius equal to the sun-star distance. The earth-sun distance forms a small portion of the circumference of this circle (see Figure A.5). We may then write

$$\frac{1 \text{ A.U.}}{d} = p_{\text{rad}}$$

where d is the distance from the sun to the star measured in astronomical units and p_{rad} is the parallax measured in radians. The relationship is quite simple; hence its great usefulness.

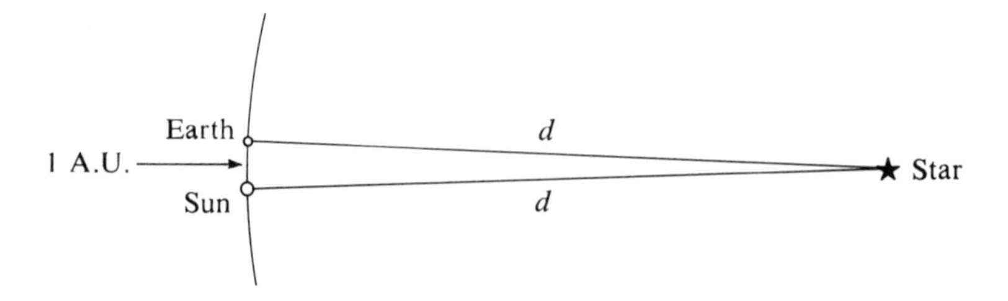

Figure A.5

Since it is usually measured in seconds of arc, the parallax must be converted to radians before the distance in astronomical units, kilometers, or parsecs may be known. One radian contains 206,265 arcsec, and 1 arcsec must be 1/206,265 radian. Thus if a star has a parallax of 1 arcsec, its distance must be 206,265 A.U. This distance is also equal to 1 pc.

It is also interesting to note that for angles of less than 1° the numerical value of the sine and the tangent are almost identical and are practically equal to the numerical value of the angle in radians. For example,

$$\sin 1^\circ = 0.0175$$
$$\tan 1^\circ = 0.0175$$
$$1^\circ = 0.017 \text{ radian}$$

A.3 Logarithms

The logarithm of a number may be defined as the power to which 10 must be raised to obtain that number. Thus the logarithm of 100 is 2 since 10^2 equals 100, and the logarithm of 1000 is 3 since 10^3 equals 1000. Similarly, the logarithm of 10 is 1. If, then, a number should lie between 100 and 1000, the logarithm of that number must lie between 2 and 3. For example, the logarithm of 657 is 2.8176. Or $10^{2.8176}$ is equal to 657. The logarithm of 65.7 is 1.8176, and the logarithm of 6.57 is 0.8176.

One of the great advantages of logarithms lies in the fact that in order to multiply two large numbers, one need only add their logarithms. To divide one large number by another, one may subtract the logarithm of one number from the logarithm of the other.

The logarithm of a number consists of two parts. The whole number to the left of the decimal point is called the **characteristic** and its value depends upon the location of the decimal point in the original number. The part of a logarithm to the right of the decimal point is called the **mantissa.** The value of the mantissa may be read from tables such as Table A.2. In the example used above,

Table A.2. Four-place logarithms.

N	0	1	2	3	4	5	6	7	8	9
10	0000	0043	0086	0128	0170	0212	0253	0294	0334	0374
11	0414	0453	0492	0531	0569	0607	0645	0682	0719	0755
12	0792	0828	0864	0899	0934	0969	1004	1038	1072	1106
13	1139	1173	1206	1239	1271	1303	1335	1367	1399	1430
14	1461	1492	1523	1553	1584	1614	1644	1673	1703	1732
15	1761	1790	1818	1847	1875	1903	1931	1959	1987	2014
16	2041	2068	2095	2122	2148	2175	2201	2227	2253	2279
17	2304	2330	2355	2380	2405	2430	2455	2480	2504	2529
18	2553	2577	2601	2625	2648	2672	2695	2718	2742	2765
19	2788	2810	2833	2856	2878	2900	2923	2945	2967	2989
20	3010	3032	3054	3075	3096	3118	3139	3160	3181	3201
21	3222	3243	3263	3284	3304	3324	3345	3365	3385	3404
22	3424	3444	3464	3483	3502	3522	3541	3560	3579	3598
23	3617	3636	3655	3674	3692	3711	3729	3747	3766	3784
24	3802	3820	3838	3856	3874	3892	3909	3927	3945	3962
25	3979	3997	4014	4031	4048	4065	4082	4099	4116	4133
26	4150	4166	4183	4200	4216	4232	4249	4265	4281	4298
27	4314	4330	4346	4362	4378	4393	4409	4425	4440	4456
28	4472	4487	4502	4518	4533	4548	4564	4579	4594	4609
29	4624	4639	4654	4669	4683	4698	4713	4728	4742	4757
30	4771	4786	4800	4814	4829	4843	4857	4871	4886	4900
31	4914	4928	4942	4955	4969	4983	4997	5011	5024	5038
32	5051	5065	5079	5092	5105	5119	5132	5145	5159	5172
33	5185	5198	5211	5224	5237	5250	5263	5276	5289	5302
34	5315	5328	5340	5353	5366	5378	5391	5403	5416	5428
35	5441	5453	5465	5478	5490	5502	5514	5527	5539	5551
36	5563	5575	5587	5599	5611	5623	5635	5647	5658	5670
37	5682	5694	5705	5717	5729	5740	5752	5763	5775	5786
38	5798	5809	5821	5832	5843	5855	5866	5877	5888	5899
39	5911	5922	5933	5944	5955	5966	5977	5988	5999	6010
40	6021	6031	6042	6053	6064	6075	6085	6096	6107	6117
41	6128	6138	6149	6160	6170	6180	6191	6201	6212	6222
42	6232	6243	6253	6263	6274	6284	6294	6304	6314	6325
43	6335	6345	6355	6365	6375	6385	6395	6405	6415	6425
44	6435	6444	6454	6464	6474	6484	6493	6503	6513	6522
45	6532	6542	6551	6561	6571	6580	6590	6599	6609	6618
46	6628	6637	6646	6656	6665	6675	6684	6693	6702	6712
47	6721	6730	6739	6749	6758	6767	6776	6785	6794	6803
48	6812	6821	6830	6839	6848	6857	6866	6875	6884	6893
49	6902	6911	6920	6928	6937	6946	6955	6964	6972	6981
50	6990	6998	7007	7016	7024	7033	7042	7050	7059	7067
51	7076	7084	7093	7101	7110	7118	7126	7135	7143	7152
52	7160	7168	7177	7185	7193	7202	7210	7218	7226	7235
53	7243	7251	7259	7267	7275	7284	7292	7300	7308	7316
54	7324	7332	7340	7348	7356	7364	7372	7380	7388	7396
	0	**1**	**2**	**3**	**4**	**5**	**6**	**7**	**8**	**9**

Table A.2. (continued).

N	0	1	2	3	4	5	6	7	8	9
55	7404	7412	7419	7427	7435	7443	7451	7459	7466	7474
56	7482	7490	7497	7505	7513	7520	7528	7536	7543	7551
57	7559	7566	7574	7582	7589	7597	7604	7612	7619	7627
58	7634	7642	7649	7657	7664	7672	7679	7686	7694	7701
59	7709	7716	7723	7731	7738	7745	7752	7760	7767	7774
60	7782	7789	7796	7803	7810	7818	7825	7832	7839	7846
61	7853	7860	7868	7875	7882	7889	7896	7903	7910	7917
62	7924	7931	7938	7945	7952	7959	7966	7973	7980	7987
63	7993	8000	8007	8014	8021	8028	8035	8041	8048	8055
64	8062	8069	8075	8082	8089	8096	8102	8109	8116	8122
65	8129	8136	8142	8149	8156	8162	8169	8176	8182	8189
66	8195	8202	8209	8215	8222	8228	8235	8241	8248	8254
67	8261	8267	8274	8280	8287	8293	8299	8306	8312	8319
68	8325	8331	8338	8344	8351	8357	8363	8370	8376	8382
69	8388	8395	8401	8407	8414	8420	8426	8432	8439	8445
70	8451	8457	8463	8470	8476	8482	8488	8494	8500	8506
71	8513	8519	8525	8531	8537	8543	8549	8555	8561	8567
72	8573	8579	8585	8591	8597	8603	8609	8615	8621	8627
73	8633	8639	8645	8651	8657	8663	8669	8675	8681	8686
74	8692	8698	8704	8710	8716	8722	8727	8733	8739	8745
75	8751	8756	8762	8768	8774	8779	8785	8791	8797	8802
76	8808	8814	8820	8825	8831	8837	8842	8848	8854	8859
77	8865	8871	8876	8882	8887	8893	8899	8904	8910	8915
78	8921	8927	8932	8938	8943	8949	8954	8960	8965	8971
79	8976	8982	8987	8993	8998	9004	9009	9015	9020	9025
80	9031	9036	9042	9047	9053	9058	9063	9069	9074	9079
81	9085	9090	9096	9101	9106	9112	9117	9122	9128	9133
82	9138	9143	9149	9154	9159	9165	9170	9175	9180	9186
83	9191	9196	9201	9206	9212	9217	9222	9227	9232	9238
84	9243	9248	9253	9258	9263	9269	9274	9279	9284	9289
85	9294	9299	9304	9309	9315	9320	9325	9330	9335	9340
86	9345	9350	9355	9360	9365	9370	9375	9380	9385	9390
87	9395	9400	9405	9410	9415	9420	9425	9430	9435	9440
88	9445	9450	9455	9460	9465	9469	9474	9479	9484	9489
89	9494	9499	9504	9509	9513	9518	9523	9528	9533	9538
90	9542	9547	9552	9557	9562	9566	9571	9576	9581	9586
91	9590	9595	9600	9605	9609	9614	9619	9624	9628	9633
92	9638	9643	9647	9652	9657	9661	9666	9671	9675	9680
93	9685	9689	9694	9699	9703	9708	9713	9717	9722	9727
94	9731	9736	9741	9745	9750	9754	9759	9763	9768	9773
95	9777	9782	9786	9791	9795	9800	9805	9809	9814	9818
96	9823	9827	9832	9836	9841	9845	9850	9854	9859	9863
97	9868	9872	9877	9881	9886	9890	9894	9899	9903	9908
98	9912	9917	9921	9926	9930	9934	9939	9943	9948	9952
99	9956	9961	9965	9969	9974	9978	9983	9987	9991	9996
	0	**1**	**2**	**3**	**4**	**5**	**6**	**7**	**8**	**9**

the table is entered with the digits 657 and the mantissa is read off as 8176. This is the decimal portion of the logarithm. The characteristic is then selected on the basis of the size of the original number. Again, since the number 657 is between 100 and 1000, the characteristic is 2 and the complete logarithm of 657 is 2.8176.

A.4 Conic Sections

The circle, ellipse, parabola, and hyperbola are four geometrical figures commonly referred to as **conic sections.** The name comes from the fact that these figures represent the following possible intersections of a plane with a cone (see Figure A.6):

1. Circle: Plane is perpendicular to the axis of the cone.
2. Ellipse: Plane is not perpendicular to the axis of the cone.
3. Parabola: Plane is parallel to the side of the cone.
4. Hyperbola: Plane makes an angle with the base that is larger than the angle between the base and the edge of the cone.

Another series of statements defines these curves in a manner that is more useful when one is actually making the geometrical construction of these figures. These definitions are as follows.

1. Circle: A sequence of points equidistant from a central point.
2. Ellipse: A sequence of points such that the sum of their distances from two points (foci) is constant.
3. Parabola: A sequence of points equidistant from a point and a line.
4. Hyperbola: A sequence of points such that the difference between the distances to two points is constant.

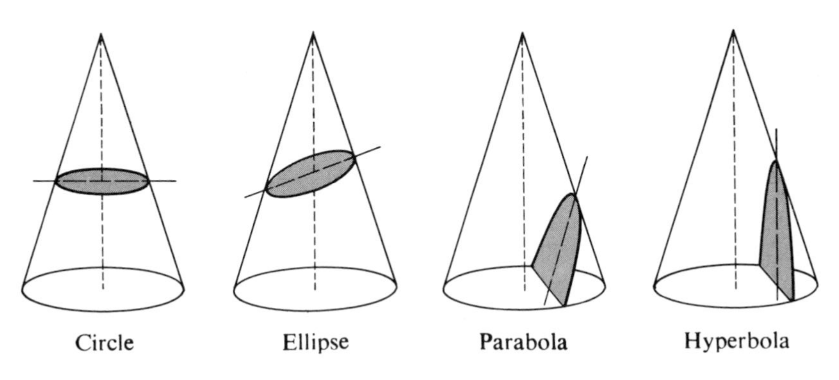

Figure A.6

Table A.3. Planetary data in miles.

	Diameter (miles)	Mean Distance from Sun (millions of miles)
Mercury	2950	35.96
Venus	7600	67.20
Earth	7913	92.9
Mars	4212	141.6
Jupiter	86,800	483.3
Saturn	75,100	886.2
Uranus	29,400	1783
Neptune	28,000	2794
Pluto	3700	3670

Table A.4. Useful constants and conversion factors.

kilometers × 0.621 = miles
meters × 3.2808 = feet
centimeters × 0.3937 = inches
parsecs × 3.26 = light-years
parsecs × 206,265 = astronomical units
light-years × 5.88×10^{12} = miles
Angstrom unit = 10^{-8} cm
1.8 × degrees centigrade + 32° = degrees Fahrenheit
1.8 × degrees Kelvin − 460° = degrees Fahrenheit
velocity of light = 300,000 km/sec = 186,000 mi/sec
constant of gravitation, $G = 6.67 \times 10^{-8}$ cm³/g-sec²
Planck's constant, $h = 6.62 \times 10^{-27}$ erg-sec
constants in Planck's law, $C_1 = 3.71 \times 10^{-5}$
$C_2 = 1.435$
Stefan-Boltzmann constant, $\sigma = 5.67 \times 10^{-5}$ erg/cm²-sec-deg⁴
constant in Wien's law, $K = 0.289$ cm-deg
mass of the earth = 5.98×10^{27} grams = 6.59×10^{21} tons
mass of the sun, $M_\odot = 1.99 \times 10^{33}$ grams = 333,000 earth masses
= 2.2×10^{27} tons

A.5 The Stellar Luminosity Function

It is instructive to note the relative numbers of stars of various luminosities or absolute magnitudes. One must begin by estimating the distances to a large number of stars in some huge volume of

Table A.5. The stellar luminosity function.

M_V	No. of Stars in 10^4 pc^3	Luminosity in 10^4 pc^3	Mass in 10^4 pc^3
−6	. . .	$1 \times L_\odot$	$0.002 \times M_\odot$
−5	0.0004	4	0.011
−4	0.0025	10	0.045
−3	0.01	16	0.10
−2	0.04	25	0.25
−1	0.2	48	1.0
0	1	96	4
1	3	138	10
2	6	91	13
3	11	71	20
4	21	52	25
5	30	30	30
6	36	14	30
7	40	6	30
8	45	3	26
9	55	1	30
10	78	1	40
11	93		35
12	115		32
13	130		28
14	140		25
15	130		25
16	100		20
17	90		18

space around the sun. One may then compute the absolute magnitude of each star and then count numbers of stars in each absolute magnitude class. The tabulation of these results is known as the **luminosity function.** Since the intrinsically bright stars can be seen at great distances, there is little problem in completing a census for them. At the other extreme, however, the census of the faint stars is very difficult for they can be seen only in a smaller volume of space near the sun.

From the tabulated numbers of stars, one may compute the amount of light radiated by each group of stars and the total amount of matter represented by each group of stars. The results of this are quite striking, for one sees that most of the light in the galaxy comes from the bright stars while most of the mass is found in the very large numbers of intrinsically faint stars.

Table A.6. Some comparative densities in the universe.

	g/cm^3	Atoms/cm^3	Relative to Physicist's Vacuum
Air at earth's surface	10^{-3}	10^{16}	10^{10}
Physicist's "vacuum"	10^{-13}	10^{6}	1
Comet tail and coma	$< 10^{-15}$	10^{4}	10^{-2}
Orion Nebula	10^{-17}	10^{2}	10^{-4}
Interstellar space	$< 10^{-19}$	1	10^{-6}
Sun: mean	1	10^{20}	10^{14}
center	200	10^{22}	10^{4}
Betalgeuse: mean	6×10^{-7}		
center	1		
Universe	$10^{-29}-10^{-31}$		

A.6 The Messier Catalogue of Nebulae and Star Clusters

Charles Messier was an eighteenth-century astronomer who was best known during his lifetime for his diligent observations of comets. We think of him today principally in connection with the famous catalogue of nebulae and star clusters which he published in the period from 1771 to 1786. Many of these objects had been discovered by Messier himself, but the list also included objects known from earlier observers and objects found by some of his contemporaries. Due to some confusion, there are a few numbers in Messier's list which do not correspond to known objects in the sky. In the list below, the missing objects are indicated with an asterisk.

Many of the Messier objects can be seen with binoculars or a small telescope of approximately 80 mm (3 in.) aperture. A careful observer using a telescope of 150 mm (6 in.) aperture should be able to find all of these objects. A very dark sky and considerable patience are necessary, however.

For a complete description of the Messier Catalogue and an interesting account of Messier's career, the reader is referred to an article by Owen Gingerich which appeared in *Sky and Telescope* in August and September, 1953. The catalogue, as included here, is from a March, 1954 article by Dr. Gingerich which also was published in *Sky and Telescope.*

The symbols in the last column have the following meanings: Pl: planetary nebula, Gb: globular cluster, Cl: open cluster, Di: diffuse nebula, El: elliptical galaxy, Sp: spiral galaxy, Ir: irregular galaxy.

Table A.7. The Messier Catalogue.

M	NGC	R.A. h m(1950)	Decl. ° ′	Const.	Size ′ ′	Mag.	Type
1	1952	05 31.5	+21 59	Tau	6 × 4	10	Pl – Crab Nebula
2	7089	21 30.9	−01 03	Aqr	12	7	Gb
3	5272	13 39.9	+28 38	CVn	19	6	Gb
4	6121	16 20.6	−26 24	Sco	23	6	Gb
5	5904	15 16.0	+02 16	Ser	20	6	Gb
6	6405	17 36.7	−32 11	Sco	26	6	Cl
7	6475	17 50.6	−34 48	Sco	50	5	Cl
8	6523	18 00.7	−24 23	Sgr	90 × 40		Di – Lagoon Nebula
9	6333	17 16.2	−18 28	Oph	6	7	Gb
10	6254	16 54.5	−04 02	Oph	12	7	Gb
11	6705	18 48.4	−06 20	Sct	12	6	Cl
12	6218	16 44.6	−01 52	Oph	12	7	Gb
13	6205	16 39.9	+36 33	Her	23	6	Gb – Hercules Cluster
14	6402	17 35.0	−03 13	Oph	7	8	Gb
15	7078	21 27.6	+11 57	Peg	12	6	Gb
16	6611	18 16.0	−13 48	Ser	8	7	Cl
17	6618	18 17.9	−16 12	Sgr	46 × 37		Di – Omega or Horseshoe Nebula
18	6613	18 17.0	−17 09	Sgr	7	7	Cl
19	6273	16 59.5	−26 11	Oph	5	7	Gb
20	6514	17 59.6	−23 02	Sgr	29 × 27		Di – Trifid Nebula
21	6531	18 01.6	−22 30	Sgr	12	7	Cl
22	6656	18 33.3	−23 58	Sgr	17	6	Gb
23	6494	17 53.9	−19 01	Sgr	27	7	Cl
24	6603	18 15.5	−18 26	Sgr	4	6	Cl
25		18 28.8	−19 17	Sgr	35		Cl – IC 4725
26	6694	18 42.5	−09 27	Sct	9	8	Cl
27	6853	19 57.5	+22 35	Vul	8 × 4	8	Pl – Dumbbell Nebula
28	6626	18 21.5	−24 54	Sgr	15	8	Gb
29	6913	20 22.1	+38 22	Cyg	7	7	Cl
30	7099	21 37.5	−23 25	Cap	9	8	Gb
31	224	00 40.0	+41 00	And	160 × 40	4	Sp – Andromeda Nebula
32	221	00 40.0	+40 36	And	3 × 2	9	El

Table A.7. (continued).

M	NGC	R.A. h m(1950)	Decl. ° ′	Const.	Size ′ ′	Mag.	Type
33	598	01 31.1	+30 24	Tri	60 × 40	7	Sp
34	1039	02 38.8	+42 34	Per	30	6	Cl
35	2168	06 05.8	+24 21	Gem	29	6	Cl
36	1960	05 32.8	+34 06	Aur	16	6	Cl
37	2099	05 49.1	+32 32	Aur	24	6	Cl
38	1912	05 25.3	+35 48	Aur	18	7	Cl
39	7092	21 30.4	+48 13	Cyg	32	6	Cl
40*							
41	2287	06 44.9	−20 41	CMa	32	6	Cl
42	1976	05 32.9	−05 25	Ori	66 × 60		Di – Great Nebula in Orion
43	1982	05 33.1	−05 18	Ori			Di – (NE wing of Great Nebula)
44	2632	08 37.2	+20 10	Cnc	90		Cl – Praesepe or Beehive
45		03 44.5	+23 57	Tau	120		Cl – Pleiades
46	2437	07 39.6	−14 42	Pup	27	9	Cl
47*							
48*							
49	4472	12 27.3	+08 16	Vir	4 × 4	9	El
50	2323	07 00.6	−08 16	Mon	16	6	Cl
51	5194–5	13 27.8	+47 27	CVn	12 × 6	9	Sp – Whirlpool Nebula
52	7654	23 22.0	+61 19	Cas	13	7	Cl
53	5024	13 10.5	+18 26	Com	14	8	Gb
54	6715	18 52.0	−30 32	Sgr	6	8	Gb
55	6809	19 36.9	−31 03	Sgr	15	5	Gb
56	6779	19 14.6	+30 05	Lyr	5	8	Gb
57	6720	18 51.8	+32 58	Lyr	1 × 1	9	Pl – Ring Nebula
58	4579	12 35.1	+12 05	Vir	4 × 3	10	Sp
59	4621	12 39.5	+11 55	Vir	3 × 2	11	El
60	4649	12 41.1	+11 49	Vir	4 × 3	10	El
61	4303	12 19.4	+04 45	Vir	6 × 6	10	Sp
62	6266	16 58.1	−30 03	Oph	6	7	Gb
63	5055	13 13.5	+42 17	CVn	8 × 3	10	Sp
64	4826	12 54.3	+21 57	Com	8 × 4	8	Sp – Blackeye Nebula
65	3623	11 16.3	+13 23	Leo	8 × 2	10	Sp
66	3627	11 17.6	+13 17	Leo	8 × 2	9	Sp
67	2682	08 48.5	+12 00	Cnc	18	7	Cl
68	4590	12 36.8	−26 29	Hya	9	8	Gb
69	6637	18 28.1	−32 23	Sgr	4	8	Gb
70	6681	18 40.0	−32 21	Sgr	4	9	Gb

M	NGC	R.A. h m(1950)	Decl. ° ′	Const.	Size ′ ′	Mag.	Type
71	6838	19 51.5	+18 39	Sge	6	9	Gb
72	6981	20 50.7	−12 44	Aqr	5	9	Gb
73	6994	20 56.2	−12 50	Aqr			
74	628	01 34.0	+15 32	Psc	8 × 8	11	Sp
75	6864	20 03.2	−22 04	Sgr	5	8	Gb
76	650−1	01 39.1	+51 19	Per	2 × 1	11	Pl
77	1068	02 40.1	−00 14	Cet	2 × 2	9	Sp
78	2068	05 44.2	+00 02	Ori	8 × 6		Di
79	1904	05 22.2	−24 34	Lep	8	8	Gb
80	6093	16 14.1	−22 52	Sco	5	7	Gb
81	3031	09 51.5	+69 18	UMa	16 × 10	8	Sp
82	3034	09 51.9	+69 56	UMa	7 × 2	9	Ir
83	5236	13 34.3	−29 37	Hya	10 × 8	9	Sp
84	4374	12 22.6	+13 10	Vir	3 × 3	10	El
85	4382	12 22.8	+18 28	Com	4 × 2	10	El
86	4406	12 23.7	+13 13	Vir	4 × 3	10	El
87	4486	12 28.3	+12 40	Vir	3 × 3	10	El
88	4501	12 29.5	+14 42	Com	6 × 3	10	Sp
89	4552	12 33.1	+12 50	Vir	2 × 2	11	El
90	4569	12 34.3	+13 26	Vir	6 × 3	11	Sp
91*							
92	6341	17 15.6	+43 12	Her	12	6	Gb
93	2447	07 42.5	−23 45	Pup	18	6	Cl
94	4736	12 48.6	+41 23	CVn	5 × 4	8	Sp
95	3351	10 41.3	+11 58	Leo	3 × 3	11	Sp
96	3368	10 44.2	+12 05	Leo	7 × 4	10	Sp
97	3587	11 11.9	+55 18	UMa	3 × 3	11	Pl—Owl Nebula
98	4192	12 11.3	+15 11	Com	8 × 2	11	Sp
99	4254	12 16.3	+14 42	Com	4 × 4	10	Sp
100	4321	12 20.4	+16 06	Com	5 × 5	10	Sp
101	5457	14 01.4	+54 35	UMa	22 × 22	8	Sp—Sombrero Nebula
102*							
103	581	01 29.9	+60 26	Cas	6	7	Cl
104	4594	12 37.3	−11 21	Vir	7 × 2	8	Sp
105	3379	10 45.2	+12 51	Leo	2 × 2	10	El
106	4258	12 16.5	+47 35	CVn	20 × 6	10	Sp
107	6171	16 29.7	−12 57	Oph	8	9	Gb
108	3556	11 08.7	+55 57	UMa	8 × 2	10	Sp
109	3992	11 55.0	+53 39	UMa	7	11	Sp

Suggestions for Further Reading

Bok, Bart J., and Priscilla F. Bok, *The Milky Way*, 3rd ed., Harvard University Press, Cambridge, Mass., 1968.

Brandt, John C., *The Physics and Astronomy of the Sun and Stars*, McGraw-Hill Book Company, New York, 1966. (For those interested in a more technical treatment.)

Fowler, William A., *Nuclear Astrophysics*, American Philosophical Society, Philadelphia, Pa., 1967. (Written on an excellent level for one who has previously read a basic astronomy text.)

Hodge, Paul W., *Galaxies and Cosmology*, McGraw-Hill Book Company, New York, 1966.

Hoyle, Fred, *From Stonehenge to Modern Cosmology*, W.H. Freeman, San Francisco, 1972.

McVittie, G. C., *Fact and Theory in Cosmology*, The Macmillan Co., New York, 1961.

Pannekoek, A., *A History of Astronomy*, Interscience Publishers, Inc., New York, 1961.

Paul, Henry E., *Telescopes for Skygazing*, 2nd ed., Chilton Books, Philadelphia, 1966.

Saslaw, W. C., and Jacobs, K. C. (eds.), *The Emerging Universe*, The University Press of Virginia, Charlottesville, 1972.

Smart, W. M., *Text-Book on Spherical Astronomy*, Cambridge University Press, Cambridge, England, 1949.

Struve, Otto, and Velta Zebergs, *Astronomy of the Twentieth Century*, The Macmillan Company, New York, 1962.

Thiel, Rudolph, *And There Was Light*, Alfred A. Knopf, New York, 1957. (Also in paperback from New American Library. An excellent history of astronomy for the general reader.)

Whipple, Fred L., *Earth, Moon and Planets*, 3rd ed., Harvard University Press, Cambridge, Mass., 1968.

Scientific American, published monthly by Scientific American, Inc., 415 Madison Avenue, New York, N.Y. 10017. (Articles related to astronomy are included almost every month.)

Sky and Telescope, published monthly by Sky Publishing Corporation, 49-50-51 Bay State Road, Cambridge, Mass. 02138.

Index

N

O